The Market Preparation of
Carolina Rice

The Market Preparation of Carolina Rice

An Illustrated History of Innovations in the Lowcountry Rice Kingdom

Richard Dwight Porcher, Jr.,
and William Robert Judd

THE UNIVERSITY OF SOUTH CAROLINA PRESS

© 2014 University of South Carolina

Published by the University of South Carolina Press
Columbia, South Carolina 29208

www.sc.edu/uscpress

Manufactured in the United States of America

23 22 21 20 19 18 17 16 15 14   10 9 8 7 6 5 4 3 2 1

Library of Congress Cataloging-in-Publication Data
Porcher, Richard Dwight, Jr.
The market preparation of Carolina rice : an illustrated
history of innovations in the lowcountry rice kingdom /
Richard Dwight Porcher, Jr., and William Robert Judd.
pages cm
Includes bibliographical references and index.
ISBN 978-1-61117-351-2 (hardbound : alk. paper)
1. Rice trade—South Carolina--History. 2. Rice—
South Carolina—History. I. Judd, William Robert. II. Title.
HD9066.S6444P67 2014
338.1′731809757—dc23
2013042683

The authors dedicate *The Market Preparation of Carolina Rice* to the late James M. Clifton, a native of Kelly, Bladen County, North Carolina. Professor Clifton was a graduate of Wake Forest College (now University) and received his masters in history from Duke University. He became a charter member of the faculty at Southeastern Community College, near Whiteville, North Carolina, when that institution opened in 1965 and taught history there until his retirement in 2004. Professor Clifton died in 2008 at the age of seventy-seven after a life marked with outstanding service to his community and profession.

James Clifton's research and writing concentrated on rice culture in the Atlantic states. He pursued both with passion and became a widely published authority in the field. His research was not limited in scope; he covered broad topics related to rice culture and history in general. His scholarship and publications provided invaluable material for this book.

Contents

List of Illustrations   xi
Foreword   xvii
   David S. Shields
Preface   xix
Acknowledgments   xxiii

Introduction   1

CHAPTER 1   A Brief History of Rice   7

African Rice   8
Asian Rice   9

CHAPTER 2   The Origins and Introductions of Rice Seeds   12

The Lords Proprietors' Search a Commercial Crop, 1663–1677   14
Seed from Madagascar and the Dubois Seed, circa 1685   15
Establishment of Rice as a Crop, late 1600s   16
Rice as an Export Crop of Carolina, 1695   17
Carolina Gold Rice, circa 1785   20
Bearded Rice, 1829   25
Long Grain Carolina Gold Rice, 1838   26

CHAPTER 3   The Culture of Carolina Rice   28

River Systems and Associated Wetlands   28
Providence Rice Culture   30
The Beginning of the Plantation Enterprise   34
The Task System   37
Reservoir Rice Culture   40
The Spread of Rice Culture   57

Tidal Rice Culture   60
  Sowing Rice Seeds   78
  Flow Culture   80
  Water Culture   82
Rice Flats   83
Canals and Floodgates   87
Deadly Rice Fields   93
Obstacles to Overcome   97
The Legacy of the Abandoned Tidal Rice Fields   102

CHAPTER 4   **Harvesting**   105

The Sickle   105
The Cradle Scythe   107
Transporting the Sheaves   108

CHAPTER 5   **Threshing**   111

The Flail   112
Seed Rice   117
The Kogar Wind Fan   118
The Andrew Meikle Thresher   120
Rolling Screens   121
The Bernard Thresher   124
The Calvin Emmons Threshing Machine   125
William Emmons's Improvement   128
Ludlow's Thresher   129
Mechanical Improvements   129
Mathewes's Machine for Threshing Rice   132
Jehiel Butts's Improved Threshing Machine   134
Rakes   137
The Chicora Wood Threshing Barn   138
The Chicora Wood Storage Barn   144
Antebellum Threshing   148
Machine Maintenance and Fire Prevention   149
Postwar Threshers   150

CHAPTER 6   **Milling**   153

Morphology and Terminology of the Rice Fruit   155
Manual-Powered Mills   157
    Mortar and Pestle   157
    The Guerard Pendulum Engine   160
    The Spring Mill   162
    The Italian Pestle Mill   162
    Deans's Rice-Pounding Machine   165
    Wooden Mill (Wooden Rotary Quern)   166
Animal-Powered Mills   170
    The Pecker Mill   171
    The Drum Mill ("Cog Mill")   173
    The Cog Mill   176
    Sieves, Screens, and Rolling Screens in Milling   177
    The Veitch Pestle Mill   181
Water-Powered Mills   184
    Sources of Water Power   188
    Construction and Operation of a Water-Powered Rice Mill   192
    Jonathan Lucas I's Water Rice Machine   202
    Mortar and Pestle Evolution   208
    Millstones   209
    Belts and Drills   214
    Jonathan Lucas I's Improved Water Rice Machine   216
    Jonathan Lucas, Jr.'s Rice Cleaner   220
    The Deforest Rice Mill   221
    Ravenel Rice-Hulling Machine   223
Steam-Powered Mills, 1830s   224
    The Early History of Steam Power   225
    The Mill Engine   231
    Portable Engines   239
    The Plain Cylindrical Boiler   241
    Rice Chimneys   248
    Carpenters, Blacksmiths and Mechanics   257
    The Napier Rice Cleaner   259
    The Williams Cam Pestle Mill   259
    McKinlay's Improvement in Cleaning Rice   260
    McKinlay's Rotary Rice Cleaner   262
    The Taylor Machine for Cleaning Rice   263
    The Lachicotte and Bowman Improvement in Cleaning Rice   265

Oliver J. Butts's Machine for Brushing Rice 266
McKinlay's Machine for Cleaning Rice 268
The Brotherhood Improvement in Rice Pounding and Hulling Mills 269
The Lockfaw Rice Mill 269
The Engelberg Huller Company, Syracuse, New York 271
The Habarnards Rice Huller 274
The Engelberg Disk Huller 274

CHAPTER 7 City Mills and Large Plantation Toll Mills 276

The Hopper-Boy 276
The Early City Mills of Jonathan Lucas I 278
Waverly Rice Mill 278
Savannah Rice Mills 281
West Point Rice Mill 282
Bennett's Rice Mill 283
Daniel Hadley's Model of a 1850 Carolina Rice Mill 285
Georgetown Rice Mill 288

CHAPTER 8 The Golden Age of Rice, 1800–1860 290

Tidal Rice-Field Expansion 293
The Labor Force 294
Mechanization and Manufacturing Expansion 294

CHAPTER 9 The Last Days of Rice Planting 297

Epilogue 313
   Richard Dwight Porcher, Jr.

Appendix 1: Original Field Research Plantations and Sites 321
Appendix 2: Museums with Originally Manufactured Products
   and/or Machinery 323
Appendix 3: United States Patents for Rice-Processing Machinery, 1829–1887 325
Notes 327
Bibliography 351
Index 363

Illustrations

Figures

1. Quarter drains in rice fields   68
2. "Planting the Rice"   79
3. "Dotterer's Patent Rice-Sowing Machine"   80
4. "Shipping Rice from a Plantation on the Savannah River"   83
5. Rice flat on the Cooper River, early 1900s   84
6. Diamond floodgate at Nieuport Plantation, early 1900s   91
7. *Bobolinks on Rice*   98
8. "Harvesting the Rice"   107
9. Hauling sheaves, early 1900s   109
10. "Threshing"   112
11. Threshing yard at Middleburg Plantation, early 1900s   114
12. Alice Ravenel Huger Smith, *Shaking the Rice from the Straw after Threshing*   114
13. Fanning rice   115
14. Winnowing platform   116
15. Winnowing house at Mansfield Plantation   117
16. Rolling screen at Cockfield Plantation   123
17. Barnyard at Chicora Wood Plantation, circa 1900   139
18. Threshing barn and storage barn on a Georgia plantation   146
19. Milling rice with a mortar and pestle   159
20. Italian pestle machine, 1578   164
21. Salzburger water mills in Ebenezer, Georgia, 1747   187
22. Rice mill at Middleburg Plantation   193
23. Stoke Rice Mill, early 1900s   196
24. Breastwork ruins at Nieuport Plantation   199
25. Polishing brush at Cockfield Plantation   218
26. West Point Rice Mill (rear view), early 1900s   228
27. Flyer for Eason Iron Works   234
28. Portable steam engine, circa 1900   241
29. Boiler at Social Hall Plantation   243
30. Rice chimney at Laurel Hill Plantation   253

31. "Rice Mills on Savannah River"   261
32. Waverly Rice Mill, circa 1900   279
33. Millstones at Savannah Rice Mill   282
34. West Point Rice Mill (front view), early 1900s   284
35. Mortars and pestles at West Point Rice Mill, early 1900s   284
36. Harvesting with a twine binder, early 1900s   308

Diagrams

1. Two types of estuaries   29
2. Lift-gate trunks   43
3. Lever-gate trunk   44
4. Swing-gate trunks   45
5. Inland-swamp rice field   49
6. Construction of tidal rice fields   66
7. Plug trunk at Social Hall Plantation   70
8. Plug trunk at Kinloch Plantation   71
9. Operation of a tide trunk   74
10. Original tide trunk   76
11. Savannah tide trunk   77
12. Rice flat designs   86
13. Typical rice field on the Combahee River   88
14. Construction of a diamond floodgate   89
15. Plan view of a diamond floodgate   90
16. One of the four gates of the diamond floodgate at Nieuport Plantation   92
17. Hopeton Plantation site plan   94
18. Operation of a typical 1800s wind fan   119
19. Interior view of Meikle's Threshing Machine, 1786   120
20. Operation of a rolling screen   124
21. Patent diagram, Benjamin Bernard's thresher, 1808   125
22. Patent diagram, Calvin Emmons's improved threshing machine, 1829   126
23. Patent diagram, William Emmons's improved threshing machine, 1831   128
24. Threshing barn   131
25. Patent diagram, William Mathewes's rice thresher, 1835   133
26. Patent diagram, Jehiel Butts's improved rice thresher   135
27. Rake at Chicora Wood Plantation   136
28. Rake at Mansfield Plantation   138
29. Threshing-barn complex at Chicora Wood Plantation   140
30. Elevated spiral conveyor   141
31. Operation of the threshing barn at Chicora Wood Plantation   143
32. Plantation sloop   145

List of Illustrations xiii

33. Storage barn at Chicora Wood Plantation   147
34. Skeleton view of the Invincible thresher   151
35. Rice fruit   155
36. Peter Jacob Guerard's "Pendulum Engine," 1691   161
37. Spring mill   163
38. Wooden rotary quern seen by Luigi Castigoni, 1785–87   168
39. Pecker mill   171
40. Drum mill (or "cog mill")   174
41. Cog mill used for a cotton gin, 1808   176
42. Cog mill for rice   178
43. Cam and spoke pestle machines   179
44. Hand-operated sieve and screen   180
45. George Veitch's pestle mill   183
46. Vitruvian water mill, circa first century B.C.E.   185
47. Rice mill powered by saltwater tides   191
48. Floor plan of a water-powered mortar-and-pestle rice mill, 1793   193
49. Typical water-powered mortar-and-pestle rice mill   195
50. Waterwheel and breastwork   199
51. Positioning of machinery within a mill   200
52. Water-powered mortar-and-pestle rice mill   201
53. Jonathan Lucas I's Water Rice Machine, 1787   203
54. Operation of millstones for hulling rice   207
55. Winnowing screen pendulum   212
56. Rice elevator   215
57. Drill conveyor   216
58. Deforest rice mill   221
59. Operation of the condensing steam engine   227
60. Oliver Evans's Columbian steam engine, 1811   229
61. Operation of Columbian noncondensing steam engine   230
62. Basic horizontal-cylinder, simple slide-valve mill steam engine   235
63. Operation of the valve stem   237
64. Operation of the slide valve   238
65. Steam-powered threshing and/or milling complex   240
66. Boilers and furnace   242
67. Boiler cap and gasket   244
68. Water-level gauges for a steam boiler   246
69. Thomas Savery's brick chimney, 1689   249
70. Rice-chimney construction   251
71. Patent diagram, Peter McKinlay's improved mill for cleaning rice, 1851   260
72. Patent diagram, Peter McKinlay's rotary rice cleaner, 1852   263
73. Patent diagram, John F. Taylor's machine for cleaning rice, 1857   264

74. Patent diagram, Philip Rossignol Lachicotte and T. B. Bowman's improved rice-cleaning machine, 1857   265
75. Patent diagram, Oliver J. Butts's machine for brushing rice, 1857   267
76. Patent diagram, McKinlay's portable rice mill, 1866   268
77. Patent diagram, Fred Brotherhood's improved mill for pounding and hulling rice, 1878   270
78. Patent diagram, John Lockfaw's rice mill, 1887   270
79. Patent diagram, Engelberg Rice Huller and Polisher, 1888   273
80. Engelberg disk huller   275
81. Hopper-boy   277

Plats

1. Col. Hezekiah Maham's rice fields   22
2. Drayton Hall rice fields, circa 1790   54
3. Woodboo Plantation, 1806   55
4. Plan of Calais Plantation, 1796   56
5. Location of a water-powered rice mill in St. John's Parish, Berkeley County, South Carolina, 1800   189
6. Plan of Limerick Plantation   190

Color Plates

following page 118

1. Small-stream floodplain on Huger Creek
2. Penny Dam, inland swamp reservoir at Fairlawn Plantation
3. Former rice fields at Drayton Hall
4. Estherville Plantation reservoir and rice fields
5. Tidal freshwater swamp on the Waccamaw River
6. Seed barn at Middleburg Plantation
7. Rolling screen at Chicora Wood Plantation
8. Rolling screen at Kinloch Plantation
9. Remains of the Butts thresher at Mansfield Plantation
10. Rake enclosure in the Chicora Wood threshing barn

following page 166

11. Threshing barn at Chicora Wood Plantation
12. Storage barn at Chicora Wood Plantation
13. Invincible thresher at Kinloch Plantation
14. Threshing barn at Kinloch Plantation
15. Ruins of the water-powered mill and millrace floor at Rose Hill Plantation

16. Ruins of the millrace at Stoke Rice Mill
17. Cast-iron waterwheel hub
18. Millstones
19. Cast-iron mortar bottom from Middleburg Plantation
20. Steam engine at Mansfield Plantation
21. Steam engine at Cedar Hill Plantation

following page 262
22. Steam engine at Middleburg Plantation
23. Smith & Porter Steam Engine
24. Boiler at Middleburg Plantation
25. Boiler at Atkinson Creek in the Santee Delta
26. Rice chimney at Fairfield Plantation
27. Rice chimney at Black Out Plantation
28. Rice chimney at Willtown Plantation
29. Chimney stabilization at Hope Plantation
30. Daniel I. Hadley's model of an 1850 Carolina rice mill
31. Abandoned rice fields on Quenby Creek
32. Abandoned slave village on Crow Island
33. Storm tower in the Santee Delta

Foreword

The Old South was generally reckoned to have been a technological backwater, with only four areas of mechanical production worth mentioning: sugar production, iron working, shipbuilding, and rice milling. Even in these circumscribed areas, commentators have reckoned that the South—with the noteworthy exception of devising submersibles—was never an innovator in the creation of products or means of production. In *The Market Preparation of Carolina Rice,* Richard Dwight Porcher, Jr., and William Robert Judd dismantle this perception forcefully and definitively, showing that in the invention of devices to harvest, thresh, mill, and polish Carolina rice, persons in South Carolina and Georgia restlessly worked at creating and refining mechanisms that could ensure the ability of the Lowcountry to supply the world with superlative rice. In rice-milling technology of the antebellum era, South Carolina was the world leader in innovation. No treatment of southern technological production in any area has chronicled developments with the care, attention to mechanical detail, and illustrative clarity of this book. Because Carolina rice was a staple grain of the American larder, and because Carolina Gold rice was reputed to be one of the great rices in world trade, Porcher's study has importance in the area of food studies as well as the histories of technology and trade.

The study details the processes involved in cultivating, harvesting, drying, threshing, screening, milling, and polishing the rice and the material means by which these labors were accomplished. Over several decades, Porcher and Judd visited important historical sites, including the difficult-to-reach marsh islands in the Santee River, to record remnants of rice fields and ruins of milling infrastructure. Some of the objects Porcher photographed and measured thirty years ago have disappeared from the landscape. In parallel with examining material evidence, Porcher explored private and public libraries and archives for written commentaries on growing and processing rice. He studied the U.S. Patent Office records for drawings of machines and sifted through the extensive corpus of agricultural literature published in the many nineteenth-century farming periodicals. He consulted the most informed contemporary commentators and used the testimonies of many experienced planters and inventors to flesh out his narrative.

The most striking dimension of this study is the richness of illustration. Early lithographs and photographs provide informative views of field culture, threshing, and milling. But the tour de force of this book is the schematic diagrams that depict the workings of the sequence of significant machines involved in preparing rice. Constructed from combinations of the material and literary evidence gathered by Porcher and drawn by Judd, these schematic drawings are a major contribution to the history of agricultural technology, and their importance cannot be overstated.

The history of technology almost invariably intersects with the history of labor. The received literature on rice culture in the Lowcountry has treated exhaustively the matter of field labor, particularly slave labor during the pre–Civil War period, focusing on mindless toil of sowing, weeding, and hand cutting. This book goes beyond this aspect of rice production to detail the labor entailed in building, operating, repairing, and renovating the machines (and in the case of threshing barns, minifactories).

The Market Preparation of Carolina Rice adds much to our knowledge of the central crop of the Lowcountry region, the distinctive technology developed to process it, and the work of at least one mechanical genius, Jonathan Lucas I, whose inventions were embraced around the world. In the process this book also memorializes features of Lowcountry civilization, providing more new information on that civilization than any book since Philip Morgan's *Slave Counterpoint* (1998).

David S. Shields

Preface

My initial introduction to the history of rice culture in the Carolina Lowcountry was field work for a study titled "A History of Land Use of Hobcaw Forest" in Georgetown County, South Carolina, which I conducted for Clemson University in 1976. This was my first major field work while an assistant professor of biology at the Citadel in Charleston. The primary land use of Hobcaw Forest was rice culture, which significantly changed its natural landscape. Continuing botanical field surveys in the Carolina Lowcountry during my thirty-three-year tenure at the Citadel (1970–2003), I spent many hours trekking through swamps and along rivers where rice was grown and on the highlands where the mills and threshing barns were located. Virtually everywhere I encountered a treasure trove of rice-culture artifacts that piqued my curiosity. I documented them with photographs and recorded their locations on maps for possible use one day on a history of rice culture. These artifacts are used throughout this book.

The brick chimneys towering above the abandoned rice fields were especially fascinating. I knew they must have played an important role in rice culture. But what? I sought information in libraries and archives about their construction, operation, and function, but I found nothing. No one I asked knew anything except that they were rice chimneys. Also scattered throughout the Lowcountry marshes and swamps were remains of millraces, steam boilers, steam engines, and mills, relics of a past industrial complex that seemed out of place in the Lowcountry swamps and marshes. Again there was little available information on how the water-powered and animal-powered threshing barns and mills were constructed and operated, or on steam engines and boilers that powered the threshing barns and rice mills.

On my initial visit to Middleburg Plantation, I stood by the abandoned millrace and tried to visualize the waterwheel—twenty-two feet in diameter and fourteen feet wide—turning in the millrace, and I decided one day to "re-create" the wonderful machines that processed rice for market. No one in the Lowcountry today has ever witnessed a waterwheel turning a Carolina rice mill or heard the pestles pounding rice and creating rhythmic sounds that reverberated over the rice fields. An entire Lowcountry industry has been lost from history. Furthermore I knew that no one would ever physically reconstruct threshing barns and mills or the machinery they housed, a neglected aspect of rice culture.

Teaching, researching, and writing two wildflower books left little time for work on a history of rice culture, but it never was far from my thoughts. After retiring from teaching in 2003 and coauthoring a book on Sea Island cotton in 2005, I decided it was time for serious work on a book about market preparation of Carolina rice, a book I was in a unique position to write. Some of the artifacts I had found had deteriorated or been destroyed since I began my field surveys in 1970. It was unlikely that anyone else would be able to document these artifacts. And I suspect that few historians would venture into the swamps and coastal riversides of the Lowcountry to look for artifacts of rice culture or even to visit the sites of those I found. I also had contacts with many of the owners of private plantations in the Lowcountry, which gave me access to sites where artifacts were located. Without these artifacts a book on the market preparation of rice could not be written. It was *now or never* for this book.

Yes, I was raised on rice, especially okra rice, more often called "okra pilau," a dish that has its origins with African slaves. Since food is history and a major component of human endeavor, the history of rice culture and its market preparation is a defining part of Lowcountry history. Americans have in the past, and I suspect today, viewed rice as a stepchild of American agribusiness. Many Americans have never eaten rice and think it is eaten only in China and other Asian countries. But here in the Lowcountry we know better; rice as food was and still is a way of life.

I met William Robert "Billy" Judd through mutual friends and was impressed with his mechanical drawings. I gained firsthand knowledge of his work when he diagramed an animal-powered roller gin for *The Story of Sea Island Cotton* (2005). We spent many days over the years documenting various artifacts in the field, Billy for his own curiosity and I for a future book. Billy made a study of the brick chimneys and steam systems that powered the barns and mills; he had an impressive portfolio of diagrams and drawings of exceptional detail and quality on many facets of rice culture. Since there are no diagrams of the machines in archival sources, I knew that Billy's mechanical diagrams and expertise would be of inestimable value to this book. He could re-create these machines in a way that I never could. (What does a field botanist know about how a steam engine operates anyway?) I asked Billy to join me as the second author, and he agreed to do so. I would not have attempted to write this book without his involvement. Billy's mechanical drawings of the fascinating machines that prepared Carolina rice for market bring to life a major aspect of the rich history of the Lowcountry.

The Market Preparation of Carolina Rice adds a new dimension to the extensive literature on rice culture. This book examines how market preparation changed over time and what motivated the changes. The answers to these questions are based on archival materials and field artifacts and are enhanced by original illustrations. The scope of this study does not duplicate earlier histories of rice culture. It is an in-depth treatment to a topic only minimally covered in many other works. *Market Preparation of Carolina Rice* documents the people who introduced, invented, and operated the implements and machines used to prepare Carolina rice for market. It describes how market preparation changed during the lifetime of the industry from the preindustrial implements used by enslaved Africans to the complex machines that came out of the Industrial Revolution and were modified and improved by artisans and manufacturers both in the North and in the Rice Kingdom.

This book is intentionally limited in scope. Recent painstaking scholarly research on plantation economy and structure, slave history, social and political history of the planter elite, and contributions of African slaves to rice culture has been published in the many books and professional journals, which are cited throughout this book. Most of these works, however, contain only brief sections on market preparation of Carolina rice, and those discussions rely on earlier historians whose accounts are often erroneous and inadequate, or at best superficial and lacking in detail. The authors of these excellent modern studies did not have access to the field artifacts and private archival materials that Billy Judd and I have been allowed to examine. One cannot fully understand the role Carolina rice played in the South and the Lowcountry without knowing how it was prepared for market, and one cannot adequately document its market preparation without these field artifacts and private archives.

One would think that this major aspect of rice culture would already have been documented; yet, although many cursory attempts were made, no one had documented the machines that processed Carolina rice for market in any detail except John Drayton in his diagram of Jonathan Lucas I's 1787 Water Rice Machine in *A View of South-Carolina* (1802). There exist only a few surviving period diagrams or written records of the operation of the animal-driven, water-driven, and steam-driven threshing barns and mills that, beginning in the early 1700s, were the main venues for preparing rice for market. Many family and company records were lost or destroyed in fires, especially during the Civil War. By examining the ruins of barns and mills and other artifacts, however, along with the scant written material in public and private archives, we have been able to diagram the implements and machines that prepared rice for market, to demonstrate how they operated, and to depict the buildings that housed these machines. Billy Judd's original diagrams and drawings show how these machines were constructed and operated and open a window to an unparalleled view of early technology in the tidewater area of the Carolinas and Georgia during the colonial and pre–Civil War periods.

Many of the artifacts documented during the fieldwork leading to this book's publication are located on private plantations and on federal, state, and county lands. In many cases landowners and agencies were either unaware of the artifacts or of their historic importance. Fortunately once their significance was made known, steps were taken to preserve them.

Our study also examines some complex patents diagrams for threshing and milling machines, many invented by Charlestonians and others in the Rice Kingdom. They are important to understanding the state of manufacturing prior to and after the Civil War. For the more detailed patents, original letters and numbers have been removed and replaced with new letters, numbers and labels. Often a patent contains several figures; only pertinent figures—usually only one for each patent—are included.

Conclusions and illustrations on some of the machines have been based on sparse material, and consequentially they may be viewed differently by future historians. We tried not to stray too far beyond the limits of what written sources and artifacts told us, but we believe it is important to apply our general knowledge of how machines work to conclusions about the remains of rice-processing machinery we found. A main purpose of original research is to provide future historians a foundation of new material and conclusions, a body of knowledge on which to build, and sometimes to revise. Many questions remain unanswered. Future historians will uncover still more new material that may well modify, add to, or change some of this book's conclusions.

Richard Dwight Porcher, Jr.

Acknowledgments

This book has been long in the making, progressing in bits and pieces until 2007, when work began in earnest. Along the way many have helped bring this project to fruition, and we are honored to acknowledge their generosity and support. We are indebted to the Georgetown Historic Ricefields Association, Inc., for helping to support publication of this book.

The Citadel Development Foundation (now the Citadel Foundation) awarded two grants for field surveys for artifacts of South Carolina rice culture. The Post and Courier Foundation provided a grant for the same purpose. The materials gathered on the surveys funded by these grants are widely discussed throughout the book, and the authors thank both organizations for their financial support.

Richard Porcher made two research trips to Great Britain, where he conversed with steam-engine expert H. Alan McEwen of Farling Top Boilerworks in Keighley, West Yorkshire, England. Alan contributed significantly to the section on steam engines, boilers, and chimneys. We viewed firsthand steam engines, boilers, and chimneys in operation. The authors thank him for his time and advice.

The valuable editorial assistance of Angela Williams, who spent many hours reviewing and editing the manuscript, is gratefully acknowledged. Jan Comfort, government documents reference librarian at the R. M. Cooper Library at Clemson University obtained copies of patents, which were of inestimable value in documenting the threshing and milling processes of rice culture. Tim Roylance and Anne Thompson of the Digital Imaging Department at the Medical University of South Carolina scanned many of the archival images used in the book. To them we owe a special thanks. We are especially grateful to Debbe Causey of the Citadel's Daniel Library, who obtained many articles and documents for us through interlibrary loan. Kevin Metzger and Mary Chapman of the Information Technology Services of the Citadel provided computer services that helped organization of the manuscript. Bud Hill of the Village Museum in McClellanville and James Fitch of the Rice Museum in Georgetown provided access to those collections, and Carter Hudgins of Drayton Hall located material on the Drayton family. Harlan Greene of Special Collections at the College of Charleston Library found many manuscripts and articles that also proved invaluable. Charlotte B. Jeffers, tourism coordinator for the City of Crowley, Louisiana, was a great help during Richard Porcher's fact-finding trip to Crowley.

Richard Porcher especially thanks Father John Seiler. We became friends when I was writing *The Story of Sea Island Cotton.* He spent countless hours on the internet tracking down manuscripts and websites that provided valuable information for the cotton book. He helped the same way with *The Market Preparation of Carolina Rice.*

We thank the private plantation owners and managers who gave us access to properties to document barn and mill ruins and former rice fields. Without their cooperation, this book would never have come to fruition. They are Jessica and Larry Loring, Old Combahee Plantation; Malcolm Rhodes, Barnhill Plantation; Marion and Wayland Cato, Jr., Cedar Hill Plantation; Marcia and Jamie Constance, Chicora Wood Plantation; Sally and John Parker, Mansfield Plantation; Suzie O'Brian, Cockfield Plantation; Jane Turner and manager Brycon McCord, Kinloch Plantation; Kay and John Maybank, Lavington Plantation; the Duell family, Middleton Place; Max Hill, Jr., Middleburg Plantation; Ken Hiller, Nightingale Plantation; the Lightsey family, Social Hall; John Nichols, Rice Hope Plantation (North Santee); Molly and Henry Fair, Rose Hill Plantation (Combahee River); Lucile Pate, Fairfield Plantation (Waccamaw River); Richard Schultz, Jr., Turnbridge Plantation; the staff of Nemours Wildlife Foundation, Nemours Plantation, Ernie Wiggers and Eddie Mills; Hugh Lane, Jr., Willtown Plantation; David Dwyer, Weymouth Plantation; Alberta Quattlebaum, Waverley Plantation; Parker and Robert Lumkin, Estherville Plantation; and Peter Manigault at Rochelle Plantation.

Thanks to the staffs of the many historical societies and libraries we consulted. We especially thank the staffs of the Charleston Library Society, the South Carolina Historical Society, the South Caroliniana Library at the University of South Carolina, Special Collections at the College of Charleston Library, and the Georgia Historical Society.

The late John Ernest Gibbs allowed Richard Porcher to copy photographs from a Middleburg Plantation family album that was subsequently lost. Copies of the photographs of rice culture will be deposited in the South Carolina Historical Society. Without John's involvement, these photographs would have been lost, and the chapter on milling would have been less developed. Thanks are also owed to the following individuals and institutions for providing illustrative materials: Alberta Lachicotte Quattlebaum, Waverly Plantation, Georgetown County, South Carolina; Arcadia Parish Library, Freeland Archives, Crowley, Louisiana; Bernard Joseph Kelley, Wadmalaw Island, South Carolina; South Carolina Heritage Trust Program, Department of Natural Resources, Columbia, South Carolina; Georgia Department of Archives and History, Athens, Georgia; Georgia Historical Society, Savannah, Georgia; Gibbes Museum of Art, Charleston, South Carolina; Historic American Building Survey, National Park Service, Washington, D.C.; Library of Congress, Washington, D.C.; Norman Sinkler Walsh, Summerville, South Carolina; South Carolina Historical Society, Charleston, South Carolina; the Charleston Museum, Charleston, South Carolina; United States Park Service, Washington, D.C.

The Market Preparation of
Carolina Rice

Introduction

Market preparation of Carolina rice falls into three distinct periods: in the first harvesting, threshing, and milling were primarily based on West African technology; in the second local artisans and plantation workers developed and used animal- and water-driven machines; and in the final period these machines were replaced by steam-driven machines that were developed during the Industrial Revolution and were used until the end of the industry in the early 1900s. Despite other advances, harvesting remained a manual process during the entire life of the industry.

The refinement of the reservoir system and the advent of tidal irrigation, which began in the mid-1700s—as well as a burgeoning overseas demand for rice—were the primary impetus for the development of efficient threshing and milling machines. With a more certain water supply, crop production increased. The preindustrial methods of market preparation were no longer able to process the larger crops in time for export before the next crop was ready. In addition using the mortar and pestle to mill rice proved wasteful. More than one planter recorded in his plantation journal or letters that much of the rice crop was broken in the mortar or only partially cleaned, decreasing its market value.

The machines that prepared rice for market resulted from repeated experiments over time and the indigenous application of borrowed technology. Each advancement led to more complex and more efficient machines until the processes reached sophisticated levels by the end of the industry. African methods of threshing and milling introduced in the Atlantic states, machines from the Industrial Revolution in Great Britain, and machines from northern American cities were borrowed and improved, or modified and incorporated into machines designed specifically for the Rice Kingdom, the Atlantic region where rice was grown: North Carolina, South Carolina, Georgia, and East Florida.[1]

Much of what we call the Rice Kingdom corresponds with the region known as the Lowcountry, the land in North and South Carolina, Georgia, and northeast Florida that lies below the zone of tidal influence of the freshwater rivers. In rough terms it is a zone about forty miles inland from the Atlantic Coast, and it corresponds to about the outer half of the geological coastal plain.

Many of the machines used in the Rice Kingdom were the result of southern invention. Iron replaced wood; then steel replaced iron. Antebellum foundries in Savannah, Charleston, Georgetown, and Wilmington produced machines to thresh and mill rice. Steam power replaced water and animal power. Technology in one area required comparable changes in other areas. Invention of a mechanical thresher spurred new methods of milling to keep up with the increased volume of harvested rice. Slaves, planters, overseers, and manufacturers all contributed to the new machines for threshing and milling.

Changes in technology took time and depended on three criteria. First, success in innovation depended on knowledge gained from prior experience. Jonathan Lucas I did not base his Water Rice Machine on totally new ideas. He constructed his mill using proven methods, but he modified them and added his own innovations. Second, a new invention had to work to the owner's advantage. If a new machine did not increase threshing and milling efficiency and produce cleaner rice for market, it was not adopted. Many threshing and milling machines, although highly sophisticated, were never commercially successful. And third, innovation had to be affordable. If a sizable number of planters did not purchase steam engines to run their threshing barns and mills, animal or water power might have remained the main power source.

Even with the explosion of technology in American farming after the Civil War, rice planters generally did not make use of the new postwar machines. Most of them were invented for grains other than rice. Machines invented for use with rice, or that could be altered for rice threshing or milling, were constructed for the Midwest rice industry, which began in 1886, and they were not suited for the rice fields of the Rice Kingdom. The rice fields of the Midwest were established on prairie lands, where the soil was firm enough to support the weight of the new machines, but the fields of the Rice Kingdom were established on swamp soil, which could not support these machines, and they quickly bogged down. With the exception of threshers, these postwar machines were generally too expensive for tidewater planters after the Civil War.

When the rice industry was fully developed in the antebellum period, its technological advances, the concomitant skills required to produce, operate, and repair the threshing and milling machines, and the knowledge and skills involved in flooding and draining the rice fields at the correct times all made the industry decidedly modern for its time. Yet the bulk of the work on a plantation was unimaginable drudgery done under extremely unhealthy and debilitation conditions. Ditching, preparing the soil, hoeing, sowing, cleaning the canals, repairing banks, and harvesting were tasks from Neolithic times. No machines replaced hand labor for these tasks. No other agricultural industry of the time was so advanced and yet so primitive.

There is a difference between the origin of rice culture and the origin of the rice industry. No matter who first began rice culture, the commercial rice industry that followed the initial success with growing rice was the creation of Europeans and European Americans. They invented and produced the machines that transformed rice culture from its preindustrial form into a worldwide industry. They financed and marketed the industry in the Atlantic world. At what date the transition from preindustrial to industrial occurred is impossible to fix because the different advances in the machinery of the industry did not occur at the same time. Harvesting was done with a sickle from the beginning until the end of rice culture; machine milling began in the mid-1700s, while mechanical threshers did not make their appearance until 1830. No one date marks the shift from African methods to European American ones, and machines that followed the early methods were based in part on African technology.

When the first colonists in Carolina transformed rice into a viable market commodity, the future of the Atlantic Coast rice industry was fixed. During the antebellum, Civil War, and post–Civil War periods, America's commercial rice industry stretched along the coast of the Carolinas, Georgia, and East Florida in a well-defined geographical region distinct from the rest of the American mainland. Although its labor-intensive cultivation resembled that used in other rice-growing areas of the world, no other slave-labor commercial crop in the British North American colonies underwent such a protracted period of experimentation and production as Carolina rice—or required so much capital.

Rice culture has been synonymous with the history of the tidewater areas of northeast Florida, Georgia, North Carolina, and South Carolina, and has shaped the course of American history. Rice produced the first generation of true wealth and created a social order different from the rest of the southeast. Rice culture forever changed the cultural, ecological, political, and economic aspects of the Rice Kingdom. Rice planters set the tone of South Carolina politics and society and put into motion forces that led to the Civil War, which ended slavery and impoverished the South for decades. The rice industry represented dramatic extremes of agricultural techniques, labor, and equipment. The knowledge and skills of experts were required to flood and drain rice fields and thresh and mill the rice; at the same time, much of the labor was simple drudgery performed under extremely harsh and unhealthy conditions. Planters helped to remove the original Native Americans from their ancestral lands and banished the original white yeomanry settlers to the upcountry, replacing their small farms with large plantations. The planter elite created a society in which they ruled socially, politically, and economically. A majority population of black slaves replaced the majority of white farmers, and thousands of acres of coastal swamps and marshes of the Atlantic states were transformed into rice fields.

The export of Carolina rice created an economic base in the tidewater area that produced the richest plantation economy in British Colonial America. Rice contributed to making Charles Towne (later Charleston) the richest city in British Colonial America in 1774. The nine persons with the greatest inventoried wealth in British America lived in Charles Towne; planters in Georgia and North Carolina also accumulated large fortunes. The port cities of Wilmington, North Carolina, George Town (later Georgetown), South Carolina, and Darien and Savannah in Georgia also prospered because of rice culture.

Rice was established as a crop by the late 1600s and became a major export crop by 1720. Because rice culture was labor intensive, it led to the import of many Africans, who became the basis for a slave-based labor system that lasted until the end of the Civil

War, when slave labor was replaced by that of freedmen working under contract to their former masters. The suitability of the tidewater area of the Atlantic coast for rice cultivation, a labor force of slaves from Africa familiar with its cultivation, the development of machines that replaced the flail and the mortar and pestle, and a growing international market demand facilitated the transition to a rice economy. Nor was any crop more profitable for capital investment than rice, justifying the long and costly period of experimentation that ultimately produced a commercial crop.

Although the final product of tidewater rice culture was uniquely American, it drew on diverse cultural experiences and technology, depended on the global integration of ecological, social, economic, and technological forces, and it did not begin or stop within the boundaries of the fields. Far from it. Light from the sun provided the energy for rice to photosynthesize; the plants used the water and nutrients that came down the river from the mountains and piedmont to produce bountiful crops. We know rice is an Asian crop cultivated in the Lowcountry using West African techniques, processed with machinery invented by Europeans as well as European Americans, and distributed, merchandised, shipped, and financed first by British practices that later developed a colonial and American aspect.

While slaves lived and labored primarily in the South, the slave trade involved northerners as well. Slave ships were based in ports along the East Coast, reaching across the ocean as part of the Middle Passage. Africans brought by slave ships worked under transplanted Europeans and grew rice as a market product that was shipped to Europe and other places. Money markets in the North provided funds to capitalize plantations in the South. As rice became an important export crop, port cities along the Atlantic Coast became part of the world trade market, expanding the Rice Kingdom's influence both at home and abroad. The integration of all these factors made possible the vast fields of golden grain that graced the former tidewater swamp lands.

South Carolina planters carried their plantation system to Georgia, northeast Florida, and North Carolina. Later, beginning in 1886, cultivation shifted to Louisiana, Texas. and Arkansas. In spatial terms, the boundaries of the South Atlantic rice-growing region was limited to the tidewater region from the Cape Fear River on the north (about north 34° 30') to the Saint Johns River on the south (about north 29° 30'). Tidewater rice plantations were unique among plantations to the Old South. Of all the major agricultural staples of the Old South, none was so restricted by geography as rice. Rice plantations were fewer, more isolated, more specialized, and, since their labor-management needs were greater, had larger slave populations than indigo, cotton, sugar, or tobacco plantations.

The tidewater Atlantic Coast and Sea Islands are a fascinating region distinct from the rest of the American mainland. Even though the landscape has now changed dramatically, much of it is still a wild and foreboding place. Those inclined to venture into the abandoned lands where rice was grown, will find a land of enchanted beauty where the past blends with the present. We hope the journey into the Lowcountry will be enhanced spiritually, culturally, and historically with the material presented in this book.

The history of rice culture left a deep imprint on the tidewater South Atlantic coast that will be with us forever. How to honor and immortalize the accomplishments and sacrifices of those engaged in market preparation of Carolina rice—one part of the history of rice culture—has been the passion that energized this project. We hope our book will add to Lowcountry history and inspire people to preserve its legacy as well as its artifacts.

Chapter 1 gives a brief history of African and Asian rices, the two domesticated species that feed much of the human race. Chapter 2 considers how Asian rices were introduced into Carolina. New heretofore unpublished material on the introduction of rice into the Rice Kingdom and the origin of Carolina Gold rice is presented.[2] We also include an important, previously unpublished map of Col. Hezekiah Maham's plantation in St. Stephen's Parish, Berkeley County, South Carolina, which may where the first crop of Carolina Gold rice may have been grown.

Chapter 3 outlines how rice was first successfully grown around 1690, and how it spread from its Cooper River epicenter in South Carolina throughout coastal Carolina and into the rest of the Rice Kingdom. The first method of rice culture, called "upland" by previous works, is more properly called "providence culture" because it depended on the chance of rainwater. The second method of culture is best termed "reservoir culture" because field studies have documented that lowland planters had other means of obtaining water for fields besides inland swamps. The chapter also discusses tidal cultivation, which replaced the reservoir-fed system, explaining how fields were formed from swamps and what methods were used to flow water onto the fields, as well as discussing the obstacles planters had to overcome to get their rice to market. Original diagrams and descriptions of the trunks that controlled the flow of water into and off the fields are included.

Chapters 4, 5, 6, and 7 are original, in-depth examinations the implements and machines used in preparing rice for market. Using field artifacts never before examined along with extensive, previously unconsulted archival materials, these chapters document, using original diagrams, the implements and machines used for threshing and milling from the beginning of the industry until its end in the early 1900s. These chapters also recount slaves' and freed people's contribution to introducing, operating, and maintaining these machines. One hundred twelve figures and plates, including the material from private and public archives and the authors' diagrams and photographs, are previously unpublished. The four sources of power—manual, animal, water, and steam—for threshing and milling Carolina rice are discussed in detail. And the machines they ran are depicted in a series of original diagrams that bring to life how these marvelous inventions prepared rice for market. For the first time, construction and operation of the steam-powered mills that cleaned rice is detailed. Chapter seven documents how the city mills, which gradually replaced the plantation mills after the Civil War, processed rough rice sent from the plantations.

Chapter 8 examines the factors that made the period 1800–60 the Golden Age of rice, the period of the rice industry's greatest production and mechanization and the codification of the task system. Chapter 9 describes the planters struggle to continue rice planting after the Civil War and why they ultimately failed.

Though the dominance of the rice industry has passed, Carolina rice culture and the former Rice Kingdom will always be part of the future.

I

A Brief History of Rice

Rice has been cultivated for so many countless ages that its origin will always be a matter of conjecture. Rice culture dates to the earliest days of humankind and eons before the time for which historical evidence documents that rice was probably the staple food and the first cultivated crop of Asia. Rice belongs to the genus *Oryza*. Of the more than twenty species of *Oryza*, only two became domesticated: Asian rice (*Oryza sativa* L.) and African rice (*O. glaberrima* Steudel). African rice was the original rice of West Africa and was cultivated in a broad variety of environments long before the first European explorers reached Africa. *O. glaberrima* closely resembles *O. sativa;* only minor differences of glume pubescence and ligule size separate the two species. Natural hybrids between the two are highly sterile.

There appears to be general agreement among scientists that *O. sativa* and *O. glaberrima* represent endpoints of independent and parallel processes of domestication. In 1976 Te-Tzu Chang, using evidence from biosystematics and paleogeology, concluded that the genus *Oryza* originated on Gondwanaland[1] and that *glaberrima* and *sativa* arose from a common ancestor that existed in the humid zone of Gondwanaland before it broke up and drifted into separate land masses about 167 million years ago.[2] Later G. Second, using modern isozyme analysis[3] together with evidence from paleogeography and anthropology, supported Chang's contention that *Oryza* had a single ancestor and origin.[4] The wild ancestors of the two species became separated during the breakup of Gondwanaland; the *sativa* ancestor separated into the Asian landmass and the *glaberrima* ancestor separated into West Africa. The ancestral plants then underwent independent and parallel evolution, ultimately producing the two modern domesticated species.

The key to the domestication of rice plants was the evolution of nonshattering inflorescences (ripe seeds are retained on the plant) versus shattering inflorescences (ripe seeds are discharged from the plant). Like wild grains in general, wild rices have shattering inflorescences. Occasionally mutations for nonshattering occurred, but they were eliminated by natural selection in the wild because nonshattering reduced the dispersal of seeds. However, ancient harvesters would have been more likely to select plants with nonshattering inflorescences because seeds remaining on the plants make harvesting easier. Once primitive tillage was practiced, a second characteristic was exposed to selection: nondormant seeds. Seeds of wild plants generally have a period of dormancy (often years) that allows them to pass through unfavorable conditions for germination and to germinate when environmental conditions are favorable. Since primitive tillage often provided favorable conditions for germination, harvesters would have selected nondormant mutants because they would be most likely to contribute to the next year's crop. These two key characteristics were selected during the domestication of both Asian and African rices and are characteristic of today's modern cultivars.

Some scientists believe that the upland culture of rice predated lowland culture. In upland culture rice grows without flooded conditions and receives only periodic rainfall while in lowland culture rice grows in more or less flooded conditions. Chang, however, suggests that the reverse occurred. Physiologically and anatomically rice is a semi-aquatic plant. According to Chang, "Observations of the very large collections at the International Rice Research Institute (IRRI) in the Philippines suggests that the greatest diversity of plant characters and the more primitive cultivars are found among varieties adapted to lowland culture."[5] Upland rices, on the other hand, often possess one or more advanced features such as "glabrous leaves and glumes, heavy grains, long and nonshattering panicles and thick roots." The culmination of these observations led Chang to accept the lowland origin of both African and Asian rices.

African Rice

David Catling has postulated the possible evolutionary pathway of African rice from its common ancestor with Asian rice.[6] The perennial *O. longistaminata* Chev. & Roer arose from this unknown common ancestor. From *O. longistaminata,* through various intermediate types, arose the annual species *O. barthii* Chev. This species occurs wild in sub-Saharan Africa from Senegal to Tanzania and is widespread in Zanzibar. Catling contends that selection from *O. barthii* by early African farmers led to the development of a great range of *O. glaberrima* cultivars about 1500 B.C.E., including floating types, swamp and upland types, weakly and strongly photoperiod-sensitive types, and short- and long-duration types.

Building on the work of these and other scientists, geographer Judy Carney has outlined additional evidence for the independent domestication of African rice in West Africa about two thousand years ago.[7] As Carney points out, some early scholars believed that African rice was of Asian origin and entered Africa with the expansion of Islam between the eighth and fourteenth centuries. Carney, however, cites a translation of Muslim documents that established the cultivation of rice prior to the arrival of Portuguese and the existence of a production system older than the tenth century period of Islamic expansion into the region. She argues that, if Muslims had introduced rice from Asia, a geological link to the Middle East or East Africa should be evident. No such evidence has been demonstrated. African rice, says Carney, was domesticated in sub-Saharan Africa in the "freshwater wetlands of the inland delta of the middle Niger River in Mali, an area where rice is grown almost within reach of the Sahara Desert."

From the primary center of domestication around two thousand years ago, two secondary centers developed. One emerged on wetlands north and south of the Gambia River; the other developed in the Guinean highlands between Sierra Leone, Guinea Conakry, and Liberia. From these two centers of domestication, indigenous African rice spread over a broad region of West Africa from Senegal southward to Liberia and inland for more than one thousand miles to the shores of Lake Chad.

Following the domestication of African rice, a succession of West African empires emerged.[8] By the first millennium B.C.E., many Mande-speaking peoples lived between the watersheds of the Senegal River and middle Niger River. From 300 B.C.E. to 300 C.E., Mande speakers turned the region into a cradle of agriculture based on *glaberrima* rice. From the eighth century C.E., the Ghana empire arose in the Soninké area between the mid-Senegal and the mid-Niger Rivers. The Ghana empire was then eclipsed by expansion of the Mande-speaking people from their linguistic heartland, resulting in a considerable expansion of *glaberrima* cultivation over a wide area of West Africa.

Berber traders spread Islam in sub-Saharan Africa as their caravans carried articles of trade across the Sahara and returned north with spices, dye, and other goods. Indigenous elites who came in contact with the traders responded by converting to Islam. The Jolof empire, founded in the thirteenth century by a Muslim dynasty, extended over what is now mostly Senegal, a prominent rice-growing area today. The state of Mali rose in the thirteenth century. Its rulers claimed a Muslim identity and descent from the muezzin. The Mali empire gradually gave way to the Muslim Songhay empire two to three centuries later. The Songhay established an empire in the inland delta centered on Timbuktu, an area of Islamic learning and brought in Arabic-speaking Muslims to help administrate the empire.

These Muslim and non-Muslim empires essentially incorporated the cultivation of *glaberrima* rice, converting more and more land into rice producing areas. *Glaberrima* rice remained the dominant rice of the region until 1950.

Asian Rice

Asian rice has been cultivated in southeastern Asia since ancient times. While the precise date or place of its domestication is uncertain, scholars have never doubted that Asia was its center of domestication. Many scientists contend that the Indian subcontinent is the ancestral home of Asian rice; yet archeological evidence goes back only to 2500 B.C.E. in that subcontinent, making India an unlikely site of domestication. As with African rice, the path to domesticated Asian rice was from a wild perennial, in this case *O. rufipogon* Griffiths via intermediates to the annual *O. nivara* Sharma & Shastry, from which Asian rice evolved.

The place of major genetic diversity in southeastern Asia is roughly an east-west belt along the Himalayas and adjoining Asian mainland from Assam and Bangladesh through Burma, Thailand, Laos, Yunnan in southern China, and northern Vietnam.[9] Asian rice was likely domesticated independently yet concurrently in multiple sites over this broad belt some seven thousand years ago. After more than two thousand years of cultivation in temperate China, Asian rice was introduced into Japan about 300 B.C.E.

Three ecogeographic races of Asian rice are generally recognized: *Japonica, Indica,* and *Javanica*.[10] The three races resulted because, during the long period in which Asian rice has been cultivated, adaptation has taken place in a diversity of climates, latitudes, soils, and water regimes. From the broad site of domestication, Asian rice was introduced into the Huang (Yellow) River valley of China where the temperate race *Japonica* evolved. The tropical race *Indica* spread southward into Sri Lanka and the Malay Archipelago and northward into central and south China. The *Indica* rices of China were initially grown in the middle Yangtze River valley basin probably before 200 C.E. From ancient India ancient farmers carried the *Indica* type westward to the Middle East, Europe, and Africa. The *Javanica* race of Indonesia appears to be a more recent product of selection from the *Indica* race. From Indonesia it spread to the Philippines, Taiwan, and Japan.

The spread of *Indica* west from its southeastern Asian center of domestication is poorly known, although Arab trade routes are considered the likely path. Andrew Watson cites sources stating that Asian rice was grown in the second century C.E. in Babylonia, Bactria, lower Syria, and Susia; in parts of Mesopotamia and Persia during the second century B.C.E.; and in Palestine at some time between the third and eight centuries C.E. Rice was known to the Greeks and Romans in classical times, but as a trade commodity from the Near East rather than as a crop. Knowledge of rice gradually spread into the classical Mediterranean after the Alexander expedition to the east, and into pre-Islamic Arabia and East Africa; however, during this period, this rice was imported into these regions for medicine and not grown there.[11]

After 700 C.E. irrigated Asian rice was grown in early Islamic agriculture wherever there was water for irrigation: in Mesopotamia, in Egyptian oases and the Nile valley, in Anatolia, and in many parts of Mediterranean Europe. It was carried to Spain with the Arab conquest in 711 C.E.

Watson argues that Asian rice first entered Africa from the eastern part of the continent, and cites the following evidence: "There is a prima facie case for Arab transmission on an important scale: in the twelfth century al-Zuhri stated that in a part of Abyssinia the people lived on rice grown along the Nile; in the thirteenth century Marco Polo said that rice was the staple of the people in East Africa; and in the fourteenth century Ibn Battuta described a rice dish that was the main food in Mogadisch."[12]

Asian rice spread easily into sub-Saharan West Africa in Islamic times by 1500 C.E. because of the irrigation and milling methods already established for African rice. Although it never became as important as wheat and sorghum, it was important in areas such as desert oases, river valleys that were flooded naturally or artificially, or swamp lands.[13]

What role the Arabs played in bringing Asian rice into sub-Saharan West Africa is not fully understood. Watson mentions three possible routes of Asian rice into West Africa: through East Africa from the tenth century onward, overland from the east coast, and southward across the Sahara.[14]

Another possible route for the introduction of Asian rice into West Africa is via Portuguese trading routes. Asian rice was carried by East Indian voyagers who crossed the Indian Ocean about two thousand years ago and colonized Madagascar. Their descendants became the Malagasy, racially and ethnically related to Sumatra rather than Africa. In Madagascar, they grew Asian rice in irrigated rice paddies. Portuguese traders could have carried this rice around the Horn of Africa to West Africa where they ported before the voyage across the Atlantic. From these ports Asian rice could have been introduced into West Africa.

The Origins and Introductions of Rice Seeds

Considerable debate exists among historians about the origins of rice culture in Carolina, and its early introduction is clouded by misconceptions and contradictions. When was rice first introduced into Carolina; was Madagascar the source of the first seeds; how many varieties were introduced; was rice introduced by chance or purposely; what was the origin of the variety Carolina Gold rice; and who first introduced rice to Carolina? These questions are unanswered today even though historians and scientists have researched these questions for years. Because of conflicting evidence and the lack of primary documents, some of these questions may never be answered with confidence.

The main problem historians have encountered documenting the introduction of rice seeds and the beginning of rice culture is the lack of primary records from those who planted rice in the late 1600s and early 1700s. For this early period, few journals and newspapers exist, and at best only fugitive items of internal communication are found. Only a few plantation records exist for the early period of rice planting and experimentation. Slaves of course left no written records. Many of the sources used in this book, although considered primary, are not from colonists and slaves who actually planted rice. Researchers are left with only the barest physical evidence from the field and narratives by noneyewitnesses who wrote many years later.

Two theories on the origin of rice culture in Carolina[1] prevailed before S. Max Edelson argued for a third view in his book *Plantation Enterprise in Colonial South Carolina* (2006). First, planters and their descendants, wishing to embellish their family escutcheons, celebrated rice as a white achievement and wrote Africans out of the story despite their vast numbers and their familiarity with rice and farming in subtropical lowland environments. Field slaves, the planters determined, were no more than "hands" and capable of no innovation. One only has to read Duncan Clinch Heyward's *Seed from*

Madagascar (1937) or Doar's *Rice and Rice Planting in the Carolina Low Country* (1936) to understand how they wrote slaves out of rice culture's origin. South Carolina planters left behind no hint that Africans added any innovation to rice culture except menial labor.

On the other hand, casting Africans as the originators of rice agriculture, Peter H. Wood (*Black Majority,* 1974), Daniel C. Littlefield (*Rice and Slaves,* 1981), and Judy Carney (*Black Rice,* 2001) have argued that African skills were critical for setting the course of the plantation rice economy. Their revisionist view points out similarities between African rice culture and the rice culture that developed in the Lowcountry. These authors have argued that early planters possessed no knowledge of rice planting or processing rice for food while Africans knew and grew rice; therefore slaves must have been the agents behind the beginnings of rice culture. (Carney's theory of rice culture's introduction is referred to by historians as the "black rice hypothesis.") "Each of these interpretations of origins," Edelson argues, "credits rice growing to a *single* [emphasis added] cultural group."[2]

Edelson notes that "how the critical materials of this plantation economy were first assembled remains this society's most compelling origin story." He argues that the origin of rice culture was slaves' use of "rice as food, [and they] valued it as a crop that linked individuals to one another in new-world families, and saw it wrenched out of this context of domestic production and consumption by an act of white appropriation." Food was the main challenge for both planters and slaves in the early settlements, "taking precedence over generating exports for Europe." From an unknown seed source and date, in 1689 and 1690 colonists "watched African slaves plant an edible species of grass [rice] that distinguished itself from the wild forage in late summer." Drawing on the slaves' success with subsequent crops of rice, European Americans planted the first crops experimentally with the intention of producing it as a commodity. By 1699 "hundreds of slaves working on scores of plantations produced the first significant crop for export." Furthermore, "within each of these moments, one can find a threshold crossed by settlers and slaves in their perceptions of rice as something to eat and something to sell. . . .Where these very different desires for the crop intersected," Edelson argues, "Carolina rice culture grew and spread."[3] While this study does not champion either version of Carolina rice culture's origins, original material included here may contribute additional material for new investigations of the topic.

Regardless of the origins of rice culture, two things are certain: rice seeds were acquired, and slaves and colonists learned how to cultivate the crop in the Lowcountry environs. While some historians argue that seeds that began rice culture were acquired by chance (some sources say accidentally), other say their acquisition was purposeful. Enough records are known today to suggest that it was not exclusively one venue or the other. Some seeds were introduced purposefully while some were introduced by chance. Both sources contributed to the origins of the industry and its expansion during the colonial and antebellum periods.

The Lords Proprietors' Search for a Commercial Crop, 1663–1677

The Province of Carolina was granted by Charles II of England, to eight of his supporters, the Lords Proprietors. At the heart of colonization was a fundamental preoccupation with land: The Crown could reward proprietors with land, and they in turn could grant land to the settlers. The proprietors had knowledge of agricultural developments in England and the West Indies, and their stated goal was to make a profit from their investment in Carolina. As early as 1663, the proprietors had laid plans for developing rice and other crops in Carolina. They were aware that rice was not climatically suited to northern Europe—as were wheat, oats, barley, and rye—and that there was a developing market in northern Europe for imported rice during poor harvest years in countries where wheat, oats, barley, and rye were grown and imported. In addition rice was an excellent food for provisioning ships on long voyages and subsistence food for the settlers.

One of the Lords Proprietors, the Duke of Albemarle, wrote to Lord Willoughby, governor of Barbados, on August 31, 1663, detailing his plans for the development of Carolina and mentioning that "the commodyties I meane are wine, oyle, reasons, currents, rice and silke."[4] In 1666 Robert Horne published a pamphlet in London advertising the natural resources of Carolina and stating: "The Meadows are very proper for Rice, Rape-seed, Lin-seed, etc., and may many of them be made to overflow at pleasure with a small charge."[5] Four years later, in April 1670, the *Carolina* landed at Albemarle Point on a bluff overlooking Old Town Creek, a tributary of the Ashley River, and established the first permanent settlement in Carolina. These first colonist numbered about 130, most of them English men and women, but with a few elite planters from Barbados.

On April 23, 1672, Governor Joseph West heralded the landing in Charles Towne of the *William & Ralph* from London. According to its bill of lading the cargo included "one barrel of rice."[6] This bill of lading, dated January 13, 1672, in London, is the earliest record of rice shipped to Carolina. Whether it was shipped for food or seed is not recorded, but knowing the proprietors' desire to initiate crop production, the seeds were likely intended for planting.

Two growing seasons after the founding of the colony, rice was supposedly shipped from St. George (the Spanish name for Charles Towne) to Barbados. In a notice dated May 8, 1674, the acting governor of St. Augustine, Florida, wrote to the queen of Spain that four Englishman had left St. George by boat because they were mistreated and arrived at the garrison in St. Augustine in April 1674. In order to learn their nationalities and their purpose for coming to St. Augustine, an Englishman was asked to take their testimonies. When asked about vessels in the port of St. George, one man, Charles Miller, replied: "The English use these vessels to take tobacco, barrel staves and some rice which is grown on the soil to Barbados."[7]

Since Charles Towne was settled in April 1670, it is highly unlikely, if not impossible, for the settlers to have started a rice crop there in 1670 (assuming they even had seeds, of which there is no record). Early settlers in America were more hunters and gathers for the first few years, and the settlers who came on the *William & Ralph* were probably no different. The earliest they could have planted rice would have been spring 1671. Since Miller landed in St. Augustine in April 1674, he could not have referred to a crop harvested in 1674 because rice was harvested in August and September. This gives a window of two growing seasons, 1671–72 and 1672–73, for production of the rice supposedly sent to Barbados. Although it is possible, it is unlikely. Rice would have been a valuable food source in the early colony and unlikely to have been exported. Miller likely mistook the rice on board the *William & Ralph* in the harbor as being *shipped* from Barbados instead of being sent there.

The proprietors wrote to the governor and council of Carolina on April 10, 1677: "We are Layinge out in severall places of ye-world for plants & seeds proper for yor-Country and for sons that are Skill'd in plantinge & produceinge vines, Mulberry trees Rice oyles & winnes and such other Comodities that enrich those other Countryes that enjoy not soe good a Climate as you."[8]

While these sources indicate the proprietors' desire to try rice, as well as other crops, but, with the possible exception of the 1672 shipment, they are not proof that the proprietors sent seeds at these early dates.

Seed from Madagascar and the Dubois Seed, circa 1685

Two introductions of rice seeds are inextricably linked with the history of Carolina rice. In 1731 Capt. Fayrer Hall published in London the first account on the beginning of rice culture in Carolina. In his article, "The Importance of the British Plantations in America to This Kingdom; with The State of their Trade and Methods for Improving It; as Also a Description of the Several Colonies There," Hall wrote:

> The production of Rice in *Carolina,* which is of such prodigious Advantage, was owing to the following Accident. A Brigantine from the Island *Madagascar* happened to put in there; they had a little Seed Rice left, not exceeding a Peck, or Quarter of a Bushel, which the Captain offered and gave to a Gentleman of the Name of *Woodward.* From Part of this he had a very good Crop, but was ignorant for some Years how to clean it. It was soon dispersed over the Province; and by frequent Experiments and Observations they found out Ways of producing and manufacturing it to so great Perfection that it is thought it exceeds any other in Value. The Writer of this hath seen the said Captain in Carolina, where he received a handsome Gratuity from the Gentlemen of that Countrey, in Acknowledgment of the Service he had done that Providence.[9]

Dr. Henry Woodward (1646?–1685) was a ship's surgeon who arrived at the Caribbean island of Barbados in 1665 at the age of nineteen. He sailed with Robert Sandford to Port Royal to explore the coast of Carolina in 1666 and remained behind after Sandford left.

After living with the Indians for about a year, he was captured by Spaniards and taken to St. Augustine. He was rescued from St. Augustine, but was subsequently shipwrecked. He finally arrived at Charles Towne in 1670 as a proprietor's deputy. His year living with Indians afforded him the opportunity to understand the importance of Indian trade, as well as their serving as an effective buffer between Carolina and the Spaniards at St. Augustine. Lord Ashley Cooper, one of the Lords Proprietors, ordered Woodward to explore the interior and negotiate alliances and trade agreements with the Indians. For his efforts on behalf of the colony, he was granted a title to two thousand acres of his choosing. He selected a tract along Abbapoola Creek on today's Johns Island. The creek is still present, but nothing remains of his plantation house or outbuildings. According to Salley, Woodward died in 1685.

If Hall's account is true, and there is no compelling evidence it is not, and Woodward was given rice seed, he must have planted it prior to his death in 1685, a date often used as the beginning of rice cultivation. It appears, then, that some variety of Asian given to Henry Woodward in 1685 could possibly have been the genesis of early rice planting in Carolina.[10]

Another introduction of rice that has been widely and often mentioned is the Dubois seed. An article in the June 1766 issue of the *Gentleman's Magazine* of London, "An Account of the Introduction or Rice and Tar into our Colonies," is often cited on the origins of rice seeds. The article contains the often-quoted source of rice from India:

> In the year 1696, my sagacious friend, *Charles Dubois,* then treasurer of the East India Company, told me often with pleasure, that he first put the *Carolinians* on the culture of rice.
>
> He happened one day, in that year to meet *Thomas Marsh,* a Carolina merchant, at the coffeehouse, to whom he said, I have been thinking, from the situation, nature of the soil, and climate, that rice may be produced to great advantage in *Carolina:* But, says *Marsh,* how shall we get some to try? Why, says Dubois, I will enquire for it amongst our *India* Captains—
>
> Accordingly, a money bag full of East India rice was given to *Marsh,* and he sent it to *South Carolina:* and in the year 1698, he told his friend *Dubois,* that it had succeeded very well.[11]

It may well be that some seeds were introduced into the colonies as Dubois related. Nothing in the article, however, identifies the variety of the rice from East India. The Dubois seeds, like the Madagascar seeds, were probably used by some planters along with other seeds that found their way to the Lowcountry.

Establishment of Rice as a Crop, late 1600s

Other sources document that rice was established as a crop by the late 1600s. John Stewart wrote to William Dunlop on April 27, 1690, from Wadboo in Berkeley County: "Our ryce is better esteem'd of in Jamacia than that from Europe sold ther for a ryall a pound its pryce here new husk't is 17/ [shillings] a hundred weight."[12] He wrote again to Dunlop on June 23, 1690, saying that "he was going to plant 9 Acres of Swamp wth 4 bushel of Rice and had done it and never had a grain wtout more drudgery and charge than it wes all worth."[13]

In 1690 several leading men of the province presented Governor Seth Sothell with a petition that the people be allowed to "pay their rents in the most valuable and merchantable produce of theire Lands . . . because [these men] are encouraged wth severall new rich Comodityes as Silck Cotton, Rice & Indigo, wch. are naturally produced here."[14] In reference to an early rice-milling engine, on September 26, 1691, the General Assembly of South Carolina ratified an act giving Peter Jacob Guerard the right to bring legal action against anyone who reproduced his engine without his consent, indicating that at this early date planters were seeking more efficient means to mill rice.[15] The 1692 will of Barnard Schencking, a prominent Barbadian planter whose home was situated along Goose Creek, is the first inventory of rice on record.[16]

Rice as an Export Crop of Carolina, 1695

Once rice was established as a crop, it quickly became an export. The first recorded shipment of barrels of rice from Carolina was included in a bill of lading to Jamaica by George Logan in 1695.[17] On March 16, 1696, the General Assembly ratified an act providing that quitrents "shall be paid to the Lords Proprietors in currant money of this part of the province, or in Indigo, Cotton, Silke, Rice, Beef or Porke." On November 19, 1698, the Commons House of Assembly requested the Lords Proprietors to "interceed with his most Gratious Majestie for ye takeing of ye Duty of Rice Turpentine Rossin Pitch and Tarr, Impor[ted] from this Province" and to "Procure and Send us by ye; first oppertunity of a moddell of a Rice Mill." In 1698 the first year records were kept, 10,407 pounds of rice were exported. Table 1 shows the exports of rice from the colonies for the years 1698–1774:

TABLE 1. **Rice Exported from Producing Areas, 1698–1774**

Year	Pounds	Year	Pounds	Year	Pounds
1698	10,407	1707	561,185	1716	4,584,927
1699	131,207	1708	675,327	1717	2,881,335
1700	394,130	1709	1,510,679	1718	2,956,727
1701	194,618	1710	1,600,983	1719	4,001,210
1702	612,646	1711	1,181,430	1720	6,485,662
1703	694,493	1713	3,850,533	1721	7,963,615
1704	759,536	1714	3,139,361	1722	9,732,377
1706	267,309	1715	2,367,605	1723	8,797,304

TABLE 1. Rice Exported from Producing Areas, 1698–1774 (*continued*)

Year	Pounds	Year	Pounds	Year	Pounds
1724	8,654,447	1742	22,706,060	1759	30,472,575
1725	7,093,600	1743	35,935,200	1760	35,327,250
1726	9,442,710	1744	39,963,630	1761	58,480,275
1728	12,884,950	1745	29,813,375	1762	47,435,325
1729	14,248,960	1746	27,335,040	1763	61,959,450
1730	18,774,900	1747	27,643,060	1764	55,907,250
1731	21,753,450	1748	28,368,550	1765	65,710,575
1732	16,886,000	1749	21,381,030	1766	48,396,000
1733	23,245,200	1750	27,372,500	1767	63,465,150
1734	13,991,850	1751	32,751,270	1768	77,284,200
1735	21,259,800	1752	42,245,850	1769	73,078,950
1736	24,804,000	1753	19,747,675	1770	83,708,625
1737	20,201,400	1754	49,179,520	1771	81,755,100
1738	16,327,350	1755	59,057,775	1772	69,218,625
1739	32,167,800	1756	45,344,250	1773	81,476,325
1740	43,326,000	1757	33,976,950	1774	76,265,700
1741	38,720,955	1758	38,527,650		

Source: Table in Henry C. Dethloff, *A History of the American Rice Industry: 1685–1985* (College Station: Texas A&M Press 1988), 10; based on statistics from U.S. Bureau of the Census, *Historical Statistics of the United States: Colonial Times to 1970* (Washington: GPO, 1975), pt. 2, ser. 481–82: 1192.

After 1698 references to rice shipped from Carolina were common. William Thornburgh, a second generation proprietor, wrote to William Popple, secretary of His Majesty's Commissioners for Trade and Plantations, on July 21, 1699: "I have herewith sent you a sample of our Carolina rice that the Rt. Hon. the Lords Commissrs. of Trade & Plantations may see what a staple the Province of Carolina may be capable of furnishing Europe withall. The Growers do assure me its better than any Foreign Rice by at least 8s." On March 8, 1700, the governor and council of Carolina wrote to the lords commissioners that Carolina "hath made fore rice yeLast Crop that wer have Ships to Transport." On July 30, 1700, Edward Randolph wrote to the same Lords Commissioners, "They have now found out the true way of raising and husking Rice there has been above 300 Tuns [tons] shipped this year to England besides about 30 Tuns more to the Islands."[18]

In 1701 Lady Rebecca Axtell and her family were growing rice for export at Newington Plantation on Dorchester Creek in the vicinity of present-day Summerville in Dorchester County, South Carolina. Entries in Daniel Axtell's account book for Newington show 3,261 pounds of rice sold in 1701 and 6,000 pounds in 1702. Subsequent entries for 1703–7 show similar sales of rice.[19]

Available evidence, though far from conclusive, suggests that the early rice grown in the colony was some type of white rice. John Lawson was the first to mention white rice. On December 28, 1700, Lawson and five other Englishmen, three Indian men, and an Indian guide's wife left Charles Towne in a large canoe bound for the mouth of the Santee River. Lawson's party arrived at present-day Washington, North Carolina, where the Tar River enters Pamlico Sound on February 23, 1701. The trip took them along the Santee River and from there along the Wateree River, leaving South Carolina in the vicinity of Lancaster. He recorded in his journal: "There are several sorts of Rice, some bearded, others not, besides the red and white; But the white Rice is the best."[20] Lawson's note is important for two reasons. First, it supports previous statements that more than one variety of rice was introduced early. Second, it is the first mention of white rice, which Robert F. W. Allston[21] mentioned later. In 1846 Allston wrote of the common white rice and said that it was "in shape and general description like No. 1 [the Gold Seed], except that the color of the husk is white or cream colored, and the grain when hulled is not as broad, nor as pearly; awn short. It is this variety that constituted the earlier crops of the country."[22]

Whether the white rice was the Dubois seed, another seed of unknown origin, or the seed from Madagascar is for future genetic research to determine. It is difficult, however, to discount Lawson and Allston's statement that some type of white rice constituted the early crop of the colony. The record allows nothing more.

By 1710 rice was not only a major export but was highly esteemed for its quality. In that year William Salmon published his *Botanologia: The English Herbal*," in which he wrote in glowing terms about the rice grown in Carolina: "[Rice] is now Sown in Carolina, and become one of the great products of the Country: I have seen it grow, and flourish there, with a vast increase, it being absolutely the best Rice which grows upon the whole Earth, as being the weightiest, largest, cleanest, and whitest, which has not been yet seen in the Habitable World."[23] John Norris, from the west of England, settled in St. Bartholomew's Parish, South Carolina, about twenty miles from Yemassee. Using the pen name James Freeman, a Carolina Planter, he wrote in "Profitable Advice for Rich and Poor" (1712), "Our chiefest Commodities sent here to England is our most excellent Rice, of which comes great Quantities."[24]

On February 16, 1715, the Commons House of Assembly voted a gratuity to John Thurber in answer to his petition seeking a reward for introducing the first Madagascar rice seed to the province. On July 19 of that year the Council of Trade wrote to James Stanhope, secretary of state for the Southern Department that "the Produce of this Colony, are, Naval Stores, vizt—Pitch & Tar in Good Abundance and some Masts, Rice of the best kind." And later, on August 26, of the same year, the Commons House of Assembly voted to reward Capt. Samuel Meade with "six Tuns of Rice" for his services rendered in the Yemassee War.[25]

A January 16, 1726, invoice from Charleston merchant Richard Splatt states: "Invoice of thirty Seven Barrells of Rice and Two Chest of Deare Skin Shipt by me Richard Splatt on Board the Lovely Polly[,] Michael Bath[,] Master[,] bound for London on my proper Account and . . . goes Consigned to Mr. William Crisp." The next day Splatt wrote on the back of the invoice: "There are a great many Vessels that there is not enough rice to load the ½ of 'em" since the demand for rice far exceeded the supply, and they will have to be "dispacht with pitch and Tarr."[26]

The archives are replete with documents dated after 1726 that attest to the status of Carolina rice as a major export crop. Production and exports increased steadily after 1700 to 76,265,700 pounds in 1774, although there were variations in production totals from year to year. Wars, hurricanes, economic downturns, and new threshing and milling machines, all contributed to a year-to-year variation in production, but over the period from 1698 to 1774, there was a steady increase in production and export.

Carolina Gold Rice, circa 1785

The name "Carolina Gold rice" comes from the golden hue of the ripe hull. Carolina Gold rice was celebrated worldwide for its quality and became the most important export crop of the Lowcountry and Rice Kingdom. Historically and traditionally, credit for the introduction of the famed Carolina Gold *sativa* rice has been given to Dr. Henry Woodward. In his *Seed from Madagascar,* Duncan Clinch Heyward (1864–1943) paraphrased the 1731 Hall article and substituted *Carolina Gold* for the generic *white rice:*

> Carolina Gold rice . . . was grown from seed brought to the province of Carolina about the year 1685. This rice had been raised in Madagascar, and a brigantine sailing from that distant island happened, in distress, to put into the port of Charles Town. While his vessel was being repaired, its captain, John Thurber, made the acquaintance of some of the leading citizens of that town. Among them was Dr. Henry Woodward, probably its best known citizen. . . . To Woodward Captain Thurber gave a small quantity of rice—less, we are told, than a bushel—which happened to be on his ship. "The gentleman of the name Woodward," to quote the earliest account of this occurrence, "himself planted some of it, and gave some of it to a few of his friends."[27]

If one reads further in *Seed from Madagascar,* it is apparent why the substitution was made. Heyward, who claimed that he was a descendant of Dr. Woodward, spent "the best years of life seeking to revive an industry in the pursuit of which four generations of [the Heyward] family had been successful." As we shall document later, the best available evidence suggests that Carolina Gold was not introduced until just after the Revolution, so Dr. Woodward could not have introduced it. One might suspect that Heyward was trying to embellish his family's status by claiming a connection to Carolina Gold.

For one thing, recent DNA analysis calls into question whether the seed from Madagascar was the famed Carolina Gold. Anna M. McClung and Robert Fjellstrom have used molecular marker technology to fingerprint historical and contemporary sources of Carolina Gold rice as well as its possible progenitors and derivatives.[28] Carolina Gold possesses a unique allele of the Waxy gene that can be traced through the last century of breeding efforts in the United States and is found in current commercial cultivars. McClung and Fjellstrom evaluated a small number of accessions from Madagascar and Indonesia and found no Waxy alleles in these samples. Since the samples were small, the evidence is not conclusive. Their study, however, produced the first scientific evidence suggesting that there is no link between Madagascar and Carolina Gold.

If the seed from Madagascar and the other earlier introductions were not Carolina Gold seed, when and where was the Gold seed first grown, and who introduced it? Two names appear in historic documents: Hezekiah Maham and Henry Laurens.

Robert Allston, an authority on rice culture, gave credit to Col. Hezekiah Maham for introducing Carolina Gold: "The Gold Seed Rice, the ordinary crop of rice most highly esteemed, and therefore universally cultivated, an oblong grain 3–8ths of an inch in length, slightly flattened on two sides of a deep yellow or golden color, awn short; when the husk and inner coat are removed, the grain presents a beautiful pearly-white appearance—an ellipsoid in figure, and somewhat translucent. This rice has been introduced into the Winyaw [Winyah] and Waccamaw region, since the revolution. It was planted by Col. Mayham [Maham], on Santee, in 1785."[29]

Allston's attribution is supported in part by archival and physical evidence that Maham owned two rice plantation though there is no record of his actually having planted rice on them. It does seem reasonable, however, to think that, if he owned rice plantations, he was engaged in rice planting.

Maham (1739–1789) worked as a plantation overseer in St. John's Parish, Berkeley County. After he obtained a grant in 1771 for five hundred acres in adjoining St. Stephen's Parish along the Santee River swamp, he became a planter. In his "Historical and Social Sketch of Craven County, South Carolina," Frederick Adolphus Porcher noted that Maham was the "most eminent military character" produced in St. Stephen's Parish during the American Revolution. Col. Maham served under General Francis Marion and retired from military life after becoming ill in 1781. He returned to his plantation along the Santee swamp near Pineville, but he struggled financially, and died there in 1789.[30]

An advertisement published in the February 21, 1793, issue of the *Charleston City Gazette,* advertising for private sale two settled plantations in St. Stephen's Parish owned by "col. Hezekiah Maham, deceased." "One plantation contains about 500 acres of high land, with about 40 acres of very rich rice swamp."[31]

Maham's two plantations had evidently been merged into one, which appears on "Map of Richmond, the Farm, Hampstead, Johns Run, Tower Hill, and part of Bluford Plantation, owned by the Est. of Robert Marion, Situate in St. Stephen['s] Parish, Berkeley County, S.C., 1928 & 29" (see plat 1). The map shows a swamp labeled "Maham Swamp," which drains into the Santee River. Lower and upper earthen banks enclosed this swamp. The upper bank held back water in a reservoir to flood the rice field. Just to the west is a swamp labeled "Old Rice Field." The lower bank apparently washed away

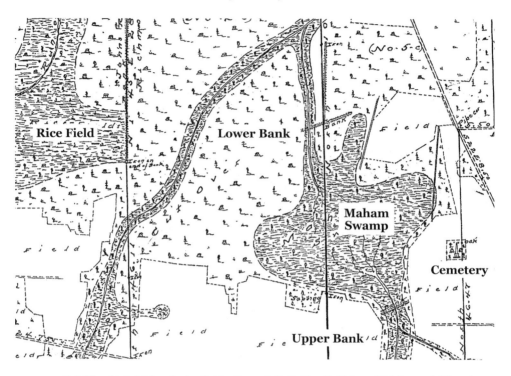

PLAT 1. Col. Hezekiah Maham's rice fields. From J. P. Gaillard's "Map of Richmond, The Farm, Hampstead, Johns Run, Tower Hill, and part of Bluford Plantations, owned by the Est. of Robert Marion, Situate in St. Stephen Parish, Berkeley County, S.C., 1928 & 29." Courtesy of South Carolina Historical Society, Gaillard Plat Collection, call number G-134.

because the map does not show it extending across the swamp.[32] The upper end of the swamp shows a bank holding back water, forming a pond that once functioned as the reservoir. Field surveys by Richard Porcher establish that both sites fit the description of inland-swamp rice fields typical around the time of the Revolution. Based on the advertisement in the *Gazette* and the "Map of Richmond," it is clear that Maham's plantations had inland rice fields, and it seems reasonable to assume he was acquainted with rice planting if not a planter himself.

To prove Allston's statement Maham did introduce Carolina Gold rice "into the Winyaw and Waccamaw region," it would help to demonstrate a connection between Maham and the Georgetown area. Such a connection is hinted at by both physical and archival evidence. Just to the east of Maham's former plantations is Maham Cemetery, the location of suggests that the adjacent land also belonged to him. The epitaph on a four-sided monument in the cemetery states in part:

WITHIN THIS CEMETERY, AND IN THE BOSOM OF THE HOMESTEAD, WHICH HE CULTIVATED AND EMBELLISHED, WHILE ON EARTH, LIE THE MORTAL REMAINS OF COL. HEZEKIAH MAHAM. HE WAS BORN IN THE PARISH OF ST. STEPHEN, AND DIED

A.D. 1789, AT 50 YEARS; LEAVING A NAME, UNSULLIED IN SOCIAL AND DOMESTIC LIFE, AND EMINENT FOR DEVOTION TO THE LIBERTIES OF HIS COUNTRY, AND FOR ACHIEVEMENTS IN ARMS, IN THE REVOLUTION WHICH ESTABLISHED HER INDEPENDENCE.

HIS RELATIVE, JOSHUA JOHN WARD, OF WACCAMAW, UNWILLING THAT THE LAST ABODE OF AN HONEST MAN, A FAITHFUL PATRIOT, AND A BRAVE AND SUCCESSFUL SOLDIER, SHOULD BE FORGOTTEN AND UNKNOWN, HAS ERECTED THIS MEMORIAL, A.D., 1845.

Joshua John Ward (1800–1853) of Brookgreen was a major grower of Carolina Gold in the Georgetown District. He was descended from Colonel Maham's sister Elizabeth, who married John Cook(e) and had a daughter, Elizabeth Cook(e), who first married a Weston and then, as a widow, married Joshua Ward (1769–1818). They were the parents of Joshua John Ward.

The Maham-Ward family connection does add some support to the idea that Maham could have brought in the Carolina Gold rice and that Maham-Ward family connection brought it to the Georgetown area, where it replaced white rice as the main crop.

If Maham first planted Carolina Gold, historians still do not know where he obtained the seeds. Maham left no records of his rice planting. Without question crediting Maham with the introduction of Carolina Gold is problematic, but the existing evidence is worthy of note.

In 1823 the South Carolina Agricultural Society gave Henry Laurens (1724–1792) credit for introducing Carolina Gold. The society had appointed a committee "to consider what beneficial effects would result to the Agricultural interests of the State, by importing Foreign Seeds, Plants and Implements of Husbandry." When the committee's report was published in the September 1823, issue of the *American Farmer,* it stated: "It is also worthy of remark that we owe, not only the original acquisition of this valuable staple to the importation of the seed, but the improved variety of it, now generally cultivated, is derived from the same source; it being understood that the late Col. HENRY LAURENS imported a small quantity of what is called the Gold-seed Rice, soon after the revolutionary war, which was found to be so far superior to the white-hulled Rice before cultivated, that the latter is now scarely [sic] to be met with."[33]

Did Henry Laurens introduce Carolina Gold? The timeline and Laurens's travel itinerary are favorable for his having had the opportunity to do so. Laurens was an American merchant and rice planter from South Carolina, who became a delegate to the Second Continental Congress and succeeded John Hancock as its president. As American minister to Holland in 1780 he was captured aboard ship by the British Navy and imprisoned in the Tower of London. Finally, on December 31, 1781, he was released in exchange for Gen. Lord Cornwallis and completed his voyage. In 1783, Laurens was in Paris as one of the peace commissioners for the negotiations leading to the Treaty of Paris. Laurens generally retired from public life in 1784 and returned to his Mepkin Plantation of the Western Branch of the Cooper River.

While in Europe Laurens continued to be active in the management of Mepkin. He instructed his overseer to resume planting rice in 1783. Laurens was interested in new rice machines and kept an active interest in agriculture, especially rice. It is possible he brought seeds of different strains of rice from Europe to experiment with in South Carolina, one which might have been Carolina Gold. Neither the Henry Laurens ledger at the College of Charleston nor the published *Papers of Henry Laurens,* however, make any mention of his bringing a new variety of rice to Mepkin Plantation, nor do any other archival sources provide evidence that the Laurens introduced Carolina Gold.[34]

Whoever introduced the Carolina Gold, where did they obtain seeds? There is no known record of their source. Could Carolina Gold have arisen as a hybrid from the many varieties that were introduced? Possibly, but there is no direct evidence of hybridization. The archival record is replete with references to different varieties of rice being sent to the Rice Kingdom. Thomas Jefferson—who said "The greatest service which can be rendered any country, is to add a useful plant to its culture"—sent upland rice to South Carolina from Italy around 1800, but the gift was not well received. Ralph Izard feared the Italian plants (which he himself had seen in Italy and had considered an inferior product) would mix with the Carolina rice and reduce its quality and yield. Later Jefferson acquired some mountain rice from Africa that the Carolina planters wanted nothing to do with, but Jefferson "understood that it had spread to upper Georgia and was highly prized by the people there."[35]

One fact is certain: tidewater planters grew many varieties of rice for which there is little or no documentation. Seeds were sent from overseas to plantations to experiment with, and no records exist to show where they came from of if they succeeded. For example Thomas Pinckney reported in 1829:

> I have for many years past, occasionally cultivated a large species of Rice, the product of which has always appeared to me at least equal to that of the Gold-seed Rice; but it has always had a capital defect in our market; which is, that it would not sell, because our merchants are ignorant what reception it might meet with abroad; I however shipped, more than thirty years ago, ten or twelve tierces of it to the house of Messers. Willink, Vanstaphorst and company, of Amsterdam, for each of which I received ten guineas; and these gentlemen assured my brother, who went soon after to Amsterdam, that such Rice would always command a good price in their market.[36]

Despite uncertainties about its origins, a variety of rice called Carolina Gold did come into being. It was probably first planted when rice culture resumed after the Revolution. Carolina Gold was the variety mainly grown in South Carolina and in North Carolina. In Georgia it was grown by planters such as the Manigaults on Argyle Island, but most Georgia planters preferred the common white rice. At present this is all historians can conclude.[37]

Genetic research may be the only way to solve the source of Carolina Gold. Hulls were collected from six abandoned rice facilities in South Carolina (Chicora Wood, Cockfield,

Chisolm's Rice Mill, Lavington, Kinloch, and Mansfield) and seven samples were sent to Anna McClung of the United States Department of Agriculture in Beaumont, Texas. According to her initial findings, "Three of these samples we have not been able to extract DNA as yet. Two of the samples we have run markers on show that there is a mixture of different rice cultivars that are evidenced by the rice hulls. One sample from 'Chicora barn trash' has markers that are typical of (and quite unique to) Carolina Gold. Thus it appears that this site may have had CGR as we know it, grown there."[38]

Yet even genetic research may not establish the source. Carolina Gold rice was exported to other countries, and its cultivation spread globally during the nineteenth century. Even if a sample of rice in another country has the Carolina Gold genetic marker, it does not prove that Carolina Gold originally came from that country because it may have gotten there from the Rice Kingdom.

Bearded Rice, 1829

William Mayrant of Stateburg in the High Hills of the Santee in Sumter County reported in 1829 that bearded rice "was brought here by Mr. Wm. Genald of this district, who, traveling to Pensacola, in its vicinity saw a small field growing on the high pine lands of that country. On his return from Pensacola he procured a parcel from the owner, who told him that it had lately introduced there, that it came from South-America, where it grew wild."[39] Genald gave some of the seed to a Mr. Span, Sr., who planted some of it in a low spot near a wetland site and harvested eight bushels of seed. After planting a second crop, Mayrant secured six bushels of the seed from Span and sent a pint each to Col. William Alston (William Alston of Clifton, 1756–1839) and a Mr. T. Ford, of Pee Dee, and four quarts to William Washington. Mayrant reported he never heard what sort of success these recipients had with the seeds. Mayrant, however, made repeated plantings with the bearded rice, generally on high land, and claimed it outperformed Carolina Gold seed. Unfortunately Mayrant did not give a date for his acquisition of the seeds, so all we know is that it was sometime prior to 1828.

Another planter, identified only as "A Black-River Planter," wrote in an 1830 article that "I have for two years planted this [bearded] rice, and feel no hesitation in saying that many objections to it are altogether unfounded. I find the bearded rice heavier and more productive than the white or gold seed."[40] He went to great lengths to extoll the virtues of bearded rice, and further said: "Nor have I been able to discover any reason or cause which would operate against the general introduction of it in tide lands."

William Washington, one of the three planters to whom Mayrant sent bearded rice seeds, stated in 1829: "To the inland planter, it offers great advantages, in-as-much, as it appears it will stand the drought much better than the Gold-seed. Another important consideration is, that it can be threshed out with a simple and cheap machine at the rate of from one hundred to one hundred and fifty bushels per day. . . . Again, it will grow well on high ground."[41]

Writing in 1830, someone identified only as "A Marsh-Planter" found that the bearded rice was not nearly as market prolific as Carolina Gold: "I consequently sent no more to mill, but fed it away, not intending again to plant any of this seed. It also shells very much in the field, and the loss, with the repeated handling is immense. Not withstanding all these losses, I threshed out 815 bushels of clean rough rice from the twelve acres. Many persons believe that the Bearded Rice and Gold Seed are of equal weight, bushel for bushel; my observation has led me to think otherwise.[42] The bearded rice, for unrecorded reasons, never was a commercial success in tidewater lands.

Robert Allston referred to "white bearded rice" under the varieties that were most common in Carolina. Whether this was the same as Mayrant's bearded rice is unknown. Allston said a sample was brought from the East Indies in 1842 by Capt. Thomas Petigru (1793–1857) of the U.S. Navy and thereafter was cultivated more or less extensively by planters for their black slaves.[43] No mention of it as a commercial crop was found.

Long Grain Carolina Gold Rice, 1838

From 1838 to 1843, Joshua John Ward of Brookgreen Plantation developed a strain of Carolina Gold called the "Big Grain Rice" or the "Long Grain Carolina Gold." In 1846 Ward described how he developed the long grain rice in a letter to the State Agricultural Society. In 1838 his overseer found in a stack of rice in the threshing barn part of an ear that was very different from the others. In the spring Ward had the overseer plant some seeds from this ear on the margin of a rice field. Because of rats and high water, only six plants survived, but the grain was once again different. The next year the overseer planted some seeds in a tub filled with swamp soil and placed the tub in a garden where it could be tended. A hog destroyed most of the shoots, but the few plants that survived were transplanted into a pond from which three pecks of "rotten light rice" were obtained. The following year, 1840, half an acre of the light rice seed was planted, which yielded forty-nine and a half bushels of clean winnowed rice. This yield was significant enough to produce seed for another year's planting.

In 1841 seeds from the previous year's crop were planted in a twenty-one acre field, which yielded 1,170 bushels of clean, winnowed rice, which in turn produced 200 bushels of milled rice. Ward said his factors sold it at a considerably higher price than Carolina Gold. The crop of 1842 was a success, the quality being better than the previous year, and in 1843 Ward planted his entire crop with the new seed. Ward stated in his letter: "I earnestly trust that this improvement in the seed, will be of incalculable benefit to the entire Rice-growing region."[44]

William M. Lawton was convinced of the superiority of the long grain, writing in 1871 that Ward's long grain "was produced chiefly in the Waccamaw area, and from its superiority of bright large grain, it usually brought in market 30 to 50 per cent. more than other fine grades, and always commanded ready sale to supply orders from Paris, where it was esteemed as a luxury for diet; and such is the taste of the French for the nutritious ingredients of Rice, that we have noticed sign plates on the cafés and restaurants of the Palais Royàle, and other resorts, *Riz au lait*—meaning that Rice, cooked in milk, would be served there."[45] Lawton added: "The long grain Seed Rice, since the war, has died out, yet orders for it from France still come to Charleston merchants." Lawton gave no reasons why planting was discontinued despite demand for it overseas.

In an 1847 letter Robert Allston wrote what he thought about why Long Grain Carolina Gold seed ultimately failed: "Many Proprietors bought this 'long grain' Seed the first year at a high price, and with good expectations of its superior productivity and quality. The greater number [of planters] were disappointed, owing to various causes, among the most efficient of which probably, is one which entirely escaped their [illegible], namely, the want of their personal attention—The new Seed, requiring a little extra preparation of the soil, and a little additional care in handling, was universally condeamned by the Overseers."[46] The problem was probably exacerbated because the planters were often absent from the plantations in the growing season and left the tedious operations with the overseers.

3

The Culture of Carolina Rice

The methods used to grow Carolina rice may be divided to three categories based on the water source: providence culture, reservoir culture, and tidal culture. Providence culture was the initial method and depended on the natural dampness of the soil from rainwater or freshets; reservoir culture relied on impounded water, springs, non-tidal river water, or bays to flood rice fields; tidal culture depended on the ebb and flow of tidal freshwater along the rivers to flood and drain rice fields.

This book uses the term "providence culture" in lieu of "upland culture," which is the term traditionally used to describe the initial experiments with rice growing, which started around 1690. The most successful early sites, however, were "low moist grounds" and not truly upland sites. Thus "providence culture" is the more apt term because acquiring moisture depended on the providential chance of rain.

The second method of rice culture is traditionally called "inland-swamp culture," based on the idea that it depended on a reservoir created by an earthen dam to trap rainwater runoff. Fieldwork, however, has documented that there were other sources of reservoir water used to irrigate rice fields besides a dammed inland swamp. Hence, the term "reservoir culture" is used.

We retain the traditional term "tidal culture" to describe the third method of growing rice. Field and archival research found nothing different from the traditional description of tidal culture that used the natural flow of the freshwater tides to irrigate rice fields.

River Systems and Associated Wetlands

A full understanding of the three methods of rice culture requires a knowledge of the river systems and associated wetlands where rice was grown. The rivers were the lifeblood of the different methods of rice culture, not only for growing of the crops, but also for transportation to the cities. The Atlantic Coast is punctuated by many estuaries, some with significant freshwater discharge (from brown-water rivers) and some with smaller amounts of freshwater discharge (from black-water rivers). The brown-water rivers—the

Altamaha, Ogeechee, Savannah, Santee, and Great Pee Dee—occupy drowned river valleys. Their brown color is caused by clay carried from erosion in the mountains and piedmont, where they originate. These rivers have broad alluvial floodplains built over eons from deposits of nutrient-rich clay sediment. They have meandering channels, oxbow lakes, and extensive natural levees along their edges, and they typically cut straight across the Pleistocene depositional terraces of the coastal plain. They produce large deltas and estuaries where they enter the ocean. The point at which their waters become salty is generally close to the ocean because of the large volume of freshwater discharge that pushes the salt water toward the mouth of the estuary. Before the swamps were cleared for fields, above the salt point and upriver for twenty or thirty miles, freshwater from the river flooded and then drained twice every twenty-four hours from the cypress and mixed-hardwood swamp forests that occupied the floodplain.

The black-water rivers, with their headwaters in the coastal plain, are generally sediment free and owe their color to organic acids derived from decaying leaves. They have narrow nonalluvial floodplains, no levees, and smaller deltas and estuaries. The Cooper River in South Carolina is an example of a black-water river. The salt point of a black-water river is generally twenty to twenty-five miles from the ocean because a lower volume of freshwater discharge allows salt water from the ocean to move farther up river. A cypress and mixed-hardwood swamp forest occupied the narrow nonalluvial floodplain above the salt point and was flooded and drained twice a day as were the brown-water river swamps.

Brown-water and black-water rivers have different types of estuaries. Brown-water rivers form salt-wedge estuaries, while black water rivers create vertically homogenous estuaries (see diagram 1). In salt-wedge estuaries the strong flow of freshwater causes a

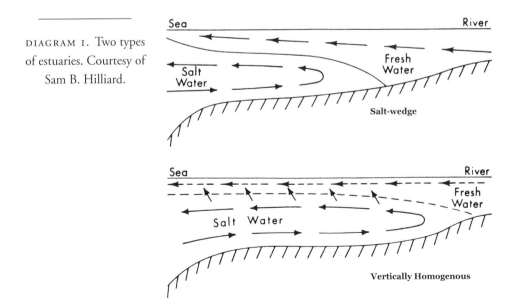

DIAGRAM 1. Two types of estuaries. Courtesy of Sam B. Hilliard.

lighter layer of freshwater to lie over the salt water. Above the salt point, these salt-wedge estuaries support cypress-hardwood swamps along their fringe. The Santee Delta and Winyah Bay in Georgetown County, South Carolina, are examples of salt-wedge estuaries. The rivers that feed these estuaries begin in the mountains of North Carolina, and so much freshwater comes down that a layer of freshwater pushes within a half mile of the ocean, allowing rice to be grown there. The Savannah, Ogeechee, and Altamaha Rivers in Georgia, with their origins high in the Georgia hills, have salt-wedge estuaries.

Fed by creeks or rivers that originate in the coastal area, tidal currents dominate the vertically homogenous estuaries; therefore, these estuaries do not have large volumes of freshwater and saltwater and freshwater mix. Salt marshes dominate the edge of the estuary near the coast, but give way to brackish marshes upriver to the salt point. Charleston Harbor in Charleston County, fed by the black-water Cooper River, is a vertically homogenous estuary. No salt wedge occurs near the ocean, and the salt point is twenty or so miles inland. Rice could not have been grown below the salt point.

Many tributaries drain into the black-water and brown-water rivers. While the larger tributaries are influenced by tides, smaller tributaries are not and have a different hydrology. They are flanked by small-stream floodplains (see color plate 1), which are fed by rainwater runoff from the surrounding highland. During heavy rains, water covers the floodplain, but it drains quickly as rains subside. Scattered swamp trees and a low ground cover of grasses and sedges or dense canebrakes dominate. The soil is continually moist except in extreme droughts. Small-stream floodplains are more common along the small, nontidal tributaries of the black-water rivers, and they were the sites where rice culture for domestic food and commodity was first successful.

Providence Rice Culture

Providence culture dates to the period when slaves and planters first experimented with growing rice. Their success demonstrated that rice could be grown in the Lowcountry strictly with moisture from rainfall. The locations where providence rice was first planted, however, are uncertain. Little documentation exists for the late 1600s, when experimentation with rice culture began, and historians only speak in general terms about early culture. Slaves left no written records, and those left by planters are sketchy as are other accounts of the period. But historians do have some information from which to hypothesize about early attempts to grow rice and where that may have taken place. They know the nature of the rice plant, what planters and slaves knew about agriculture, and the natural setting of the Lowcountry around the Cooper River. They also have clues from the letters of Scottish agricultural innovator John Stewart and promotional literature, travel descriptions, and official reports of the Lords Proprietors and the Board of Trade.

Most rices worldwide are wetland plants that need large quantities of water to grow well and generally grow best in standing water. Growing in standing water, however, presents a problem with getting oxygen to the roots since roots cannot extract enough dissolved oxygen from water. A morphological and physiological adaptation evolved in rice plants that solved the problem of growing in standing water. As oxygen is used by the roots, oxygen from leaves flows down through special ducts in the plant body to the roots while the carbon dioxide produced by root respiration is given off in solution into the surrounding water. For plants capable of this gas exchange, growing with their foliage in air and their roots in water permits enormous productivity. Wetland rices are such plants. Upland rice varieties can grow without standing water but require an assured rainfall of at least 5 to 6.5 feet a year over three or four months during the growing season and cannot tolerate desiccation. They can be grown without irrigation in a sufficiently rainy climate, but they are low yielding.

If, as S. Max Edelson argued, African slaves were the first to plant rice and their knowledge was appropriated by planters who turned rice into a commodity, historians have no primary records to support his argument. It is unlikely any exists. A case can be built in other ways, however, to support his argument that African slaves were likely the first to experiment with rice planting. Rice was a staple food for Africans in their homeland. Some of the slaves the Goose Creek Men brought with them from Barbados in 1670 probably knew of rice culture from their African origins, even if secondhand. Once they were here in the Lowcountry and once a source of rice seeds became available, it would have been natural for them to experiment with rice growing for food. Rice brought on early slave ships as food for the long voyage could have been used by slaves as a seed source to begin planting. The early shipment of rice on the *William & Ralph* in 1672 could have supplied seed for beginning a plot of rice, and slaves may have had access to the seed from Madagascar that Henry Woodward received. And the Lowcountry was certainly conducive to rice growing. Given these factors, it is logical to conclude that slaves probably experimented with growing rice for food, and their success led planters to turn rice into a commodity.

John Stewart may have been one of planters who appropriated slaves' knowledge of rice culture. The letters Stewart wrote to William Dunlop in 1690 are the most important primary record of rice growing in early Carolina. Stewart was managing Gov. James Colleton's Wadboo Barony at the confluence of Biggin and Wadboo Creeks, which form the headwaters of the Western Branch of the Cooper River in present-day Berkeley County, South Carolina. Stewart's letters cover a wide variety of early agricultural topics, including his experiments with rice. Stewart added agricultural science to early rice growing, and his letters documented that the mysteries of rice culture were not easily solved. Stewart was not the only one experimenting with rice growing; the Goose Creek Men were also active. Their contributions were not documented, however, so Stewart's experiments are the only preserved evidence relating to early rice culture.

Stewart knew well that crops required moisture no matter how the seeds were sown. Rice was no exception. It would have taken him only a few upland plantings to realize these sites were not ideal for rice. Stewart rated "Marsh, Swamp and Sevano [savanna] Land" the "best" lands for rice.[1] All three are common in the Wadboo area and in the

Lowcountry. Stewart's swamps were probably flat, small-stream floodplains flanking narrow waterways not subject to tidal waters and dominated by swamp trees with herbaceous ground cover or canebrakes. These waterways received runoff from the surrounding highlands and emptied into larger creeks and rivers. Small-stream floodplains are common flanking the smaller tributaries of the Cooper River, so the fertile floodplains had adequate moisture. Here Stewart, and probably other planters in the Cooper River area such as the Goose Creek Men, found that rice grew well in the cleared small-stream floodplains.

Savannas have a herbaceous ground cover and scattered trees. Two types of savannas in the coastal region could have supported early rice culture: longleaf-pine savannas and pond-cypress savannas. Both have a sparse tree cover making it easy to convert them to crop land without using a large labor force, either with or without clearing the trees; both types of savannas are underlain by a clay pan that keeps rainwater close to the surface and holds it there for extended periods; and both are generally infertile because they receive their moisture primarily from rainfall, which contains few nutrients. Stewart may have tried these savanna habitats but ultimately realized that the low soil fertility precluded a good growth of rice, abandoned them, and concentrated on the swamp lands, where he met with success.

Stewart's marshes are harder to document. Many of today's marshes are the result of so many years of disturbance that early natural marshes are difficult to describe. Stewart's marshes, however, may have been freshwater marshes that flank the edges of the small-stream floodplains and depend on rainwater or freshets for their main moisture supply. They were incorporated into early rice fields along with the small-stream floodplains.

Rice grown on marshes, swamps, and savannas depended on direct rainwater or runoff from the adjacent uplands—hence this type of culture is aptly called *providence culture*. The planter simply sowed his seeds and hoped for enough rain to make the seeds sprout. Some cultivation was supplied with a hoe. There was no provision for water control on the fields. Providence rice was subject to two factors that determined the success of a crop: droughts or freshets. A drought meant failure of the crop because there was no means to supply water from elsewhere. Freshets flooded the land at critical times and were equally devastating. Adequate moisture in the soil supported a crop of rice between these two extremes.

A document written by Thomas Nairne in 1710 supports the notion that initial success with rice culture was on swamp lands. He explained that rice was "vry much sow'd here, not only because it is a valuable Commodity, but thriving best in low moist Lands, it includes People to improve that Sort of Ground, which being planted a few Years with Rice, and then laid by, turns to the best pasture."[2] Nairne's "low moist Lands" were probably small-stream floodplains, or perhaps marshes, more than likely not savannas, and certainly not highlands.

By the end of the first decade of the 1700s, colonists had perfected rice growing. As John Norris wrote in 1712, "As to the Manner of Planting our Rice, after the Land is clear'd or chenged, as aforesaid, we, with Hoes, trench the Land something like Furrows made with a Plough, but not so deep, and about a Foot Distance, between each Trench; and when the Land is so Trench'd, in the Month of April we seed it, carefully, within each Trench, and cover it thin with earth, one Peck and a Half is sufficient for to seed an Acre, then, with narrow Hoes made for the Purpose, about Five or Six Inches broad in the Mouth, we Hoe, Weed, or cut up the Grass, or other Trash, growing between the said Trenches of Rice."[3]

English naturalist Mark Catesby, who toured Carolina in 1712 and 1726, writing decades after rice had become established, added additional information on the sites where rice was initially grown:

> The Soyl of Carolina is various, but that which is generally cultivated consists principally of three kinds, which are distinguished by the names of Rice Land, Oak and Hiccory [hickory] Land, and Pine Barren: Rice Land is most valuable, though only productive of that Grain, it being too wet for any thing else. The Scituation of this Land is various, but always low, and usually at the Head of Creeks and Rivers, and before they are cleared of Wood are called Swamps, which being impregnated by the washings from the higher Lands, in a series of years are become vastly Rich, and deep of Soyl, consisting of a Sandy Loam of a dark brown colour. These Swamps, before they are [prepared] for Rice, are thick, over-grown with Underwood and Lofty Trees of mighty Bulk, which by excluding the Sun's Beams, and preventing the Exhalation of these stagnating Waters, occasions the Land to be always wet, but by cutting down the Wood is partially evaporated, all the Earth better adapted to the Culture of Rice; yet great Rains which usually fall at the latter Part of the Summer, raises the water two or three Feet, and Frequently covers the Rice wholly, which nevertheless, though it usually remains in that state for some Weeks, receives no Detriment.[4]

Catesby's swamps "at the Head of Creeks and Rivers" were small-stream floodplains, and his description supports statements by Nairne and Norris that early rice crops were successfully grown on swamp lands.

Several references state that rice was first grown as an upland crop, then shifted to "low wet spots" as planters realized rice thrived best with adequate moisture. Although no primary records close to the time of initial planting record upland planting, one later document is probably the basis for the upland version in the literature. In *An Historical Account of the Rise and Progress of the Colonies of South Carolina and Georgia* (1779), published in London in 1779, Alexander Hewatt, who first treated the historical development of rice growing, claimed that it was "certain that the planters long went on with this article [rice], and exhausted their strength in raising it on higher lands, which poorly rewarded them for their toil." It is reasonable to accept that some planters, including Stewart, tried early on to grow rice on highlands, as did their British friends, who grew a variety of crops in the highlands of their homeland.

No records reveal whether the first rices planted were *upland* or *wetland* varieties or a mixture of both. If Hewatt's "higher lands" were oak-hickory forests or longleaf-pine flatwoods, upland varieties would have received enough rainfall to survive in sites where rainfall averaged fifty to sixty inches for the year, well below that required during the growing season. It could be, however, that Hewatt's highlands were pond-cypress savannas and longleaf-pine savannas, where both slaves and Stewart may have attempted to grow rice. Both these landforms occur in Wadboo Barony. If so upland rice might have grown in the cypress savannas or longleaf savannas, but they did not yield good crops and were quickly abandoned for the more productive swamp lands.

From a simple beginning along the swamps of the Cooper River and its tributaries, slaves and then planters, solved the mysteries of growing a species of grass in Lowcountry swamps, first for food and then for export, setting forces into play that would forever change the Carolina Lowcountry in ways they could not have possibly envision.

Although Stewart's letters place him as one of the earliest rice planters, it was the Goose Creek Men who brought rice into the plantation enterprise and made it a major export crop.

The Beginning of the Plantation Enterprise

Barbados, settled in 1627, became the model for early Carolina rice plantations. Early on the settlers of Barbados struggled to survive in what appeared to be a tropical paradise, but developing agriculture there proved more formidable than expected. Once land was cleared, the colonists tried a variety of crops, but with little success. When sugarcane was introduced in 1638, the nature of the struggling colony changed. The world had longed for a cheap sweetener, and sugar from the cane satisfied that desire. African slaves replaced white indentured servants, and an exploitative and materialistic society based solely on becoming rich emerged in Barbados. Barbados became the richest, most highly developed, and most populous English colony—fifty thousand lived on the small island, thirty thousand of them Africans. Removed from the restraints of the mother country by an ocean, the planter elite focused on the pursuit of wealth and social trappings.

After Barbados became densely populated and new land for sugar production became scarce, one Barbadian, Sir John Colleton, saw an opportunity for reward if a royal colony could be established along the southeastern coast of North America. Though Jamaica lured some English West Indians, it was surrounded by Spanish-held islands and had a reputation for sickness. Other West Indian islands posed problems for settlement. The southeast coast, however, had commanded the attention of Europeans from the beginning of the sixteenth century. Spain and France were the first to evince interest long before England started to consider colonization of the southeast coast in the 1580s; however, the southeast lacked the spices and gold sought by Spain and France. France abandoned its interest, and Spain retreated to St. Augustine and some Indian missions and was never a serious presence in the Southeast again.

Though the proprietors' first two colonies, located at Cape Fear in present-day North Carolina failed, the next one succeeded. The 130 settlers aboard the *Carolina* when it landed at Albemarle Point on the Ashley River in 1670, included people from all social classes, the majority from small planter and freeman classes of families. One of three ships that left England and reached Charles Towne in April 1670, the *Carolina,* however, had stopped in Barbados because the proprietors had hoped that some wealthy Barbadian planters with expertise in agriculture would join the settlers. Because all the better agricultural land in Barbados had been claimed, land for expansion of the planters' holdings had become nonexistent, and some of them did indeed join the settlers. The Barbardians were experienced planters and well educated members of the Church of England and brought investment capital with them. They also brought with them three important factors in establishing rice culture: a well-developed plantation culture, slaves (some perhaps with knowledge of rice culture), and a labor system that gave slaves free time to grow provision crops and tend to other duties or wishes. Over the next two years, half of the white settlers who came to the new settlement came from Barbados, and brought with them Africans and a Barbadian plantation cultural that became the initial model for the planters of the fledging colony.[5]

Before the Carolina colony was established at Albemarle Point, the English had already experimented with rice in Virginia, as had the Portuguese in Brazil, the Spanish in Central and South America, the Dutch in the West Indies, and the French along the northern coast of the Gulf of Mexico. By the time the *Carolina* landed, the proprietors had laid plans to introduce crops. During the first years settlers experimented by planting an array of exotic crops such as grapes, sugarcane, olive trees, and citrus. The cold Carolina winters dashed hopes for these crops. Tobacco proved successful, but it could not compete with the already-established Chesapeake tobacco. Indian corn also did well, as did a variety of root crops. The colonists added cattle and hogs. By 1674 Carolina had achieved self-sufficiency in food production.

The Barbadians settled on Goose Creek, a tributary of the Cooper River. This region was attractive because it could be reached by water from Charles Towne; the land supported a lush growth of oak and hickory indicating fertile ground suitable for crops; it backed up against longleaf-pine forest that provided lumber for building and grazing for cattle; and the river swamps contained bald cypress, which was also a good building material and, along with pine, valuable for export. It was only later that these swamp lands proved suitable for rice cultivation. By the time Carolina became a royal colony in 1719, a rich plantation society was in place, and Goose Creek had become a prosperous corner of Carolina.

The Goose Creek Men were often at odds with the proposals and goals of the Lords Proprietors. They came to Carolina for one reason: to make money and did not look favorably on interference in their business activities. At first the Goose Creek Men dealt mainly in illegal Indian slave trade; later deerskin trade and naval stores dominated. By the 1690s, they held many important offices in the colonial government, forming the dominant political faction in the first generation of the settlement. Seven early Carolina governors had Barbados backgrounds. The years of 1670–1712 became the "Age of the

Goose Creek Men," a period that coincided with the introduction and development of rice as an export crop.

The Goose Creek Men brought African slaves with them, making Carolina the only English colony in the New World in which black slavery was introduced at the beginning of settlement. The typical arrangement between slave and master that existed in Barbados was transplanted to the new colony. A slave in Barbados was defined as "freehold property," which entitled its holder only to service, not to absolute ownership. In Barbados a slave was allocated a small plot of land to grow provisions and raise livestock and was given time to tend the crops. Slaves could sell excess provisions to their masters, who in turn could sell the excess in the market. This system, which has been aptly called the "provision ground system," resulted in more food production for the plantation. Slaves had a vested interest in the land and were less likely to run away.[6] Elsewhere in the Caribbean, a similar system prevailed. British West Indies crops that required little supervision—notably coffee and pimento—were grown by slaves under a similar labor system.[7]

This provision ground system of labor was transferred to the Lowcountry with the Goose Creek Men and other early plantation settlers from the Caribbean. A casual and open exchange between master and slave existed in the pioneer years with the slaves being given wide latitude over their lives. Master and slave often worked side by side in the fields in the early days of the colony. This system of labor led to the task system, which prevailed until the Civil War.

Were the Goose Creek Men the first to experiment with rice culture? It would have been in their interest to develop a suitable staple crop to replace sugarcane. The truth may never be known. An early letter and does indicate, however, that they were either the first to experiment with rice culture or among the first. John Stewart wrote to William Dunlop on June 23, 1690: "The Govenor both in Sevanoh [savanna] and swamp sow'd his Rice thin after the Goose creek philosophers' old measurs, and when it was 6 inchs above ground, I advised him to plow it all up and sow at least 2 bushel and a half on an acre not in Rows or planted bot as barly."[8] A second reference to early rice planting is the previously mentioned 1692 will of Barnard Schencking, a Barbadian planter whose home was situated along Goose Creek. Both these references indicate that the Goose Creek Men were involved in rice planting prior to 1700 and created the first plantation enterprise devoted to rice culture.

It seems plausible to view the Goose Creek Men as early experimenters with rice. They were near the port of Charles Towne, where goods and seeds arrived from overseas. The port was an avenue to the promising English market and its reexport market to continental Europe. The Goose Creek Men had acquired large property holdings suitable for rice agriculture. They were driven to make money, and rice was a crop that promised to do that. They could afford the high cost and lengthy time it took to develop a sustainable crop like rice. They had commercial connections necessary to become exporters. They had agricultural background from their West Indies heritage, they had a slave force of first- or second-generation Africans, some of whom may have been familiar with rice cultivation. Finally they were located in the coastal area of South Carolina, which was bathed in a humid subtropical climate, and had abundant rainfall, surface water resources, and loamy Ultisols,[9] which were conducive to the growth of Asian rices.

The Task System

As these planters were experimenting with methods to grow rice and create a plantation system, the task system of labor was essential to this success. The task system originated in the early days of experimentation with rice growing and was the central feature of the slaves' daily routine. During the colonial period, increased productivity had little to do with improvements in market preparation; rather it depended on adding more fields to one's cultivated land and developing the task system. In a 1982 article on the task system that evolved in the Georgia and South Carolina rice region, historian Philip Morgan wrote "that a particular mode of labor organization [the task system] and a particular domestic economy [of the slaves] evolved simultaneously in the colonial and antebellum lowcountry."[10]

Morgan argued that the task system began in the first decade of the 1700s, when slaves were planting staple crops such as corn, rice, potatoes, tobacco, peanuts, sugar, pumpkins, and watermelons for themselves and that "the opportunity to grow such a wide range of provisions on readily available land owed much to the early establishment and institutionalization of the daily work requirement." This daily work regime, which was essentially the provision ground system, evolved into the task system. "By midcentury the basic 'task' unit had been set at a quarter of an acre," a square 105 feet on each side. Tidal culture made the development of the task system highly effective because irrigation imposed an orderly grid on the landscape. Small quarter drains in the fields delineated quarter-acre units of work, each quarter-acre containing seventy-eight trenches that constituted a task, or a quarter acre to be hoed as a task. Since rice was a hardy plant and required just a few straightforward operations and little supervision, once it was established as a commercial crop, it lent itself easily to a system in which tasks could be easily measured on various projects in its culture. Initially planting and weeding stages were brought under the task system and provided the rationale for its development. Planters benefited from the system. Letting slaves work on their own and be held accountable at the end of a day meant they could work under less supervision, and the planters could vacate the plantations in the malarial season and leave the supervision to an overseer.

By midcentury, the basic task unit had been set at a quarter of an acre for field work. According to J. Motte Alston, "For every kind of work there was a set task, and so, according to ability, there were full-task, half-, and quarter-task hands. When two tasks were accomplished in one day by any hand, he was not expected to work the next, and these tasks were *never* increased."[11] Those not able to do any field work—the ill, the very old and the very young—were classed as nonhands. When a slave's ability to perform work changed, his classification was changed. For example an old man might be dropped from a full hand to a three-quarter hand. Throughout the evolution of the task system, planter and slave disputed the fairness and uniformity of task units, and planters came to recognize slaves' power in defining and redefining terms of labor.

The equivalent of this measurement was expected whether the work was planting, harvesting, or preparing rice for market. Once tasking became firmly established and routine, tasks were established for other plantation operations, including ditching new land; harvesting with the sickle, one quarter of an acre to cut and haul to one hand; sowing by the best hands; cleaning canals; weeding; trenching; digging stiff land; seven mortars of pounding per day; and leveling fields. When tidal culture replaced reservoir culture at the turn of the century, the task system was transferred to tidal culture. By the early nineteenth century there had evolved a standard method that was followed on most plantations, and the task system reached its full flowering.

The work day began at sunrise. In the summer months this allowed the most industrious slaves to miss some of the midday heat. Each morning a slave was assigned a task by the overseer or plantation owner. Tasks were calculated to last a day, but once the task was completed the slave was free to pursue his or her own interests. Obviously the reward of free time was an incentive for slaves to work hard and well. The more industrious and active hands often finished their tasks by two o'clock and had afternoon time to tend to their gardens. Once slaves produced their own crops, they found ways to market them, either to fellow slaves, planters, or anyone else they came into contact with. Free time also allowed them to grow plants of medicinal value, or collect such plants in the wild, and a slave medicinal cornucopia developed that was for a time the only real medicinal alternative.

The many descriptions of the task system in historic documents are in general agreement on the operation of the system. Written by planters, however, they do not portray how difficult working conditions were under even the most liberal system. No matter the task system or the plantation, workers toiled under appalling conditions in the rice fields regardless of time of year or individual tasks (see "Deadly Rice Fields" later in this chapter). In the following four descriptions of the task system, only one writer, Frances Kemble, touches on the bleak conditions enforced upon the workers.

James Ritchie Sparkman described his task system in a letter to Benjamin Allston:

> The ordinary plantation task is easily accomplished, during the winter months in 8 to 9 hours and in summer my people seldom exceed 10 hours labor per day. Whenever the daily tasks is finished the balance of the day is appropriated to their own purposes. In severe freezing weather no task is exacted, and such work is selected as can be done with least exposure. During heavy rains and in Thunder showers, my people are always dismissed and allowed to go home. The task is allotted to each slave in proportion to his age and physical ability. Thus they are considered ¼, ½, ¾, or full task hands. Men and Women are all engaged together in the planting,

cultivation and harvesting of the crop, but in the preparation of the Rice Lands, as ditching, embanking etc. the men alone are engaged with the spade. It is customary (and never objected to) for the more active and industrious hands to assist those who are slower and more tardy in finishing their daily task.[12]

James C. Darby, who planted on the North Santee, described his operation in 1827: "From the fifteenth to the twentieth of March is the most advisable time to commence Rice planting, and all the planters should have their land nearly in order by that time, and ready for planting, and the task in trenching is three quarters of an acre to each hand, the sowers nine quarters each, and those that cover, each three quarters."[13]

Age, gender, and physical ability were taken into consideration in assigning tasks and most planters instructed their overseers not to assign a task beyond reasonable means to complete. A decrease from the "quarter acre" occurred when more strenuous assignments were performed (like clearing new land), or increased when less strenuous tasks were assigned. Women slaves were excluded from the more strenuous tasks such as construction of banks and ditches. Pregnant women were a special case, although every planter did not adhere to special conditions employed by others. James Ritchie Sparkman wrote to Benjamin Allston describing his policy toward pregnant slaves: "Allowance is invariably made for the women as soon as they report themselves *pregnant* they being appointed to light work as will insure a proper consideration for the offspring. No woman is called out to work after her confinement, until the lapse of 30 days, and for the first fortnight thereafter her duties are selected on the upland, or in the cultivation of the provision crops, and she is not sent with the gang on the low damp tide lands."[14]

On the other hand, Frances Kemble reported that on Butler's Island in Georgia, Mr. Butler "was called out this evening to listen to a complaint of overwork from a gang of pregnant women. I did not stay to listen to the details of their petition, for I am unable to command myself on such occasions, and Mr. [Butler] seemed positively degraded in my eyes as he stood enforcing upon these women the necessity of their fulfilling their appointed tasks."[15]

Only half-tasks were done on Saturday on most plantations, and "no work of any sort is to be permitted to be done by Negroes on Good Friday, or Christmas day, or on any Sunday. . . . The two days following Christmas day; the first Saturdays after finishing threshing, planting, hoeing and harvest, are always to be holidays, during which the people may work for themselves."[16] Women with six children alive at any one time were allowed Saturdays off. Because of the heat in the summer months, the work expected on a task might be decreased.

Although a standard for tasks developed throughout the industry by the 1840s, some planters varied from the standard. Gov. William Aiken of South Carolina required only a third-acre per worker on his Jehossee Island plantation, thus allowing his slaves more time to raise small crops or hunt in the woods or fish the surrounding waters of the island. "Aiken reasoned that such incentives as short workdays and the opportunity to make some money for themselves in their free time would produce better discipline among the slaves than frequent use of the lash."[17]

The task system was not applied to all work on a plantation. Certain types of work were not easily divided into tasks and were better suited for gang work instead. Gang work—from sun to sun—included of burning field stubble, loading schooners with rice, emergency harvesting when weather conditions seemed ominous, and repairing breaks in banks. "During the rice harvest time there could be no tasks. From early morn, til late in the night—when torch lights were burned in the barn yard to enable the hands to see how to put the newly cut sheaves into racks—the work went on."[18]

The mechanical improvements in market preparation, which began in the mid-1700s, changed the plantation task system very little. Operation of the animal-, water-, and steam-powered machines did not require a large labor force and was not suited to a task system. Water-powered barns and mills depended on tides not under control of the work force. When the tides were right for operation, hands had to work for the duration of a tide cycle. For steam-driven threshing barns and mills, it was impractical to shut down the boilers and engines after half a day or so of work. Workers trained to run the steam engines stayed as long as the engine was in operation, which was often late into the day.

Perhaps six to eight laborers could run a steam-driven mill. On Jehossee Island, Governor Aiken's steam-driven threshing barn "required the labor of only a few slaves, two or three to feed the rice stalks into the mill and two or three to supply the feeders, plus one to fire the boiler."[19] But improvements in market preparation did create a special class of skilled workers. Overseer Jesse T. Cooper on Argyle Island wrote to Charles Manigault on October 20, 1849, that "Williams has attended faithfully to his duties in the mill, & have done all this time necessary to be done—& without a trial warrants it. He has fitted up the Elevators, rigged the upwright shaft, put in all his pullies—made & put down a new sluice Gate, & several minor jobs in the mill."[20]

The main types of labor on the plantation remained the same: maintaining the fields, repairing banks, cleaning the canals, sowing, planting, and harvesting all continued under the task system until the Civil War.

The task system on the rice plantations was better than the gang system on the cotton plantations or other labor systems. Rice slaves enjoyed considerably more freedom and mobility to pursue a variety of activities. Most important, their labor system allowed them to preserve much of their African culture and to blend it with the culture of the white slaveholders. The task system helped slaves survive as a community in the "deadly fields" until emancipation. The early task system also made possible the success of reservoir culture.

Reservoir Rice Culture

Reservoir rice culture depended on impounded water, artesian springs, nontidal river water, or bays to flood rice fields. At some point planters realized that control of the water supply for rice crops was more efficient than relying on direct rainwater (as in providence rice planting), but historians are not in agreement on an approximate date when water control began. No primary source records when the first dams, canals, or trunks were used to control the movement of water on and off Lowcountry rice fields. No plats exist for the early period of water control, and physical evidence in the field is not conclusive.

Water-control measures have been known since the dawn of agriculture. As early as 722–481 B.C.E., the Chinese harnessed rivers, developed water-conservation projects and irrigation systems, and constructed artificial levees that divided a river's water into inner and outer streams. The irrigation works of ancient Sri Lanka, the earliest dating from about 300 B.C.E., were one of the most complex irrigation systems of the ancient world. In addition to underground canals, the Sinhalese built completely artificial reservoirs to store water. In South America the preconquest Aztecs had several main methods of agriculture, including irrigation farming, which they used in the valleys. Dams diverted water from natural springs to the fields, allowing harvests on a regular basis. Canal systems diverted a portion of a river to provide irrigation to large areas of fields. Between 1300 and 1150 B.C.E., the Incas mastered high-altitude agriculture, building tiers of agricultural terraces on mountain slopes and irrigating the fields.

In 1006 there were more than five thousand water mills in England, many of them tide mills situated on the coast. Tide mills operate by using flap-valve trunks to trap tidal salt water in a pond at high tide. At low tide, when a sufficient water head is created, the flap valve is opened, and the flowing water turns a waterwheel.[21] Water mills driven by streams were also common in England.

Since the fifteenth century, rice has been grown in the piedmont area of northern Italy, where the growing conditions necessitated summer cultivation and a system of canals to deliver river water through gravity flow, not unlike delivering water from reservoirs. Englishmen traveled to Italy often, and Italian agricultural practices would have been observed and taken home.

As Judy Carney has noted, "Descriptions of West African agriculture in the early period of contact indicate that rice cultivation was well established along the Upper Guinea Coast prior to the arrival of Portuguese mariners in the mid-fifteenth century." Furthermore "environments identified as being planted to rice in the earliest contact period include the use of tidal floodplains for both wet- and dry-season cultivation, inland swamps, and the rain-fed uplands."[22]

Writing about fens, bogs and quagmires in England in his classic book *The English Improver Improved* (1653), Walter Blith stated, "These Lands thus perfectly Drayned, will return to be the richest of all your Lands, and the better Drayned the better Land."[23] By the second half of the 1600s, the English at home had perfected large-scale drainage and floating (flooded) meadows, and the use of sluices, floodgates, weirs, and dams, all to bring land under farmer's control.[24] Draining and floating dominated seventeenth-century British agriculture. John Mortimer's *The Whole Art of Husbandry,* (1708), reads almost like a guide for early rice planters. He discussed topics such as enclosing land with ditches and banks, methods to overflow fens with water, using cow dung to fertilizer fields, and a machine that could lift water from a river to elevated fields. He even discussed ideas about how to farm upland lands.

The continuous influx of people from England to Barbados would have spread such agricultural information to the island. Like John Stewart, the English Barbadians who settled in the Goose Creek area likely had knowledge of British agricultural practices, especially water control, and would have understood how to apply them to rice growing. Although rice was not grown on Barbados, and water control was not implemented in sugar growing, most books on agriculture coming from England during this time discussed the topic of water control, and it is hard to imagine that the planter elites did not have agricultural libraries or would not have felt it prudent to obtain agricultural books after they settled Carolina. Slave knowledge was also a possible source. If only one in a hundred slaves brought to the Lowcountry had African knowledge of rice growing using water control, this knowledge could have easily spread to the various plantations through the close-knit plantation society. In short at least some methods of water control were probably known to the early rice planters. Without water control on fields, the water regime for the benefit of the plants was ineffective, and weed and insect control was limited to manual methods. Trunks, in progressive designs, solved these problems.

The origin of the term "trunk" for the conduit to direct water into and out of rice fields is uncertain. A romanticized explanation of its origin is that a hollow tree trunk was used. Judy Carney has documented that hollow logs were traditionally used in Africa, suggesting an African connection.[25]

The earliest rice trunk may have been a *plug trunk,* nothing more than a hollow log with a plug in one end. The worker either reached down into the water and pushed in the plug or removed it; the plug might have been on the end of the log or in a hole on the top of the log.

No remains of plug trunks have been found in reservoir-fed rice fields. Physical remains of three other types of trunks do survive from such fields. These lift-gate, swing-gate, and lever-gate trunks are certainly not the only early types that were used, but they are the only ones documented through archeological remains. There were many reservoir fields, and the three designs more than likely represent only a fraction of trunks that were made and employed, far too few to develop a definitive chronology of trunk designs. Plantation slave craftsmen—working under direction of the planters or basing their designs on their own knowledge and ingenuity—undoubtedly fashioned a variety of trunks that were not based on previous plans or concepts. If a particular design worked, it was kept, no need to come up with another. Yet these different designs have been lost forever because no written records were made of them and no surviving physical structures are complete enough for us to understand their construction.

Lift-gates trunks were found in two separate reservoir rice fields. A lift-gate trunk found by Drayton Hall staff flowed water in one direction from the reservoir to the rice fields (see diagram 2A). Similar trunks might have been placed in the banks between

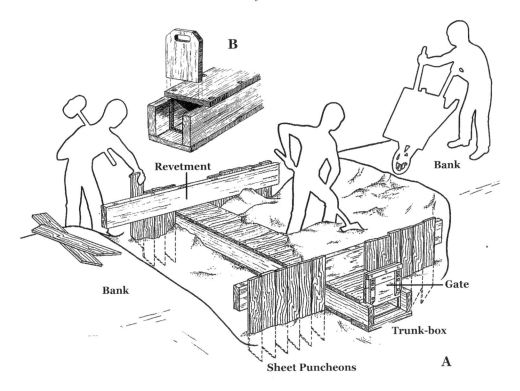

DIAGRAM 2. Lift-gate trunks. Illustration by William Robert Judd.

fields and flowed water from one field to the next one lower down. The design was simple and varied according to the builder's preference and available materials. The Drayton Hall lift-gate trunk is no more than a rectangular wooden trunk box approximately twelve inches high, two feet wide, and sixteen feet long. The sides are constructed of two sixteen-foot planks. The cover and bottom have cross planks, allowing construction of a wider trunk box. A hand-operated board-type gate on the reservoir end was lifted or lowered to control water flow. A vertical groove on the inner sides of the trunk box guided the movement of the gate. A horizontal groove in a bottom plank held the bottom of the closed gate to insure a tight seal.

A lift-gate trunk found at Caw Caw County Park in Ravenel, South Carolina, is similar in design to the one at Drayton Hall, but made from four planks that run lengthwise (see diagram 2B). The gates of both operated in a similar manner.

The partial remains of what appears to be a lift-gate trunk were found on the Combahee River. This trunk was fashioned from a carved log topped with a dressed board; then the trunk was turned over and installed with the board side placed down. Like the one at Drayton Hall, the gate fits into a groove. This trunk was put together with pegs, indicating that it is significantly older than the Drayton Hall trunk, which was put together with nails.

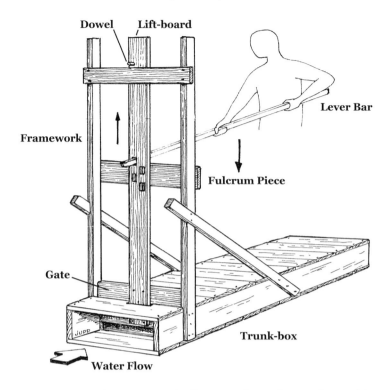

DIAGRAM 3. A lever-gate trunk. Illustration by William Robert Judd.

The wider trunk box of the *lever-gate trunk* gave planters the ability to flow a greater volume of water than they could with the narrower lift-gate trunks. With a wider trunk box, however, the gate had to be wider, and its extra weight, as well as the greater water pressure exerted against the gate, might have made it too heavy for a worker to lift manually. A lever solved this problem.

A trunk with a mechanically operated lever gate (see diagram 3) was found at Bonneau Ferry, a South Carolina Department of Natural Resources Heritage Preserve on the Cooper River in Berkeley County. This lever-gate trunk is located in an earthen bank that separated an inland reservoir from a rice field. Water flowed from the reservoir into the rice field. The larger and heavier gate required a centered, vertical lift board, which operated within a vertical framework attached to the reservoir end of the trunk. A vertical groove cut in each inner side of the trunk box guided the gate's movement. The first top cross plank was notched to clear the lift board. A space was left between the first and second top cross planks, creating a slot for the gate to move up and down. Staggered rectangle openings were placed midway in the lift board to receive a lever bar, which pried against the fulcrum piece of the framework and lifted or lowered the gate. At the top of the lift board were staggered one-inch adjustment holes through which a wooden dowel was inserted to hold the gate open.

The lever-gate trunk at Bonneau Ferry has not been dated by archeologists; however, its location in an obvious inland-swamp rice field suggests it is a heretofore unrecorded step in the evolution of rice trunks.

Inland rice fields often bordered on and emptied into salt marshes. If the planter wanted to drain a field over a few days using either a lift-gate or lever-gate trunk, to prevent salt water from entering the field, a worker had to man the trunk and close the gate each time the tide rose. A possible solution was a *swing-gate trunk* from England. In 1708 John Mortimer described a trunk in Essex, England, that worked automatically with the change in tide level (see diagram 4A). The Essex trunk was used to drain fens that bordered tidal creeks: "only according to the Quantity of the Water that they have to vent, they lay in it several square Troughs which are composed of four large Planks of the same length that they design the Thickness of the Head to be, and towards the Sea is fitted a small Door which opens when the fresh water bears upon it, and shuts when the Door is being put next the salt Water."[26]

A portion of a swing-gate trunk was found in 1993 at Popular Grove Plantation on the South Carolina shore of Back River in Jasper County by Tidewater Atlantic Research, Inc. during a survey for the U.S. Army Corps of Engineers. Its similarity to the Essex trunk suggests that the South Carolina version was derived from the English model. A conceptual drawing was made from the remains (diagram 4B). Disruption of the site made it impossible to determine how this trunk was used. Found on the edge of a canal, it was likely employed to drain water from a field but not to allow water to reenter. A swinging gate on one end indicates it was not manually operated but subject automatically to action of the tides.

Only measurements of exposed portions were documented. No digging or uncovering were attempted; therefore much information about this trunk was unobtainable. The trunk was buried in mud with only a portion of the gate visible. The trunk box measured

A. Essex Swing-gate Trunk

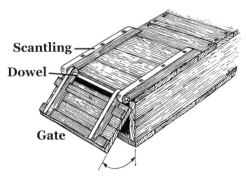
B. Rice Field Swing-gate Trunk

DIAGRAM 4. Swing-gate trunks. *A:* Essex swing-gate trunk from John Mortimer, *The Whole Art of Husbandry* (2nd. ed., 1708). *B:* illustration by William Robert Judd based on remains found on the Back River, Jasper County, South Carolina.

3' 4" wide and approximately 1' 6" high; its length was undetermined. There was unlikely a gate on each end since it appears the trunk flowed water in one direction. The pivoting gate on the river side was suspended on a horizontal, 1" diameter wooden dowel fastened to the ends of two short pieces of scantling. Each scantling was secured to the top of the trunk box on each side. This assembly functioned as a hinge and allowed the gate to swing.

The swing-gate trunk may have acted as a check valve similar to riser boards installed in rice trunks today. They were placed in a rice field bank at a higher elevation than the tide trunk. A sudden storm could have flooded the rice field while the inner door of the tide trunk was closed, the rising water possibly breaking the bank. Excess water could flow automatically through the swing-gate trunk into the river.

With the evolution and employment of trunks, greater control of water allowed the development of the major method of reservoir culture, inland-swamp culture.

With the beginning of inland-swamp rice culture, planters started to use slaves to impose artificial order on the natural landscape on a grand scale unlike anything else in the colonies, and this imposition proceeded on an even grander scale with the advent of tidal culture, which completed the transformation of the tidewater land to an artificial landscape. So when did planters first turn to controlling to flow water onto inland-swamp rice fields? This book suggests a time around the late seventeenth century or the early eighteenth century, a date earlier than historians have previously argued. The first mention of using water control in the colonies may have been in 1690 when John Stewart wrote: "The discovery of pine land to excel far our oakground either for graine Englysh of Ryce yea for flax too and especially for Cotton The proposing our dry swamps to be clear'd befor all other Vineyards and Ryce and with drains for graine one acre to yield [illegible] with four of hickory land never to wear out."[27] His "drains for graine" may or may not be a reference to water control, but it certainly is possible given that he was an agricultural innovator and surely would have had knowledge of water-control methods. His drains may have been nothing more than drains for removing excess water from his fields. Nonetheless this passage is the earliest known record of an attempt to control water.

Early rice planters and field workers would certainly have observed that rice fields on small-stream floodplains flooded during heavy rains. Flooding was critical to keeping the soil moist and the main reason rice was planted on small-stream floodplains. If flooding occurred during harvest time, however, it destroyed the crop. A means to prevent flooding during such critical times was needed. Certainly someone realized that a dam constructed at the head of the tributary that flooded the crop would back up the water, preventing flooding. Building dams high enough to hold back all water during heavy floods would not work, so they fashioned primitive overflows or bypasses to reroute excess water past the dam and around the fields. This technology might have been intuitive to the planters and perhaps part of a slave's African past. Again intuition or past knowledge could have made someone realize backed-up water could also serve as a reservoir, and the water could be let into the fields during a drought. A primitive plug trunk (or another unknown kind of trunk) would have provided a means to flow the water as needed from the reservoir to the fields. This initial system was not perfect, however, for during heavy freshets, water would have flowed over the dam and flooded the fields at inopportune times anyway. But a beginning of water control was established, and this method of growing rice became known as inland-swamp rice culture.

An early precedent was set for using dams in the Lowcountry. In 1701 Daniel Axtell and his partners on Newington in Dorchester, South Carolina, dammed a creek to create a reservoir that could power a sawmill and possibility flood a rice field. They used a water-control structure of unrecorded means to control the flow of water from the reservoir to the sawmill. This knowledge of using dams to hold back water would certainly have been known as well by the Cooper River planters only a few miles distant.

Rice planters and workers must have noticed that, when rice grows in water, it flourishes and produces more fruit than when it grows in soil that is only wet. They would have observed that rice suffered no ill effects during floods that occurred during the growing season and left fields underwater for extended times. On the contrary, the rice flourished. Hence a second dam was constructed at the other end of the swamp to enclose the field, and periodic floodings were practiced from the reservoir. A trunk in the lower bank drained the field for harvesting. This early controlled flooding was only for the benefit of the plant. Using water to control weeds came later. With reservoirs, banked fields, and bypass measures in place, the basic inland-swamp method of growing rice was in place. In time, it became more sophisticated and spread throughout the Rice Kingdom.

Naturalist Mark Catesby recorded observations that have been interpreted in different ways. Commenting on rice growing, he wrote: "It agrees best with a rich and moist Soyl, which is usually two Feet under Water, at least two Months in the Year. It requires several Weedings till it is upward of two Feet high, not only with a Hough, but with the Assistance of Fingers."[28] Does Catesby's "two Feet under Water, at least two Months in the Year" refer to intentional water control on the rice fields or flooding by a freshet? He also recorded, "At the latter End of July or August it rains in great Quantities usually a Fortnight or three Weeks, overflowing all the *Savannah* and lower Ground,"[29] and "great rains, which usually fall at the latter Part of the Summer, raises the water two or three feet, which nevertheless, though it usually remains in that state for some weeks, [rice] receives no Detriment."[30]

The "two Feet under Water, at least two Months in the Year" is his most intriguing observation. No inland swamp would have flooded every year. Rainfall from year to year was not the same, and many years could have passed before a rain heavy enough to flood a swamp occurred. So why did Catesby record this statement? He may well have been describing a rice field that was under water control and was flooded at his visit. His second observation, perhaps at another plantation's field, that "great rains "cover the rice wholly" may refer to a period of flooding from heavy rains when the banks were overtopped. His omission of any reference to banks does not necessarily mean that they were not there.

It is still a stretch of the imagination to interpret his comments as proof of water-control structures, but such a reading is not unreasonable. In light of the history of water control since the dawn of agriculture, these early water-control structures were not new technologically, and knowledge of them was certainly not beyond the reach of early planters or slaves. Even if they had no prior knowledge of water control, it was not a great leap of ingenuity to realize that a dam thrown up at the head of the tributary feeding the fields could have temporarily served this purpose. In fact it stretches credulity to think that no one would have reached this conclusion as they viewed their rice under water.

It took years for water-control practices to be implemented fully throughout the fledgling Rice Kingdom. The first archival reference that unequivocally documents inland-swamp rice culture using water control was a legislative act of May 29, 1744. Although the reference might be used to place the beginning of inland-swamp culture close to this date, the evidence suggests instead that water control had become widely used enough to require regulation and that it had reached a complex level. The act's preamble begins: "Whereas, it hath (of late) been frequently the practice of many persons to make dams or banks for the reserving of water, whereby they sometimes prevent their neighbors planting, and at other times let off their reserved waters and overflow the lands of their neighbors at their manifest prejudice; to prevent, therefore, the like evil for the future, we hereby pray."[31]

With new water-control methods in place, crop production increased. Rice became not only a domestic crop but a major export crop. By 1730 South Carolina was exporting eighteen million pounds annually (see table 1).

Inland swamps constituted the greatest acreage devoted to reservoir-fed rice fields. Small waterways flanked by small-stream floodplains flowed into virtually every river or creek in the Rice Kingdom. Inland-swamp fields and reservoirs required a large slave force to clear the swamp vegetation and construct the elaborate system of banks and canals necessary to control water. A slave population, expanding both through natural increase and importation, provided the labor force. As the demand for labor grew with inland-swamp rice culture, the colonists became more active in the slave trade from West Africa.

The inland-swamp system depended on rain-fed reservoirs (also called "reserves" or "backwaters") to supply water for the rice crop. Although each field varied depending on topography, a description of the inland-swamp rice field at the Bluff Plantation on the Western Branch of the Cooper River in Berkeley County documents the basic design of the inland-swamp rice field (see diagram 5). Earthen banks (1 and 2) were constructed across two inland swamps. Bank 1 was on the down-flow end of the swamp and trapped water above the dam. (The upper bank is not shown.) Bank 2 is the upper bank of the other reservoir, and the water was held in the swamp between the two banks. (The lower bank is not shown.) Further down the swamp, at a lower elevation, an upper bank and a lower bank were constructed from upland to upland. The area between these two banks was then cleared of trees and ditched. This was the inland rice field. A trunk, probably a lever-gate type, was installed through the upper bank to control water flow from the

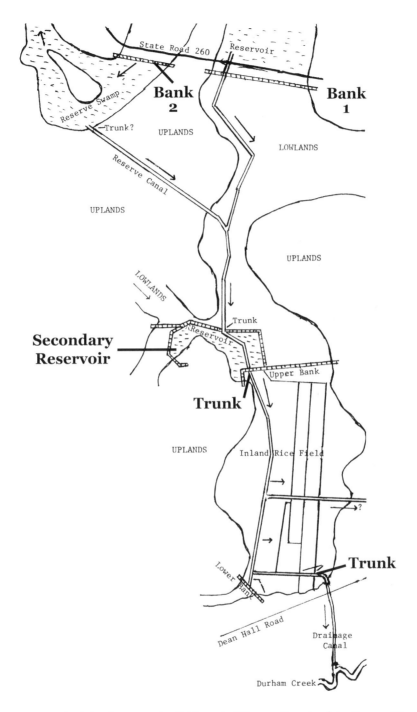

DIAGRAM 5. An inland-swamp rice field on the Western Branch of the Cooper River, Berkeley County, South Carolina. Numerical and letter labels have been added to the original map. The inland rice field at the Bluff was recently converted to a waterfowl impoundment, and all traces of the canals, banks, and trunks were destroyed. Illustration by Richard D. Porcher, Jr., originally published in "Rice Culture in South Carolina" (1987).

reservoirs to the fields. Another trunk installed in the lower bank drained the water from the fields into an adjacent waterway. The fields were now ready for planting. A secondary reservoir was constructed just above the upper bank of the rice field as an extra supply of water. (Secondary reservoirs were not made on all inland rice fields.) Water flow was one way: reservoirs → rice field → drainage canal → adjacent creek.

The Bluff had an unusual system to prevent the fields from being flooded during heavy rains. The inland swamps were the main reservoirs. Excess water from heavy rainfall could be let out of the reservoir swamps downstream through trunks, thereby diverting the water from the fields.

Other inland fields had a different method for flood control. A flanking canal was constructed from the reservoir down the side of the rice fields and emptied into an adjacent river below the lowest field. A trunk (probably a lever-gate type) controlled water from the reservoir into the flanking canal. During freshets, when water threatened to overtop the bank, the trunk's gate was opened, and excess water drained from the reservoir into the flanking canal. The flanking canal also served to flood and then drain rice fields down the canal. Placing a trunk in the middle of the dam opening into a major canal was another method to let off excess water from the reservoir. Rice fields were placed successively down both sides of the canal. Excess water from the reservoir was let into the canal and emptied into a river below the last field. The canal was also the source of water for the rice fields.

Flanking canals also solved or lessened a problem that developed when an entire stream or tributary with multiple owners was developed into rice fields. Planters on the upper end of the waterway controlled the water flow. They quickly filled their reservoirs, sometimes taking so much water that the planters lower down had little or no water for their reservoirs. With flanking canals, overflow water could be sent down the flanking canals to the reservoirs downstream. Everyone benefited. Planters with reservoirs in the upper end of the waterway could prevent their fields from being flooded by freshets by sending the excess water around their fields; reservoirs lower down could be filled with this excess water.

Often an inland swamp ran for a mile or so through the Lowcountry before it emptied into a river or creek. Along its way it was fed by rainwater from tributaries and/or springs. The swamp traversed several plantations, and each planter converted his section of swamp to inland fields. Small tributaries that emptied into only one planter's field posed no problem; however, water passing down the swamp from higher up was used in succession by the various planters, and they had to work together to ensure that each planter had enough water for his fields. Problems often arose; either planters settled then as gentlemen, or they settled in courts. The few lawsuits between planters testify to how well they worked together to distribute the water supply.

The inland-swamp system had one main drawback in producing a marketable crop: the variable water supply. Droughts, which often occurred, meant crop loss. Despite flanking canals, freshets often overtopped reservoir banks and flooded crops at inopportune times. Planters especially feared freshets. Peter Manigault reported in 1766 that South Carolina had "such incipient Rains that all the Rice Lands are under Water and numbers of people will not be able to plant this year."[32]

Weeds had been a problem from the beginning of rice culture. John Stewart advised fellow planters to sow rice "thick as barley [as] in britain its roots chokt the weeds" and they would "have my way at least 600 bushel of Rice without howing on stroak or pulling up one weed." Once cleared of trees, inland swamps received more sunlight, which proved an ideal environment for opportunistic weeds. Weedy species such as barnyard grass, *Echinochloa crusgalli* (L.) Beauvois, and others lay in wait on the nearby uplands to invade the rice fields. Weed seeds quickly spread into the cleared swamp, responding to the increased sunlight and moisture-laden fields and little competition, and by late spring the rice fields were a carpet of weeds. Stewart's method of choking weeds with thick rice plants would not have worked in the new fields, for the weeds would have developed height so quickly that rice would not have choked them out.

Slaves had to pull weeds by hand and hoe endlessly in knee-deep muck and summer heat. The toil of weeding prompted slaves to abscond, eroding the planter's authority over their labor. Weeding became a serious point of contention among slaves, overseers, and planters, and spurred the development of the task system. Writing in 1774, Josiah Smith, fearing his slaves would run away when the weeds were thickest, expressed the hope his overseer's moderate treatment and supplemental rations of beef and rum would keep them in the fields.[33]

At some point inland-swamp planters resorted to irrigation to control weeds, a practice called *water culture*. When water culture began is not clear. What may be a reference to water culture appears in James Glen's 1761 *Description of South Carolina*: "The Country abounds every where with large Swamps, which when cleared, opened and sweetened by Culture, yields plentiful crops of Rice: Along the Banks of our Rivers and Creeks, there are also Swamps and Marshes, fit either for Rice, or, by the hardness of their Bottoms for pasturage."[34] Since his book was published in 1761, his observations must have been made earlier, but how much earlier is unknown. His "sweetened by Culture" undoubtedly refers to water control, but it is a stretch to say it refers to water culture to control weeds. It is also not clear if Glen was referring to inland swamps, but at this period water control for weeds in tidal swamps was in the future. The mention at the end of swamps and marshes on riverbanks seems to imply that the first reference was to inland swamps.

William Gerard De Brahm, the German-born surveyor general of the Southern District, made the first unmistakable reference to using water to control weeds. His general survey of the Southern District, which began in 1765 and ended in 1770, was published in 1773 as *Report of the General Survey in the Southern District of North America*. In his chapter on South Carolina, he wrote that the country "produces also many ingenious Stamp Machines, or rather Rice Mills, and has erected innumerable Dams for the Use of reserving Water to inundate the Rice Fields, after the first hewing or weeding, in order to hinder the Roots of Grass and Weeds from sprouting again."[35] This statement clearly refers to inland-swamp culture and water control of weeds. This reports pushes the advent of water culture for weed control to a few years prior to 1770, possibility as early as 1765, when he began his tour.

The weed control De Brahm described relied on timely flooding. Initial hoeing beat back the first crop of weeds that had sprouted among the rice plants, especially grasses, which were the main weeds competing with rice. The fields were flooded, keeping the water level below the rice heads but above the grass tops. Rice plants kept growing, but the weeds could not stretch under water. Water culture was not perfect, and slaves often had to return to the fields to hoe or remove weeds manually.

Historian David Ramsay (1749–1815) gave credit for the introduction of water culture to Gideon Dupont, Sr. (1712–1788), saying that South Carolina is indebted "to Gideon Dupont [Sr.], of St. James Goose Creek, for the water culture of rice; he was an experienced planter of discernment and sound judgement, who after repeated trials ascertained its practicability." Ramsay reported that Dupont petitioned the legislature in 1783 for compensation for his discovery, but his resolution of petition was turned down because the state treasury was empty. Furthermore, according to Ramsay, Dupont's "method is now in general use on river swamp lands." Was Ramsay correct in giving Dupont credit? If Dupont did indeed introduce water culture, he has to have done it before 1770, the year of De Brahm's report, seemingly a long time before he petitioned for compensation.[36] Another problem with Dupont's claim and Ramsay's support is neither said whether the new method was for inland-swamp or tidal culture.

James Louis Petigru gave strong evidence that Dupont was not the innovator of water culture. In an 1843 letter to Robert Allston, who had asked Petigru to investigate the veracity of Ramsay's claim concerning Dupont, Petigru replied that "I am much disposed to think it fabulous. The water culture of Rice must have been more or less understood from the beginning and the additions that were made to the stock of Knowledge among those who cultivated the grain, were likely to be the gradual results of experience, rather than the sudden accession of a discovery." Petigru went on to say that he could not find the 1783 resolution and could not trace the Dupont family.[37] If there is other evidence that Dupont was the innovator of water culture, it has not surfaced.

From 1770 on, references to using water culture to control weeds were common, and the practice became widespread. "An American" reported in 1775 that "the great object of [water] culture is to keep the land clean from weeds, which is absolutely necessary, and the worst weed is grass: if they would say a man is a bad manager, they do not observe such a person's plantation is not clean, or is weedy, but *such a man is in the grass;* intimating that he has not enough negroes to keep his rice free of grass. This is the only object until it is reaped."[38]

Using reservoir water to control weeds, however, was not always reliable. Often there was not enough water in the reservoir for the two flooding necessary to control weeds adequately. Techniques developed in inland-swamp culture were later refined for tidal culture. River water was more reliable to control weeds and became a mainstay of tidal rice culture, where water culture became highly technical and reduced the toilsome manual labor needed to bring a crop to market.

Inland-swamp culture persisted until the Revolution, after which it was replaced as the major method of rice growing by the tidal method. Inland-swamp culture did not cease entirely, however, and there is no one date historians can point to as its end. Plantations without tidal river swamps to convert to tidal fields retained their reservoir-fed fields until the Civil War and afterward if the planter could still plant rice profitably. This was true also in North Carolina and Georgia.

At Limerick Plantation on Huger Creek in Berkeley County, South Carolina, inland-swamp fields expanded from four fields in 1786 to twelve fields by 1797. This increase in inland-swamp fields occurred because Limerick had fewer than ninety-five acres of tidal marsh.[39] On Jericho Plantation, along Nicholson Creek in Berkeley County, Mathurin Guerin Gibbs (1788–1849) planted inland-swamp fields from 1845 to 1849 and drew water from Hell Hole Swamp. He stated in his plantation register on November 27, 1845, that he had harvested a total of 270 bushels of rice; he saved 50 bushels for seed and kept 220 bushels for "family use and market."[40] Duncan Heyward wrote that his father moved slaves in 1858 from his Combahee River plantation to Goodwill Plantation on the Wateree River twelve miles south of Columbia, where they expanded the acreage of rice fields in the river swamp, indicating that inland-swamp rice was grown at Goodwill a century after tidal rice commenced.[41]

Today inland rice fields and reservoirs lie abandoned throughout the Carolina Lowcountry and the rest of the Rice Kingdom, and miles of canals and earthen banks traverse swamps and woodlands, seemingly out of place in dense Lowcountry swamps. Most of the inland-swamp fields have been reclaimed by forest, and some have been timbered more than once.

The reservoirs (see color plate 2) are a timeless legacy of inland-swamp cultivation. Many reservoirs still have intact banks and hold water most of the year. They provide habitats for wading birds and other waterfowl. The abandoned fields and reservoirs are a permanent, silent testimony to an earlier time, when the land was used to begin and sustain rice culture and slaves were introduced to the Lowcountry.

Although inland swamps were the main locations for reservoir culture, planters created reservoir-fed fields elsewhere as well. While these fields were minor in acreage and production compared to the inland-swamp fields, their designs, however, were ingenious.

A salt marsh at Drayton Hall in Charleston County may be one of the earlier sites where rice was initially grown using reservoir water. A small finger of salt marsh penetrating the uplands (see color plate 3) is crossed by the remains of four earthen banks. The outer bank (see plat 2) was constructed at the outer fringe of the marsh, where it was continuous with the salt marsh of the Ashley River, thus preventing salt water from entering the field. Once the outer bank was constructed, and the three inner banks were built to divide the marsh into individual fields (see plat 2, numbers 2, 3 and 4), a combination of

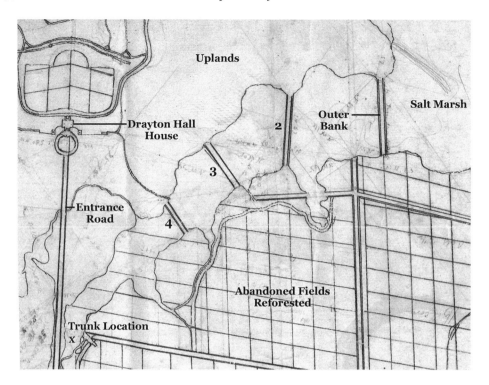

PLAT 2. Drayton Hall rice fields, circa 1790. Map by Charles Drayton. Courtesy of Drayton Hall National Historic Trust. The banks that separated individual fields are labeled "Outer Bank" and numbered 2–4. The *x* marks the site of the lift-gate trunk.

rainwater and water from the reservoirs flushed the salt water from the fields. Then rice was planted and supplied with water from the rain-fed reservoirs. A swing-gate trunk in the outer bank may have been installed to prevent salt water from entering the fields while allowing water to drain from them.

This salt-marsh field is the location of the lift-gate trunk discussed earlier (see diagram 2A). Because it was constructed with iron nails, it is probably not the original trunk. It may have replaced a similar trunk made with wooden pegs.

The main advantage of using salt marsh for rice fields is that the marsh did not have to be cleared of woody vegetation, so converting it to rice fields could have been accomplished with a small labor force, making it cost effective in the early years in the rice industry, before large numbers of enslaved Africans were imported.

Artesian springs served as water source for rice growing, but only to a limited degree. A spring at Indianfield Plantation was typical of those in Middle St. John's Parish (see plat 3, insert). On Woodboo Plantation, Big Spring fed water directly into a rice field. Other Woodboo springs fed water into a reservoir adjacent to the rice fields, where the water was held and released into the fields at planters' discretion. Col. John Christian Senf commented on the spring at Woodboo and documented that it was used for the

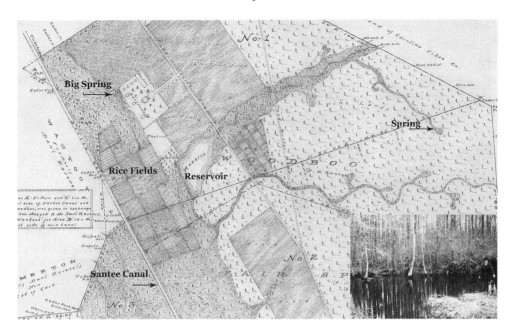

PLAT 3. Woodboo Plantation, 1806. Map by J. O. Palmer: "A Plan Exhibiting the Shape, Marks, Buttings, and Boundaries of the Woodboo Plantation, Situate in St. Johns Parish—Berkeley County and Charleston District.—Belonging to Stephen Mazyck," certified March 22, 1806. Courtesy of the South Carolina Historical Society. *Insert:* Limestone Spring in St. John's Parish, Berkeley County, South Carolina, 1921. From "A Souvenir of a Hunting Trip at Northampton, Bonneau, S.C., February 1921." Photograph from the collection of Richard Dwight Porcher, Jr.

rice fields of Woodboo: "—Towards the Rice field four yards back, through the left Bank enters the Canal the Drain from the large Spring of Stephen Mazycke [Mazÿck], from which the Rice field has at all times a sure supply of Water, without recourse to Biggin creek."[42]

Edmund Ravenel described a rice field in adjacent Wantoot Plantation that was supplied by water from an artesian spring: "The water here issues from the marl which is about two or three feet below the surface at this spot. This water passes South and is carried under the Santee Canal in a Brick-Aqueduct, to be used on the Rice-Fields of Wantoot Plantation."[43] Wantoot and Woodboo were two of the many plantations located in Middle St. John's Parish that had limestone springs. Both plantations were flooded by Lake Moultrie in 1942.

Artesian springs were used on Cumberland Island, Georgia, to grow rice. Great Swamp Field, which occupied one-third of the island's twenty-four thousand acres, contained several artesian springs that provided a flow of freshwater north through the swamp until it met salt water. Rice cultivation on Cumberland probably began in earnest after 1770, when wealthy South Carolina planter Thomas Lynch and his agent and factor Alexander Rose bought the swamp.[44]

On Hobcaw Barony along Winyah Bay in Georgetown County, an area labeled the "1000-Acre Rice Field" on an 1840 map was used to grow rice (see plat 4). Whether the original vegetation was swamp or marsh is unknown. The southern area of the "1000-Acre Rice Field" borders Winyah Bay, where the water is salty and unfit for rice culture. An ingenious plan was devised to bring freshwater to the field. A canal was constructed from the field northward through the uplands to Winyah Bay above the salt point. Fresh water from Winyah Bay, which functioned as a "reservoir," was flowed through the canal into the rice field; water was drained off the field lower down into Winyah Bay at low tide.

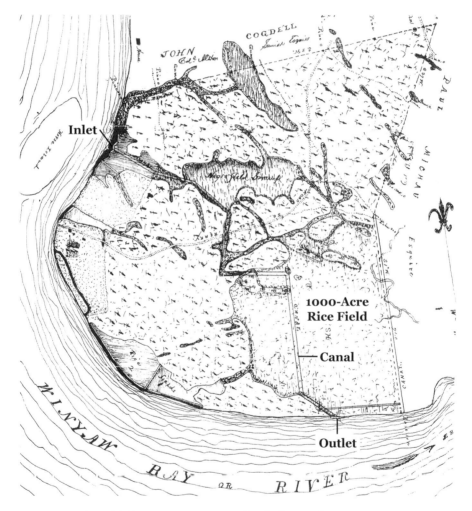

PLAT 4. Plan of Calais, 1796. From J. Hardwick, "PLAN of—CALAIS—A PLANTATION belonging to the REVd Mr HUGH FRASER. Situated in All Saints Parish Georgetown, District State of So Carolina, Having such Courses, distances, bounds, and marks as are represented and expressed in the PLAN From an actual Survey taken in July 1796." Plat book B, pp. 114–15, Georgetown County Courthouse, Georgetown, South Carolina. Calais was located on the southern tip of Hobcaw Barony on Winyah Bay.

Rice could also be grown on inland ridges and swales. Across Winyah Bay from the "1000-Acre Rice Field" is Estherville Plantation. The uplands of Estherville exhibit the ridge and swale geological formation created as the ocean receded during the late Pleistocene epoch (see color plate 4).[45] The swales supported swamp vegetation; rainwater and runoff from the adjacent upland ridges drained down the swales into Winyah Bay. Two of these swales were dammed near their confluence with the bay, forming reservoirs.

Land suitable for rice fields occurred along the southeast of Estherville along Winyah Bay. Because the bay was salty at that point, however, a freshwater supply had to be procured elsewhere. A canal was constructed from the reservoirs extending south to the rice fields and supplied the freshwater necessary for rice culture. At harvest time, the water was drained through trunks into Winyah Bay.

Rice could also be grown on barrier island ridges and swales. It was grown on at least one barrier island along the South Carolina coast, Hilton Head Island in Beaufort County. Ridge-swale formations similar to the one at Estherville, but formed during the Holocene epoch, comprise the larger coastal barrier islands along the South Carolina coast. On Hilton Head the upper sections of several of these swales were banked to form a reservoir. At the lower end another bank was constructed from highland to highland, and the vegetation was cleared for rice growing. Water flowed from the reservoir into the fields, then drained into the adjacent bay to allow harvesting. Within the sounds of waves breaking on the seashore, rice was grown.

With methods in place to grow rice successfully and with means of water control established, rice culture spread from its Cooper River epicenter throughout what became the Rice Kingdom.

The Spread of Rice Culture

From a beginning along the Cooper River and its tributaries, rice culture spread north and south through the Carolina Lowcountry. Entire watersheds were converted to reservoir-fed fields. In 1704 Parliament added rice to the enumerated goods that could be shipped only to England, Wales, or Berwick-upon-Tweed, where it was repackaged and sent to European and other foreign ports. The act did not significantly reduce the expansion of the international rice trade, and in 1731 Parliament relaxed the act and opened certain regions to direct colonial trade. By 1720 rice had emerged as a major Carolina export crop and continued to be the principal staple throughout the rest of the colonial era. In 1720 more than half the value of Carolina's exports came from rice.

Starting around 1738, when planters realized tidal water was a more certain water supply, they began to experiment with tidal culture. Tidal culture either spread or was developed in place north to the Santee Delta and Winyah Bay, where five freshwater rivers joined. The tidal swamps of four of these rivers—the Waccamaw, Sampit, Black, and Great Pee Dee—were suitable for tidal rice culture. Southward, every freshwater river had tidal swamps that could support rice culture was claimed, and rice plantations developed on the Edisto, Ashepoo, Combahee, Coosawhatchie, New, and Wright Rivers.

Commercial rice culture began in what would become North Carolina around the 1720s, when a group of wealthy Goose Creek planters settled in Lower Cape Fear. These planters went to the area for the naval-stores industry—tar, pitch, and turpentine from the abundant longleaf pine forests of the coastal region. A cessation of the bounty on naval stores in Carolina was one reason they migrated to Cape Fear. In essence the Cape Fear plantation system was an extension of the plantation enterprise of the Goose Creek Men and by extension that of Barbados and the West Indies.

Naval stores remained the principal economic interest in the Cape Fear region throughout the colonial period. Because the settlers had knowledge of rice culture and a contingent of slaves, rice developed into a secondary export in the region, which had an environment similar to the Cooper River area. As early as 1731, travelers in the Cape Fear region reported rice culture.[46] In the same year the North Carolina assembly established rice as one of the official commodities of the colony, an indication that rice had become an export crop.[47]

The many small tributaries of the Lower Cape Fear were ideal sites for inland-swamp fields, but because they were subjected to droughts like their neighbors to the south, rice culture was shifted to the tidal swamps of the Cape Fear River. Cape Fear was the northern extent of tidal rice culture in North Carolina. The river is the only one along the state's entire coastline that empties directly into the Atlantic Ocean, creating tidal dynamics that produced a freshwater system capable of supporting rice cultivation. North Carolina rice culture never reached the extent it did in South Carolina. Of the five hundred to six hundred tidal rice plantations in the Rice Kingdom, South Carolina had more than four hundred, Georgia about one hundred, and North Carolina, some twenty-five.[48] Tidal rice was grown until rice culture ended in North Carolina in 1906.

By the time the Georgia colony founded by James Oglethorpe was settled in 1733, South Carolina was already an important rice-producing colony. Although Carolina planters were aware of the potential of the Georgia swamps and tidelands for rice growing, the ban against slavery and limitations on fee-simple land acquisition implemented by the trustees of the Georgia colony discouraged them from migrating to Georgia. Eager to exploit the Georgia tidelands if the trustees' experiment failed, Carolina planters fostered discontent among the Georgia colonists. By the mid-1700s lands suitable for planting rice in South Carolina were at a premium. Since eldest sons usually inherited the estates (by the law of primogeniture), like their ancestors who left Barbados seeking new lands to plant, the last-born sons of Carolina planters had to look elsewhere for lands suitable for rice cultivation.

Georgia's social experiment ended in 1750, when the trustees dropped the ban on slavery, and the state reverted to the English Crown in 1752. Carolina planters rapidly settled on Georgia's coastal plain, where they found a climate and environment of inland and tidal swamps similar to the Carolina coastal area. They brought their slaves to Georgia, where cheap and fertile land, the promise of lower taxes, and increased credit quickened their expectation of success as rice planters. They began cultivation in the late 1750s. Surveyor De Brahm commented that "many rich Carolina Planters . . . came with all their Families and Negroes to settle Georgia in 1752; the Spirit of Emigration out of South Carolina into Georgia became so universal that year near one thousand Negroes were brought in Georgia, where in 1751 were scarce above three dozen."[49]

Carolina planters established reservoir and tidal culture along the major rivers, first on the Savannah and Ogeechee, which were close to Savannah. Charles Colcock Jones (1831–1893) described a typical migration to Georgia: "After laying by their crops in Carolina in the fall of that year, the planters came with able-bodied hands, and during the winter, cleared land and built houses. In a season or two having thus sufficiently prepared the way, they brought their families and servants in the early spring and at once entered upon the cultivation of the soil. Thus was the removal rendered as safe and comfortable as the nature of the case permitted."[50]

John Potter of Charleston settled on the Georgia side of the Savannah River. The Midway District between the Ogeechee and South Newport Rivers was settled next. Stephen Elliott, Daniel Blake, and Arthur Heyward, sons of wealthy Carolina planters, settled plantations along the Ogeechee. Among the Carolinians who settled on the Altamaha and Satilla Rivers were Duncan Lamont Clinch, William Gignilliat, Hugh Fraser Grant, and T. Pinckney Huger. When the boundary of Georgia was extended south from the Altamaha River to the Saint Marys River, Carolina planters extended their new rice frontier southward. By the end of the century, just as in Carolina, most of the reservoir fields were abandoned, and until the end of rice production in Georgia, tidal rice was grown along the Georgia freshwater rivers in a narrow band of tidewater counties, including Chatham, Bryan, Liberty, McIntosh, Glynn, and Camden.

The availability of credit, slaves, and large land grants all stimulated Georgia's economy, and by 1775 slavery had established itself as Georgia's most important social and economic institution. During the Revolutionary War, Georgia tidewater rice plantations of both loyalists and patriots fell into disrepair, as they also did in South and North Carolina. After the Revolution the tidal rice fields were slowly rehabilitated. The plantations of Georgia loyalists were claimed by the new state government and distributed to new landowners. The state favored Revolutionary War heroes, two of whom were Generals Nathanael Greene and Anthony Wayne. Rice production reached even greater heights after the Revolution and became Georgia's first important staple commodity. Rice was the unchallenged market staple in tidewater Georgia; thirty thousand acres were planted along the coastal rivers. By 1860 Georgia tidewater plantations were producing 28 percent of the total rice yield in the United States. The day of the great Georgia rice plantations, modeled after those in Carolina, had arrived. It was only natural that the Georgia plantations were similar to those of the Carolina coast; they were owned by sons or relatives of the Carolina planters.[51]

St. Peter's Parish on the Carolina side of the Savannah River was the last area to be converted to tidal culture. When planters realized the potential in the lower Savannah

River swamps for tidal culture on the Carolina side, they made plans to settle there. Charles and Jermyn Wright, who obtained land adjacent to Purrysburg in St. Peter's Parish, were first to plant the Carolina side, starting in the late 1750s. Henry Laurens purchased the most valuable of the Wright property in 1768, and until the Revolutionary War, he made continual progress in rice planting. Around 1800 John Williamson moved his slaves to his land opposite Argyle Island, and John Rutledge, Jr., developed rice fields on the Carolina side opposite the northern end of Hutchinson Island. The settlement of tidal rice lands in South Carolina was complete.[52]

Commercial rice culture spread to Florida on a relatively small scale along the Saint Marys and Saint Johns Rivers in East Florida. In 1763 Spain gave up Florida to ransom Cuba and the Philippines from the British. Spanish, Indians, and free black people loyal to the Spanish evacuated Florida, taking all the infrastructure they could load onto ships. The British controlled Florida from 1763 to 1784, and during this period, attempts were made to develop a plantation system similar to those in Georgia and Carolina. When Britain acquired Florida in 1763, Governor James Grant encouraged colonization by offering political offices and generous land grants to Georgia and Carolina planters, who came with their hundreds of slaves to break in their new estates. By the sweat and blood of African slaves, impressive plantations were established along the Saint Marys and Saint Johns Rivers.

As the American Revolution closed, loyalists fled Charleston and Savannah for East Florida with more than eight thousand slaves. They intended to transplant their plantation enterprise to Florida, but the disruption caused by the war made plantation operations almost impossible. Florida was retroceded to Spain in 1784 by the Treaty of Paris, and East Florida was disrupted by a colonial transfer once again. Nearly all British planters evacuated the province. On hearing Florida was being ceded back to Spain, major British planters such as John Moultrie shipped their slaves back into South Carolina, Georgia, or the West Indies.

Spanish rule lasted until 1821, when the United States took control of Florida, but rice cultivation never regained prominence there.

Tidal Rice Culture

The vicissitudes of reservoir rice culture were many. The water supply was too difficult to control; there was too much water during heavy rains and not enough during droughts for the benefit of the plants and for weed control. The fields were scattered, and there were no new inland swamps to cultivate. They created a high labor demand, and their shallow soils and decreasing fertility resulted in diminishing yields. All these factors prompted planters to seek better lands for rice growing. Beginning as early as 1737, planters found a more certain supply of water in the tidal swamps bordering freshwater rivers that traversed the Lowcountry. For the most part, their problems were solved.

The rivers of the Lowcountry were bordered by fringes of freshwater tidal marshes and wide expanses of low-lying, freshwater tidal swamps (see color plate 5). As the salt tide rose in the coastal estuaries into which the rivers emptied, freshwater twenty to thirty miles inland from the ocean backed up and rose. When the salt tide ebbed in the estuaries, it allowed the freshwater to flow into the estuaries, and the water level in the rivers dropped. Twice a day the freshwater ebbed and flooded, draining and then flooding the adjacent marshes and swamps. Swamps that could be used for tidal culture were limited to a coastal strip no more than fifteen to eighteen miles wide. The inland limit for culture was fixed by a diminished tidal range, and the seaward limit was created by encroaching salt water.

Planters developed an ingenious system to apply this diurnal rhythm of nature to rice growing, thus ensuring a more consistent and reliable supply of freshwater for benefit of the plants and for weed and insect control. On every freshwater river and estuary along the coast of the Rice Kingdom above the influence of salt water and upriver to the point where at least a three-foot difference in low and high tide occurred, slaves converted freshwater tidal swamps and marshes to rice fields. In some deltas, fields were created upriver thirty to thirty miles from the coast. Entire deltas and floodplains were converted to tidal rice fields, with fields extending several miles across and ten miles upriver (as in the Santee and Altamaha deltas), and entire islands within deltas were banked and planted in rice (as was the Argyle Island in the Savannah River). The advantage of tidal irrigation over reservoir irrigation was striking: inland-swamp fields yielded six hundred to one thousand pounds of rice per acre; tidal fields yielded twelve hundred to fifteen hundred pounds per acre. The labor of one slave could produce five or six times more rice with tidal fields. With the system in place, tidal irrigation led to a more calibrated flooding and draining, creating optimum growing conditions for rice and reducing the need for hoeing to control weeds.

The origins of tidal culture are uncertain. The *South Carolina Gazette* for July 16, 1737, carried an advertisement for the second Landgrave Thomas Smith, who was selling land on the Black River near Winyah Bay, "part of which is good rice swamp, that the Spring Tide flows on." On August 19, 1738, the *Gazette* printed a notice for a sale of land by Wm. Swinton of Winyah saying that "each contains as much river swamp, as will make two fields for 20 Negros, which is over-flow'd with fresh water, every high tide, and of Consequence not subject to the Droughts."

In the January 22, 1741, edition of the *Gazette,* Joshua Sanders offered for sale fourteen acres of land on the Combahee River in Colleton County, South Carolina, which he described as "in a good Tide's way." Whether these swamp lands were actually converted to rice fields in 1737, 1738, or 1741, or soon afterward, is unknown. The ads show, however, that at least as early as 1737 planters were cognizant that tidal irrigation was possible.

Who first developed tidal culture and where is also uncertain. William Ashmead Courtenay, mayor of Charleston in 1883–87, offered one possible answer: "I have made extended inquiries as to the date of the earliest successful experiment in reclaiming river swamp land for the culture, and find that Mr. McKewn Johnstone the Elder, raised a crop on such land at the 'Estherville' plantation on Wanyah Bay as early as 1758."[53] The date of 1758 seems reasonable given that it would have taken time to master the hydraulics of tidal culture if planters started with the system around 1737. The fact that Courtenay said the earliest *successful* experiment indicates that others had tried before but perhaps did not meet with success. It must be borne in mind, however, that Courtenay was writing in 1883, more than a century after Johnstone reportedly planted at Estherville.

Lord Adam Gordon, a British officer sent to gather information on the colonies in 1764–65, wrote from near Charles Towne: "The Tide Swamp in these Southern Provinces is by much the most valuable, since, when they are properly banked in, and your trunks and dams in perfect order, by a judicious use of these advantages, it is alternately equally capable and fit to produce . . . Rice."[54] Traveling near Savannah, he said that the soil "is light and Sandy, which is troublesome in windy weather, the Tide land on both sides the river, is mostly Cypress Swamp, and the fittest for Rice since, when properly banked in, and drained, you need have no dependence on Seasons, but either overflow or keep dry your fields, just as you please."[55] Courtenay's and Gordon's narratives are strong evidence that tidal culture was in place a decade or so before the Revolutionary War.

More than likely the development of tidal culture was a gradual evolution of complexity and sophistication starting around 1737–41 with contributions from many individuals, perhaps including Johnstone, some working together, others working independently, with knowledge of English farming, and perhaps with slave knowledge. There were many who had knowledge of inland-swamp culture and would have noticed the ebb and flow of tidal water into swamps and realized that here was a means to irrigate rice fields.

If a date of around 1737–41 represents the beginning of tidal culture, it was not until after the Revolution that tidal culture became the primary method of rice growing. The delay was probably influenced by several factors: the Stono Rebellion slave uprising of 1739; the time it took to solidify the task system of slave labor; the Spanish disruption of sea trade that provoked the War of Jenkins Ear (1739–42), which expanded into the War of Austrian Succession (1741–48) and brought rice culture to its lowest ebb; the long time it took to develop a commercially feasible method of processing rice for market; the time it took to master the complex system of gates, canals, and trunks to irrigate fields; and the devastating effects of the Revolution, which contributed to widespread economic distress in the Rice Kingdom.

The Revolutionary War interrupted the colonial rice trade. Before the war, England oversaw and supported production and transfer of rice to Caribbean and Old World markets. Based on the Navigation Acts, British economic policy dictated that all enumerated products from the American colonies must clear customs in England and be repacked and shipped to other ports in British ships. Closure of commerce with the British empire during the war ended the main market for rice. With the advent of the Revolution, the British stopped shipping slaves to the American colonies. Then British troops landed in Georgia in the fall of 1778 and captured Savannah. A continuous parade of British ships and Tory privateers from St. Augustine virtually halted rice trade from the Beaufort-Savannah area. Henry Laurens, who owned three rice plantations in Georgia, which he administrated from his Ansonborough residence in Charleston as part of his multiple-plantation enterprise, knew well the situation along the coast. With "Picaroons from St. Augustine" cruising the coast, Florida raiders "all over the Country," British men-of-war "sweeping" plantations along the Savannah River of slaves, and "a Number of Indians . . . coming against the Frontiers of Georgia," the "whole will either be destroyed stolen or lye [lie] . . . to perish by time & Vermin[,] no small sacrifice to the shrine of Liberty."[56] Property was stolen. Slaves fled to British camps seeking freedom. Disease and privation increased among slaves because medical help, although rudimentary at the time, was even scarcer during the war. On May 12, 1780, Charlestown, the main port city for rice exports, capitulated to Sir Henry Clinton and was occupied until the end of the war.

In North Carolina the Revolutionary era brought a significant change in the economy of the Lower Cape Fear. "No longer was the British bounty available for naval stores as an inducement for imperial production; consequently, Cape Fear exports declined sharply." Indigo production ended completely after the war, and planters then turned to rice. Since tidal culture had proved successful, it marked the postwar beginning of a large-scale commercial rice industry in the Lower Cape Fear.[57]

Anarchy reigned throughout much of the rice-growing area. The planters' community was broken apart, and their prosperity unraveled. White planters were drawn into the conflict as either patriots or tories. Overseers served in the Continental Army or local militia leaving little authority behind on plantations. Slaves seized the opportunity to flee or rebel against planters' authority. Other slaves volunteered or were forced into the American armies as common laborers or skilled workers helping army engineers build military fortifications. The English enticed slaves to fight for the Crown with the promise of freedom. Roughly one-third of the slaves were lost to the planters by the time the war ended. Many planters were loyalists, their lands confiscated by rebels who did not always know how to operate a rice plantation. Estates of leading revolutionaries were confiscated as Clinton's armies plundered the Lowcountry. Fighting between patriots and loyalists took its toll on resources and people throughout the tidewater area, making it difficult to plant and harvest crops. Reservoir-fed fields without continuous management quickly reverted to woody vegetation; banks and canals fell into disrepair from neglect. Without labor, banks broken by freshets could not be repaired. A broken bank not quickly repaired damaged the rest of the irrigation system. Crops were lost because few workers were available to harvest and prepare rice for market.

The war over, planters returned home to contemplate resumption of rice production. Specie was in short supply, and the state had burdened the people with debts for wartime loans. Did rice culture have a viable future? Could the plantation infrastructure be rebuilt? In 1784 Henry Laurens of Mepkin Plantation described the efforts of rice planters:

"All are busy in their respective vocations, covering as fast as they can the marks of British cruelty, by new Buildings, Inclosures, and other Improvements, and recovering their former State of happiness and Prosperity."[58] Some slaves who had stayed away several years returned to their masters' plantations, joining those who had stayed; together they provided the necessary labor to begin planting again.

And rebuild they did—but it took time and a herculean effort. Although tidal culture had begun before 1776, many planters were reluctant to abandon reservoir culture because of the necessity to increase labor and raise capital for tidal culture. Georgian George Baillie wrote that "the resettling of plantations that are so intirely [sic] gone to ruin, must be attended with nearly as much expense and difficulty, as the first settling of them."[59] The major problem planters had before the war, especially inland-swamp planters, was flooding. After the war, the state and local planters constructed a series of public canals or drains as water-control devices to prevent flooding. This work undoubtedly made it possible for some inland-swamp planters to resume planting.[60]

But for most planters, the war changed their outlook. Because they were aware that planters who had tried tidal culture before the war had reaped greater profits and because they would have had to rebuild prewar inland fields before they could use them anyway, most planters decided to abandon reservoir culture and switch to tidal culture. Inland rice fields had also experienced decreasing soil fertility. Unlike the tidal fields, which were replenished with the flooding from river waters, inland fields were fed by rainwater and local runoff, which did not carry sufficient nutrients to replenish what was lost through a hundred years of crop harvests. Planters had no feasible means of soil amelioration for the inland fields. The rich soils of the tidal fields were their salvation. The planter elite rebuilt the infrastructure of tidal culture on a massive scale. They employed the most advanced threshing and milling machines available, cleared the last remaining tidal swamps, and ultimately abandoned the inland fields. By the second decade of the 1800s, most of the available tidal swamp had been converted to rice fields, and for practical purposes labor requirements had been met. In the post-Revolution era, a concentration of plantations into larger ones by the most wealthy produced some of the richest planters, and rice agriculture was profitable and productive until the Civil War. The "Golden Age of Rice" had begun.

The planters' former enemies helped financially to restart the industry. Despite detrimental commercial legislation passed at the end of the war as reprisal for rebellion against the mother country, London remained the new colonies' financial capital for credit and insurance throughout the early years of the Republic. London needed trade with its former colonies, and merchants continued to handle the market in Europe for American tobacco, rice, and cotton. In 1787 moves were made to establish a direct trade link between the United States and France. Once the link became operable, new trade venues were possible. The slave trade reopened in 1783 and provided the labor to increase the production of rice. Even though the U.S. Constitution prohibited foreign slave trade after 1808, illegal importations continued after that date. Georgia disallowed the importation of slaves in 1799, but the law was not strictly enforced, and its slave population increased.

Wealthy rice planters and their families continued to sojourn in England and Europe, buying British goods and remaining part of the social and business established before the war. Only after the War of 1812 did American manufacturing reduce the demand for overseas goods. By then manufacturing had reached a state of independence from England.

Tidal culture represented in part a culmination of long-term trends in the tidal region of the Rice Kingdom. Agriculturists consider the use of tidal dynamics to flood the rice fields of the Lowcountry without parallel elsewhere in North America. As more planters switched to tidal irrigation, they created an increasingly artificial landscape throughout the Rice Kingdom. With the shift from reservoir to tidal culture and its enormous profits, the elite and wealthy planters ruled politically and socially after the Revolution as they did before. Along with shaping a tidewater region marked by an ecologically distinct crop grown with tidal irrigation, they recognized and accepted a formal autonomy in the slave community in the form of the task system, and set the tidewater economically apart from the rest of the slave South. The planters' leisured provincial culture set the tone for the social, cultural, and political environment. Their reign lasted until the Civil War.

The construction of tidal rice fields was far from easy. The conditions that made tidal floodplains ideal for rice growing were also terrible obstacles to converting those swamps into rice fields. The dark muck soil was unstable. To avoid sinking, one had to step from root to root or onto a dubiously stable tussock. Twice a day the land was flooded at high tide. Throughout the swamp, thickets of cat briar impeded progress and tore at flesh and clothing. The heat and humidity in the summer was oppressive; cottonmouths and alligators also must have caught the attention of workers, as well as swarms of mosquitoes. The work went on. African slaves working with oxen, shovels, and axes transformed the swamps into the richest agricultural land in the nation, but the "deadly fields" extracted a horrendous cost to life and health.

Although rice-field construction varied from one tidal swamp to another, even on adjacent plantations, variations were relatively minor. There were more similarities than differences. The lay of the land of each plantation did have to be taken into consideration and might have required a different method from that used on a neighboring plantation. The plantation owner generally hired someone knowledgeable in rice-field construction to oversee the work on a new field. A familiarity with the science of hydraulics and the ability to manage the slave force were both critical to construction, and the lack of one or the other could spell disaster. If a plantation owner was versed in engineering, as was Robert Allston, a West Point graduate, he probably served as his own engineer. Otherwise the owner had to hire outside the plantation.

The following description of construction of a tidal field from swamp forest was drawn from several sources and is probably close to a standard method that evolved over time (see diagram 6). The proposed rice field was first surveyed. The boundary of a typical field ran from the highland through the swamp to the river's edge, along the river's edge, and then back through the swamp to the highland. Next the inner and outer margins of the proposed permanent outer bank were established. The outer margin of the permanent bank followed the survey line of the rice field boundary where it ran through the swamp; however, along the river's edge, the outer margin of the permanent outer bank was set back fifty to eighty feet from the water's edge. A right-of-way was then cleared of vegetation along the course of the proposed permanent outer bank to a distance inward of the inner margin in order to prevent the subsequent cutting down of the remaining large trees, injuring the bank. The trees cut from the right-of-way were moved inward among the standing trees out of the way of the bank and main ditch constructed later. The margins of the permanent outer bank were then surveyed precisely twelve feet wide at the base, and then along the center of the base a temporary ditch (step 1) three feet deep and three feet wide was constructed. The earth from the temporary ditch was closely packed alongside the outer margin of the ditch to form a temporary bank, which functioned to keep out water from the river so as not impede the main operation that followed. At suitable locations tide trunks were placed through this temporary bank. Serving to drain the work site of water from rain, and springs or seepage in the temporary bank, these trunks would later be placed in the permanent outer bank.

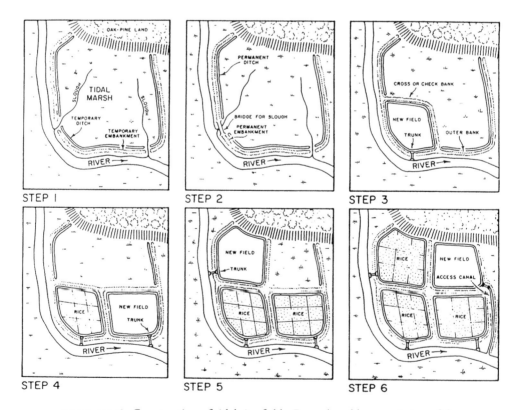

DIAGRAM 6. Construction of tidal rice fields. Reproduced by permission of the Institute of Archeology and Anthropology, University of South Carolina.

During construction of the temporary bank, all slues that serve as outlets of small creeks were omitted at first. When the temporary bank and ditch were complete, a bank was made across the slues to the same height as the temporary bank. Next, from the inner margin of the proposed permanent bank, a line was staked off parallel to it and fifteen feet from the inner margin. Along the inside of this line, the main ditch was dug, eight feet wide and five feet deep. The earth from this permanent ditch was thrown into the temporary ditch, first to fill it up and then to build the permanent outer bank. The permanent outer bank (step 2) was made two feet higher than the highest spring tide. The second advantage of constructing the temporary ditch (the first was for the fill to build the temporary bank) was to remove stumps from the bottom of the permanent outer bank; if left, the stumps would decompose in time, causing the bank to collapse and leak.

Bank construction was taken extremely seriously by some planters. Mr. Myrick of Richmond Plantation on the Cooper River mixed highland earth with the mud of the swamp soil, which, he professed, contained so much organic matter that a bank made entire of swamp soil would contract and separate and settle. He placed the highland earth *"in the middle of the bank"* from the foundation to the top. His bank not only did not crack but kept crayfish from burrowing and could not be washed away by spring tides or when the fields were flowed.[61] The hundreds of miles of abandoned banks still intact today are a testimony to how well planters and slaves constructed the banks.

Next the task of clearing began. Using primitive hand tools and oxen, slaves felled, piled, and burned underbrush and small trees. The largest trees were killed by girdling, and after they died they were cut at ground level. These trees were removed the following year(s) and used for timber or burned. The stumps of the largest trees were left, and today they can be seen in abandoned fields at low tide.

Once vegetation had been cleared, slaves divided the large banked area into smaller individual fields of about twenty acres each separated by check banks to ensure an even flow of water over the entire crop so it would ripen uniformly (step 3). The land sloped from the highland to the river, and if the entire area was treated as one large field, rice flooded to the correct depth at one end of the field would not be flooded at the same depth at the other end. Each field was constructed and leveled in a step-down fashion from the highland toward the river.

Next tide trunks made by plantation carpenters were installed in the bank surrounding each individual field. Placement of the tide trunk through a bank was a labor-intensive task requiring precision and planning. Unlike the low-profile banks of the inland-swamp system, the banks of tidal fields required a higher elevation to hold back the flood tides of the rivers. To put down a tide trunk, an opening in the bank was created at ebb tide and an area called the "trunk dock" was leveled to receive the tide trunk. The area was inspected to insure it was free of sand, which if not removed, could cause the dock to vent, undermining the tide trunk and allowing water pressure from the river to blow it out of the bank. A tide trunk was towed to the site, and at slack tide, floated into the opening of the bank and leveled. A log was thrown across each end to keep it down. Working quickly to complete the task before the tide rose, slaves used carts to haul fill from the highland and dumped fill on either side of the tide trunk. Fill was quickly packed underneath and on either side, then the fill was raised to the level of the bank. The tide trunk needed to be sealed on bottom and on either side to keep water from seeping through an unfilled space and undermining it. The doors to the tide trunk were hung last. At flood tide the correct level of water could be let onto and maintained in each individual field, or a particular field could be flooded or drained without affecting water levels in other fields.

Flowing a newly installed trunk could prove hazardous. In a March 4, 1845, letter Manigault's overseer, James Haynes, cautioned: "I would like too to let the trunk that has recently put down in No. 15 to settle and become more firm in its bed and the earth settled more that is upon it before the field is flowed for fear it may from the heavy press of water on the inside cause it to blow."[62]

After the check banks were constructed, parallel quarter drains (see figure 1), so called because they were a quarter of an acre apart, were constructed throughout the field; they emptied into the main drainage ditch (step 4). All planters knew that perfect drainage was necessary throughout cultivation of the rice crop. Not only would seeds rot in poorly drained soil, but when it was time to drain the fields to hoe weeds, no implement would work in wet soil. The fields also had to be dry at harvest time and spring planting time. Ultimately all the swamp land owned by an individual planter was converted to rice fields

FIGURE 1. Quarter drains in rice fields. Reproduced from *Harper's Weekly*, January 5, 1867.

(steps 4–6). Each field was flooded or drained by its own tide trunk. A canal was dug from the river to reach back fields that could not be flooded from the river.

Virtually all tidal freshwater swamps along the black-water and brown-water rivers from Cape Fear, North Carolina, to the Saint Johns River in Florida were converted to rice fields. In essence the planters created an artificial landscape. This conversion, which began with reservoir irrigation, was the culmination of their regard for the natural world as a source of personal wealth. They succeeded beyond their initial aspirations. In South Carolina alone planters converted approximately 150,000 acres of tidal freshwater swamps and marshes to tidal rice fields. In Georgia approximately 25,000 acres were transformed.

Two types of trunks are known from the period of tidal rice culture: plug trunks and tide trunks. Whether the plug trunk used in the tidal fields is a carry-over from reservoir culture is unknown because no remains of reservoir plug trunks have been found. More than likely tidal plug trunks were similar to earlier reservoir plug trunks but modified to better suited conditions in tidal fields.

The earliest mention of the operation of a tidal plug trunk is a letter from Louis Manigault to his father, Charles, in 1854: "The rule is to have ditches always perfectly Clean & allow fresh water from the river to flow in & out at times through little plug trunks."[63] This reference undoubtedly refers to water flowing in two directions—that is back and forth between a river and a field.

David Doar mentioned another function of a plug trunk: "Very often, in order to pass water through the fields, there were put little trunks leading from field to field. They were locally called 'plug trunks.' For years the origin of this name bothered me. I asked every old planter I knew, but no one could enlighten me. One day a friend of mine who planted on one of the lowest places on North Santee River said to me with a smiling face: 'I have solved that little trunk question. In putting down another one, I unearthed the granddaddy of plug trunks made before I was born.'" Doar then went on to explain that the plug trunk "was simply a hollow cypress log with a large hole from top to bottom. When it was to be stopped up a large plug was put in tightly and it acted on the same principle as a wooden spigot to a beer keg."[64] Doar indicated the plug trunk was used to flow water from one field to an adjacent field. From these two references, we can deduce that plug trunks were evidently used in more than one way (and possibly others not recorded).

The plug trunk that Doar described was not preserved. The remains of only two tidal plug trunks are known, both from South Carolina: one from the Ashepoo River in Colleton County and the other from the North Santee River in Georgetown County.

In 1969 a plug trunk was uncovered accidently by a drag line at Social Hall Plantation on the Ashepoo River (see diagram 7) and is housed at the Lightsey Brothers office in Miley, South Carolina. The plug trunk was buried within a bank separating two canals. It was partially destroyed by the drag-line bucket, but about five feet of the plug end with the plug was recovered intact. The trunk measured two feet square with an estimated

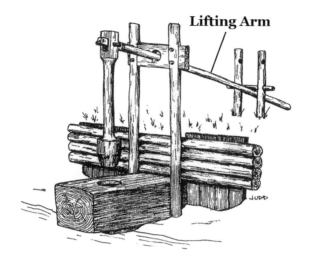

DIAGRAM 7. The plug trunk found at Social Hall Plantation.
Illustration by William Robert Judd.

length of thirty feet. It was made from a single hollowed-out and squared-off log, with its inner portion carved out resembling an old style watering trough. The carved-out portion was covered with cross planks fastened with wooden pegs, installed with the planked side down. A tapered hole was carved in the top near one end.

The lifting apparatus (see diagram 7) is conjectural. Some type of securing device for the lifting arm would have been needed to prevent the pressure from any retained water from blowing out the plug. To operate the trunk, the plug was lifted, allowing water to flow through the hole, down through the trunk, and out the opposite end. To stop the flow of water, the plug was lowered into the tapered hole and secured in place. To reverse the flow of water, the plug was lifted, allowing water to flow through the tapered hole.

At Kinloch Plantation on the North Santee River, a mostly intact plug trunk was excavated from a tidal rice-field bank (see diagram 8). The trunk box was fashioned from a cypress log 17' long, 16" wide, and 11" deep. Its bottom was left rounded, its carved sides taper inward toward the top. The inner portion was carved 6½" deep and 8" wide from end to end. Each end cants inward toward the top. A recess 12" from one end is 8" × 12" × 1½" deep with a centered 6" diameter cylindrical hole to receive a plug.

The carved portion of the trunk is covered with two planks of different lengths, one butting against the other, both secured to the trunk by wooden pegs. The cover plank toward the lift-gate end was 10' long. A shorter cover plank (missing) on the plug end would have had a plug-shaft opening directly above the hole in the bottom of the trunk to guide the plug shaft (also missing), which operated in a vertical up and down motion inside the trunk box. A loose clearance was provided in the opening in the cover plank for the up-and-down movement of the plug shaft. When seated in the cylindrical hole,

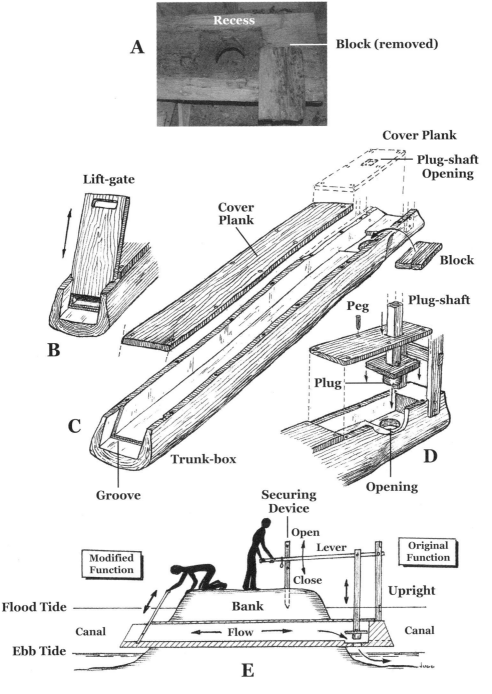

DIAGRAM 8. The plug trunk found at Kinloch Plantation on the North Santee River in Georgetown County, South Carolina. Illustration by William Robert Judd. Photograph (A) by Richard Dwight Porcher, Jr.

the plug stopped the flow of water, but it probably would have allowed some leakage. As with all trunks, however, minimal leakage was unpreventable.

Originally located on each side at the plug end was a vertical upright housing a pivoting lever, which operated the plug to control the water flow. This assumption is based on existing pegs and peg holes on the side of the trunk box. What the actual upright looked like is unknown, but a conceptual design is presented. Raising the lever would raise the plug, allowing water to flow from whichever side of the bank held the higher level of water to the other side. Lowering the lever sealed the hole, stopping the flow of water. The lever could be set in an open or closed position by a securing device.

The trunk box appears to have been modified. It originally functioned as a plug trunk; however, at a later date a permanent wooden block was wedged in the recess area, blocking the hole (see diagram 8A). Then, at the opposite end (see diagrams 8B and 8C), a canted, one-inch-wide groove in both inner walls and across the carved bottom was constructed, suggesting a hand-operated lift gate was installed. It was cantered to allow easier lifting by a worker on the bank. The intact end of the trunk was cut off allowing water to pass directly through and the trunk became a lift-gate trunk.

The two lift-gate trunks previously described were associated with inland reservoir rice fields, where the volume of water was less and water flow was only in one direction. The two plug trunks were found in tidal fields.

The origin of the *tide trunk* may have been England, but there is no direct evidence. It seems unlikely that it was entirely a local innovation since rice planters during the colonial period had access to books and manuscripts on farming techniques in England. Controlling water flow with automatic floodgates was known to the English before settlement of Carolina. In seventeenth-century Cornwall, engineer Richard Carew of Antony (1655–1720) described a floodgate used to create tidal power for a corn mill: "Among other commodities afforded by the sea, the inhabitants make use of divers creeks for grist-mills, by thwarting a bank from side to side, in which a floodgate is placed with two leaves; these the flowing tide openeth, and after full sea the weight of the ebb closeth fast, which no other force can do, and so the imprisoned water payeth the ransom of driving an undershot wheel for his enlargement."[65]

The term "two leaves" implies that the floodgate had two pivoting adjustable leaves (gates) at each end and worked automatically with the change in tides, not unlike the tide trunks. This leaves little doubt that the tide trunk was based, at least in part, on the automatic floodgates used in the tide mills of England.

When the tide trunk first made an appearance in Carolina is uncertain. The following lines in George Ogilvie's *Carolina; or, The Planter,* possibly refer to a tide trunk:

> Unless some sluice, or floodgates, valves allow
> The waters, to their parent stream to flow;
> Till rising tides, by Cynthia backwards roll'd,
> With outward weight th' obedient gates refold.[66]

The gates of this tide trunk close automatically with the tide ("th' obedient gates refold [close]"). Ogilvie sent letters back to Scotland between 1774 and 1789. If his reference of a tide trunk is accurate, it was in use at least from 1789 and possibility as early as 1776.

Whatever its origin, the tide trunk was an ingenious adaptation to tidal cultivation. Based on field surveys of the remains of tide trunks that have survived in the abandoned fields, cypress, the "wood eternal," was the preferred wood because it withstood decay, was strong, and was easy to work. Trunk components were fastened together with wooden pegs. Trunks were submerged on the flood tide and partially exposed on the ebb tide. The wooden pegs would expand and contract at the same ratio as the wood of the trunks, thereby maintaining a tight seal. "Trunks [the trunk box (see diagram 9)] were built with two sides of varying width, planked top and bottom, generally twenty or thirty feet long so as to extend through the bank."[67] Trunk boxes were long rectangular-shaped wood structures of varying lengths and width, from two feet to eight feet wide, approximately two feet in height, and as long as forty feet in the extreme. Trunk boxes eight feet wide or wider had a longitudinal center brace to support the weight of the bank. Tide trunks were installed through the banks surrounding the rice fields—one tide trunk for each field—set deep enough in the bank so their bottoms were a little below low-water mark. Two uprights eight to nine feet high were attached to a trunk box to support the gates. A pivoting gate that fitted tightly against the opening of the trunk box hung from two uprights on either end. The opening at each end of the trunk box was canted inward (set at an angle) so gravity acting on the weight of the gate would seal the opening. Each gate could be raised and secured in the open position or lowered to swing freely, either sealing the opening of the trunk box or allowing water to pass. As the fields were prepared for planting (step 1), the outer gate was lowered to its closed position. Pressure from rising tides against the gate kept it tightly closed, and no water could enter the fields. To flood the field from the river or canal, the outer gate was raised and secured open at ebb tide (step 2). The inner gate was lowered to swing freely. The rising water flowed through the trunk box, pushed the inner gate open, and flowed into the field. As the river ebbed, the pressure from the higher water level within the field pressed the inner gate closed, trapping water in the field (step 3). It generally took three high tides to flood a field.

To drain a field, the inner gate was raised and secured open on the flood tide. The outer gate was lowered to swing freely. As the tide ebbed, water flowed from the field through the trunk box, pushing the outer gate open as it flowed into the river (step 4). When the tide rose, water pressure closed the outer gate, preventing water from flowing back into the field. Three ebb tides were generally enough to drain a field. As long as the drains and tide trunks were kept in good order, water control of the fields was automatic and complete. Once the trunk minder set the gates in the correct configuration, his job was finished until it was time to change the direction of water flow or adjust the water level in the field.

The trunk minder was the key player in operating the tide trunk; he had to know the proper times to open or shut the gates, a skill gained through years of trunk minding.

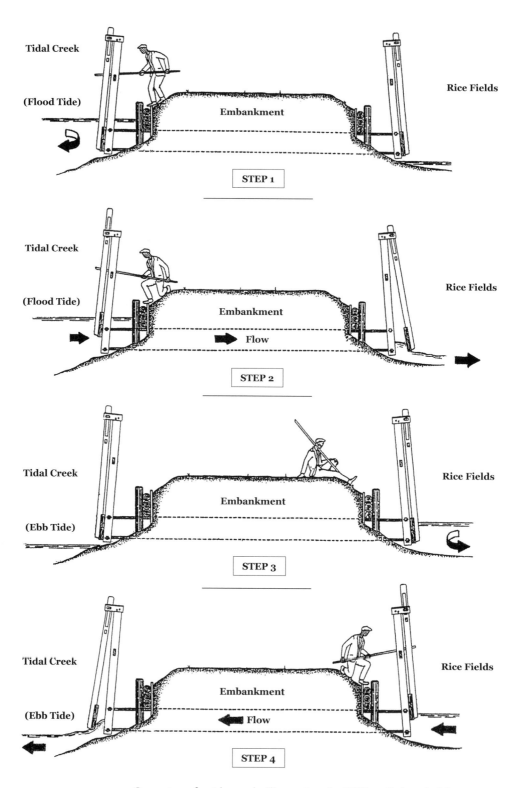

DIAGRAM 9. Operation of a tide trunk. Illustrations by William Robert Judd.

He had to know when to flood and drain the fields and maintain the proper water level, day or night. To determine if the water was fresh, he had two methods. When the water tasted "sweet"—that is, fresh, or when the water would permit soap to lather, the trunk minder considered the water fresh enough to flood the fields. In 1802 planter Nathaniel Heyward instructed his trunk minder to keep both gates slightly propped open when the water was fresh to allow constant freshening of the water in the ditches.[68] The trunk minder could control within inches the depth of water on a field at any stage of the plants' growth. Gauge marks or notches were sometimes etched on the trunks or elsewhere to denote the level of water to be maintained during different stages of the rice plant's growth. As retained water in the fields evaporated or filtered into the soil, the trunk minder used the marks a guides while restoring the proper water level at each tide change.

Hurricanes, other storms, or freshets were always a threat to the valuable floodgates and tide trunks. Keeping both gates of the tide trunks partially open during a hurricane or storm prevented them from being damaged or blown out, thus creating a costly break or washout in the bank. On February 28, 1854, Charles Manigault wrote to his son and plantation manager Louis Manigault about an expected freshet coming down the river: "Take great care of the Flood gates, or they will be torn to pieces. Fasten the gates open, & brace them open with pieces of Scantling. And take off such trunk Doors as are liable to be injured. Put them in a flat & carry them to a place of safety."[69]

There were two basic tide-trunk designs, each with variations: the original (see diagram 10) and the Savannah (see diagram 11). The difference in the two was the design of the hanging gates and pivoting components; the trunk boxes were similar. During active rice culture, the original tide-trunk design dominated north of Charleston while the Savannah River design dominated rice fields to the south. In modern times the two tide-trunk designs came to be called the Santee and ACE Basin types, the names most often used today. The Santee did not have a designation when it first came into being, so in this book it is referred to as the original design. The ACE Basin trunk was originally called the Savannah River trunk when it was used in the antebellum period. The term "ACE Basin" is a recent designation created by the South Carolina Department of Natural Resources when the basins of the Ashepoo, Combahee, and Edisto Rivers became a focus of land protection.

A February 10, 1859, letter from Charles Manigault to his son helps to establish the notion that the Savannah area was the origin of the Savannah tide trunk, when it mentions "4 New Trunks to be made on the Savannah River Plan, with Stuff from Town on the spot, & John Izard who has assisted in making so many Trunks knows just as much about it as Billy."[70] He recorded in his plantation journal when he visited Gowrie and East Hermitage in March 1867: "The Trunks built by Jack Savage after the 'McAlpin pattern,' with Arms to slip in and out, were perfect."[71] These statements suggest that the Savannah design was different from the original.

Each hanging gate of the original tide-trunk design was suspended from a horizontal pivot pole mounted at the top of the uprights of a vertical framework (see 10A and

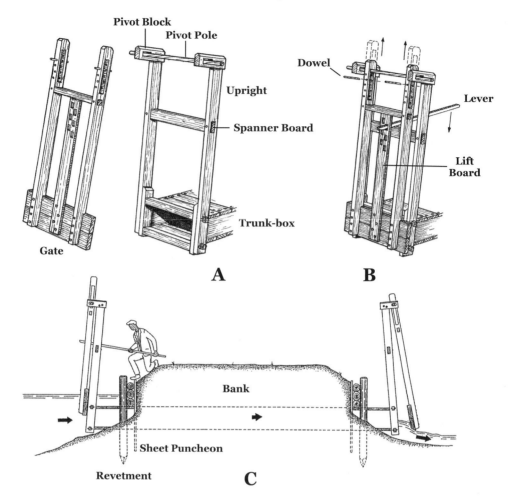

DIAGRAM 10. The original tide trunk. Illustrations by William Robert Judd.

10B). The uprights were aligned flush with the cant of the opening of the trunk box and fastened to each side. A spanner board fastened midway up the upright boards acted as a fulcrum. A pivot block projecting forward was mortised on top of each upright and housed a pivot pole from which hung the gate assembly. A high water level pressing on the gate from either end increased its seal against the trunk-box opening, restricting water flow through the trunk box. To raise the gate, a lever was inserted into one of the rectangular openings in the lift board. Pressing the lever downward against the spanner board lifted the gate (see 10C). The gate assembly slid upward along the face of the uprights, its movement guided by the pivot pole. At the end of each downward action of the lever, the gate was temporarily "pinned" at that height with wooden dowels. The lever was then placed into a lower rectangular opening, the pinning dowels removed, and the lever pushed downward. The process was repeated until the gate was raised to the desired height and pinned open by the dowels.

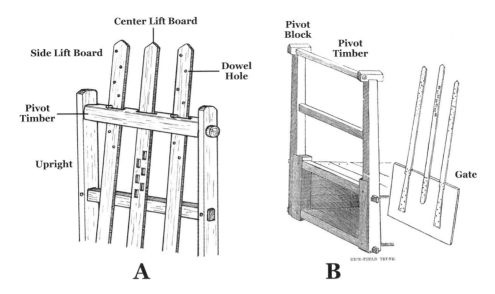

DIAGRAM 11. The Savannah tide trunk. Illustration 11A by William Robert Judd.

Louis Manigault recognized a flaw in the tide trunk's pivot-pole design. He wrote to his father that "I have not ordered the iron bars [pivot poles] to hang the doors upon, but 'maske' that, the wooden ones [pivot poles] are easily made, & as we expect to put the trunks down on Monday or Tuesday next we shall not wait for them, but as soon as the iron bars are made the wooden ones will be taken out." Manigault feared that the doors were so heavy they would put too much a strain on the wooden pivot poles. He was also concerned that the wooden pins [dowels] were too slight and should be replaced by "six iron pins to poke through the stem [lift board] of the door to rest on the iron bar."[72]

The pivoting mechanism of the Savannah design appears to be structurally more sound than that of the original design. The Savannah tide trunk had a square pivot timber (see diagram 11A) pivoting in round openings cut in the uprights of the framework. The lift boards connected to the gate passed through slots in the pivot timber. The center lift board had rectangular holes for the lever to lift the gate. The two side lift boards had dowel holes. The lifting method was similar to the original truck.

One interesting early diagram of a tide trunk appears in A. M. Forster's "Rice Culture in South Carolina," published in 1860.[73] Forster's diagram appears to be a Savannah design, but perhaps an earlier model (see diagram 11B). A pivot block atop the uprights of the framework housed a squared pivot timber. The gate is undoubtedly the Savannah design. Perhaps the blocks housing the pivoting timber were short-lived because of structural stress and were later replaced by the pivot timber pivoting in round openings cut in the uprights. Forester's sketch does contain one obvious flaw: no slots are shown in the pivot timber for the boards connected to the gate. No remains of a trunk design resembling Forster's have been found. Forster's diagram is important, however, because it is the earliest archival diagram of a tide trunk—of either design.

Years after the end of rice culture, the tide trunk persists in performing its original function. Private landowners and state and federal organizations managing old tidal rice fields for waterfowl throughout the former Rice Kingdom still use tide trunks that are virtually unchanged in design from the ones fashioned by slaves more than two-hundred years ago. The simplicity of the tide trunk's design and operation is a testimony to past technology. One can float down a Lowcountry river and pass remnants of old and new tide trunks, side by side, bridging the past and present.

Sowing Rice Seeds

Sowing rice seeds came in late March to early April, or in late May to June in order to miss the spring migration of bobolinks, which could wreak havoc on a newly planted rice crop. Some planters timed their sowing to catch the first run of high tides in March, when the difference in high and low tides was the greatest, allowing better control over flooding and draining the fields. This was risky because of the possibility of cold weather retarding seed sprouting, or a late harvest exposing the rice to bobolinks during their fall migration.

Methods of sowing seeds changed over time. Early on seeds were broadcast because stumps prevented digging trenches. This method sufficed in the early days but produced small crops. As stumps rotted or were removed manually, slaves began to cut trenches to sow seed. When this transformation occurred is unrecorded, but by the time tidal culture was in place, sowing in trenches was the norm. Two main methods of sowing seed in trenches evolved.

One method was *covered seed sowing*. As planting time approached, the land was thoroughly chopped or broken crudely, all clots pulverized, and the surface smoothed and leveled with the hoe. On old and well-cleared plantations, the plow and the harrow often replaced the hoe. Sowing came next. Early on sowing was done by barefoot slaves who pressed the seeds into the waterlogged muck with their heels, a method practiced by their African ancestors. Trenching replaced the foot method. Trenchers, using roughly four-inch-wide hoes made for the purpose, made trenches of the same width. They cut trenches twelve, thirteen, or fourteen inches apart depending on individual preference, and two to three inches deep, about 100 to 125 trenches per half acre. An alternative method was to use an animal-pulled drag that marked where the trenches were to be excavated. Close upon the heels of the trenchers were sowers, generally women, who scattered or strung out the seeds in the trenches (see figure 2). The common practice was to sow about 2¼ bushels to an acre. Next came slaves who covered the seeds lightly in the trenches with about two inches of soil using the universal hoe or light wooden bats. Covering the seeds prevented birds from consuming the seeds and prevented them from floating away when the field was flooded.

In 1826 Capt. John H. Allston of Prince George Winyah introduced an alternative method of sowing seed called *open-trench planting*.[74] This method probably originated in South Carolina during the 1820s to solve the problem of declining crop yields on largely

FIGURE 2. "Planting the Rice" using narrow trenching hoes. Reproduced from *Harper's New Monthly Magazine* (November 1859).

worn-out soils. According to planters who used open-trench sowing, the method saved labor, yields increased, the growing season was shortened, and the fields were kept cleaner of weeds and red rice.

In open-trench planting seeds were stirred in clayed water and dried, or slaves stomped the seeds in a bed of clay. The clay coated the seeds so they adhered to the soil, preventing them from floating away when the field was flooded. "Claying" saved labor by eliminating covering the seeds with a harrow. Water was held on the seeds after they germinated until a leaf and a branch were produced; then the water was drained off the field to allow the embryo to set root. How widespread was claying is not recorded, but Theodore DuBose Ravenel stated the "open-trench method is not commonly planted, not over 5 per cent of the entire acreage was planted open."[75] Whether he referred to the 5 percent of the entire Rice Kingdom or a particular plantation's crop is uncertain. Archival records do not contain much mention of claying seeds. Few Georgia planters used open-trench sowing because their soils were not exhausted.

Robert Allston reported that, on his return from his native Scotland, Dr. Robert Nesbit introduced a "Drill Plough," "which excited no little interest among the planters in his neighborhood." It was designed "to open the trenches and deposit the seed, which it was found to do very well when managed with care, to the saving of so much labor." The drill plough had two wheels and was drawn by a horse. Firmly fixed to the machine were rods with shoes on their ends that opened the trench for the seed. Regulators distributed seeds from a box into tubes that directed the seeds into the trench. "Drawn by a good horse, over ground in high tilth, and managed by a skillful and judidicious [sic] hand, the drill plough would trench and seed from eight to twelve acres of ground in a day," Allston reported. It was used successfully by planters in the Pee Dee and Waccamaw region but was later abandoned because, in Allston's view, the field laborers did not have the judgment to use the machine. Nonetheless Allston expected that the drill plough would be introduced again and be successfully used in rice planting.[76]

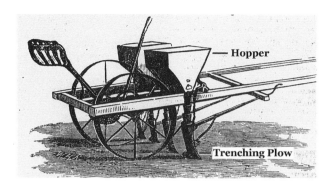

FIGURE 3. "Dotterer's Patent Rice-Sowing Machine." Reproduced from the *Rural Carolinian* (1870).

In 1870 Thomas D. Dotterer, Jr. (1832–1894), of Charleston, South Carolina advertised his "Patent Rice-Sowing Machine," manufactured at the foundry of William S. Henerey, 314 Meeting Street, in Charleston (see figure 3). One horse pulled the machine, which as operated by one driver. Trenching plows cut four trenches. A hopper containing seeds distributed the seeds via screws in an even flow into the four trenches. A laborer followed behind and covered the seed. The screws were operated from gearing attached to one of the wheels. Dotterer's machine was undoubtedly based on Nesbit's drill plough. How he claimed to improve on Nesbit's machine and whether it was successful is unrecorded. Drill ploughs, however, became standard machines in agriculture throughout the country, and Dotterer's machine, manufactured locally, must have gained some use.

Flow Culture

The flow culture of rice is probably the standard that evolved in the early 1800s and was used until the end of the industry. It consisted of three steps:[77]

1. The first flow of water (the sprout flow) started seed germination.
2. The second flow (the stretch flow) caused the plants to increase in height rapidly and kept down grasses and weeds since rice grew much faster than the other species; this second flow also controlled insects.
3. The last flow (the harvest flow) supported the plants as the grain ripened; the weight of the ripening grain caused lodging (falling down) of the plants if not supported by water.

Planters generally agreed on only the sprout flow and the harvest flow. Between these flows there was considerable variation in the number, duration, and types of flowings as well as in the drying periods and hoeing.

The *sprout flow* was imposed as soon as the seeds were sown and covered. A field could be flooded the same day it was planted. Depending on the temperature, which determined the time it took seeds to germinate, the sprout flow lasted from four to six days. The sprout flow had three purposes: to promote germination of the seeds, to prevent germination of weedy seeds and the growth of weeds already established, and to kill grubs and maggots already in the soil. The rice seeds germinated under water without oxygen (anaerobic germination), but the seeds of the weedy species required oxygen for germination (aerobic germination). Under flooded conditions, there was not enough dissolved oxygen in the water for aerobic germination. After the rice seeds germinated, or "pipped" as it was referred to locally, the water was drained so the young embryos could set root into the soil. During this dry period, workers entered the fields to hoe whatever weeds had developed while the field was flooded and pull grasses from around the roots of the rice plants.

The *stretch flow* followed after the new plants developed two leaves. At first the water was flowed deep enough to cover the rice completely (generally from ten to twelve inches); then it was gradually drawn down to about six inches, where it was held for about two weeks. The different levels of water flowed on the fields were maintained by the trunk minder, who determined them by "notches" or "marks" on the trunk structures. The stretch flow killed grasses and other weeds that had germinated after the sprout-flow water had been drained. Grasses in particular were killed or their growth was restricted because they could not stretch under water, while rice plants stretched rapidly to about six or eight inches. Grubs and maggots, the most serious insect pests, were controlled by flooding. The stretch flow lasted for about two weeks. Then fields were drained again, and the crop passed into the dry-growth stage, which lasted for about forty days.

Dry growth was the critical growth period, for branching occurred, and every branch that developed produced one ear containing several hundred grains of rice. Here is when it was necessary to have the quarter drains in good shape, so the fields could be drained completely. When a field was sufficiently dry, cultivation with a hoe removed weeds and volunteer rice. After hoeing, the crop remained without irrigation until jointing.

At the end of the dry-growth period, the trunk minder opened the trunks for the final flow, the *harvest flow,* also called the "layby flow," which lasted until eight days before harvest (about sixty to seventy days). The harvest flow was only for the benefit of the rice plants since any weeds not hoed during dry growth did not affect the crop at this stage. During the harvest flow, the water was raised continuously as the plants grew. Water supported the plants, which were over-burdened with rice, and prevented the plants from lodging. While water was held on the field, it was changed at least once a week to keep it from becoming stagnant. As quickly as possible after the harvest flow was taken off and the soil dried, workers began harvesting with sickles.

Some planters divided the stretch flow into two separate flows, the *point flow* and the *long-water flow*. After the sprout flow, when the plants had pipped and the young plants could be seen the whole length of the row, water was again applied with the goal of stimulating the young plants and bringing them up early together while destroying any young grass that might have vegetated, controlling insects, and securing the field from bird depredation. This was the point flow.

A few days later, when the plants were strong enough not to fall when the water was removed, the fields were drained. As soon as the plants were strong enough to bear the hoe, one or two hoeings were given and the water again put on. This was the long water, and the rice was overtopped for three or four days. Workers removed trash that floated to the banks, and insects died. Then the water was gradually drawn down to about six inches and kept that way from twelve to twenty-three days. The water was then gradually slackened until no water was on the surface. When the field was again dry, the soil was broken with the hoe to enable the plants to spread, and a light hoeing was done. During this period the plant underwent jointing. The harvest flow, the last step of flooding, was then applied.

Water Culture

Another effort to increase yields was the use of *water culture,* not just as flooding to control weeds, the sense in which the term is employed by modern scholars, but as a means simplifying and reducing the labor required for producing a rice crop. In this sense water culture involved only two floodings of the crop. The method had limited success. Water culture required very level fields, which many plantations did not have.

Apparently William Butler on the Santee River developed water culture in 1786 and claimed that it was superior to "the slovenly [careless] method of flowing fields, and hoeing or chopping [grasses and weeds] thro' the water."[78] According to John Drayton, "It is said, that by this process, a greater quantity of rice is made to the hand, although less be made to the acre, than in the first method [flow method]. Some planters have adopted it; but the other mode is most generally pursued."[79] Drayton did not say why most planters did not adopt water culture. No record was found to indicate that water culture was practiced in Georgia or North Carolina, and in South Carolina it had limited success. The various descriptions do not reveal how it was used to control weeds.

Water culture involved only two floodings and one dry period. The sprout flow was kept on the field and slowly raised to keep up with the growth of the plants until the crop had shoots with three leaves (about twenty days after planting). Next the fields were drained for about three weeks while hoeing was done. The fields were then reflooded, and the water kept on them until they were drained for harvest. Because water culture required only one draining and hoeing before harvest, it reduced the labor effort. Planters kept the floodgates open during flooding to circulate the water so it would not stagnate. Water culture required replanting by hand from a nursery where plants had died in low spots.

Thomas Pinckney conducted elaborate experiments comparing water culture with flow culture in 1810 on the Santee River. He reported that he made 1,069 bushels on twenty-two acres using water culture and only 990 bushels on the same amount of land using flow culture. This method required less diligence and decision making on the part of the trunk minder, and as Pinckney pointed out, slaves would have more time to perform other work because they spent less time tending the rice fields. He also noted that shifting the water would reduce the incidence of diseases in the fields and that two crops might be obtained in one year, although he did not say how this might be done.[80] Despite the apparent advantages of water culture, the method was not widely used.

Rice Flats

Throughout the colonies and the new republic, the primarily means of transportation was by watercraft along natural waterways and manmade canals. For rice planters the watercraft of choice was the rice flat. As tidal culture expanded and became the dominant method of growing rice, rice flats became the indispensable workhorses of the rice plantation and continued to be used until the end of the industry (see figure 4).

Rice flats were entirely a product of tidal rice culture, and their design cannot be understood apart from their function in the rice fields. The rice flat was an ingenious solution to a transportation problem unique to the rice fields. This flat-bottom vessel was well adapted to working in and around the plantation's shallow rice fields and narrow canals and ditches. Flats were easily poled through the larger canals and carried workers and materials to and from the fields as well as to and from the city markets (see figure 5). Flats varied in length from small to extralarge, depending on their use. The average flats used in the canals surrounding the rice fields were thirty-five to fifty feet long and eleven to fourteen feet wide with a depth of two to three feet. These were flat-bottomed, rectangular-shaped, shallow boxes with ramped ends capable of carrying a cargo of two

FIGURE 4. "Shipping Rice from a Plantation on the Savannah River," with rough rice being poured into the flat to be shipped to a Savannah mill. Reproduced from *Frank Leslie's Illustrated Newspaper,* September 7, 1872.

FIGURE 5. A rice flat on the Cooper River in Berkeley County, South Carolina, early 1900s. From the Middleburg photograph album (now lost), courtesy of John Ernest Gibbs. Richard Dwight Porcher, Jr., copied some photographs from the album before it was lost.

to three tons. According to Elizabeth Pringle, her largest flat was expected to carry nine acres of threshed rice. To support this weight, multiple stringers (keelsons), providing longitudinal support, and stretchers (frames), providing lateral support, were placed in the bottom of the vessel. The rice flat had shallow draft that made it ideal for navigating the shallow rice-field canals.

Flats were open vessels; the whole of their interior was dedicated to cargo. When decking was present, it was usually limited to the ramped ends. At least one planter ordered a flat built with a deck and hatches, but when Charles Manigault did so in 1851, overseer K. Washington Skinner responded: "I write you immediately to inform you that you had better not have the Deck & Hatches put on or fastened on until after the Harvest."[81] Skinner was concerned that it would be too difficult to store the load of rice sheaves through the hatch and that removing sheaves through the hatch would reduce the sheaves to straw. Skinner's solution was to suggest that the decking be removable so it could be added only when the flat was to be used for a purpose other than transporting rice.

Flats were decked when they were used as platforms for field workers or pile drivers. In another letter to Manigault, Skinner stated, "The pile-driver is on the new flat, which is fastened at the wharf in front of the pounding mill. I had it carried there to drive several large piles on the side of that wharf, where it is giving way."[82]

Only essential items were kept aboard a rice flat—oars and push poles, a sweep for steering, a small wooden hand pump for bailing, a grapnel hook for snagging the banks, and a ladle and small cask of drinking water for the crew. Walk boards were sometimes placed along each side the whole length of the flat for hands to walk on as they pushed the vessel through the canals and along the river's edge when it was not moved by oars. Most plantations owned only one or two flats and hired or rented additional flats when they were needed. The flats transported rice from the fields to the threshing yard and milled rice to town. Although many plantations along the Savannah River had their own threshing barns, the proximity to the city meant those who chose to could avoid the cost of a mill by sending rough rice to the city for milling. Flats were operated by all-slave crews of plantation "sailors," one of whom commanded the vessel.

The care and handling of flats was a constant concern to a plantation overseer. Neglect or unforeseen circumstances could result in a flat's removal from service for repairs. Elizabeth Pringle shared in this concern: "I have sent for a tug to tow the two flats up on the flood-tide this evening—just now it is dead water, and the flats are aground, which always scares me; for, if by any chance they get on a log or any inequality, they get badly strained and often leak and ruin the rice."[83]

Caulking the seams of the flats was a never-ending task, usually done by plantation carpenters. When the carpenters were overburdened with other work, the flats were sent to the city boatyards for repairs. No matter how careful the flat operators were, submerged obstacles and swift currents created unpreventable accidents, such as the one described by Charles Manigault's overseer: "On Monday the new flat inclined side-wise and turned one third of the produce of 13 acres into the river. I borrowed a flat from Mr. Haynes' immediately and saved it, some dry, but mostly wet."[84]

The rice flat was so important to the operation of a plantation that a building called a *flat house* was built over a basin to store the flats at night and during storms, affording them shelter from the elements and security from theft. Charles Manigault's flat house at East Hermitage was sixty feet long and twenty feet wide with a gable roof. Its two outer walls were sided with one-inch boards.[85] One planter, Plowden Weston, posted a set of rules to protect his flats: "All the flats, except those in immediate use, should be kept under cover, and sheltered from the sun. Every boat must be locked up every evening and their keys taken to the Overseer. No negro will be allowed to keep a boat."[86]

Archival construction plans of rice flats are nonexistent, and historical descriptions are scarce. Unlike well-documented designs of larger sailing vessels, designs of small vessels such as rice flats were scratched on available paper at a city boatyard or by plantation carpenters, perhaps on the ground, and not thought important enough to record. However, enough photographs, newspaper illustrations, physical remains and plantation journal descriptions of rice flats have survived to allow us to diagram the two types of flats. The shape of flats depicted in nineteenth-century drawings and photographs remains the same, but a variety of construction methods were used to achieve that shape. The types of flats are named for the way in which the chine and side were constructed. A "chine" is the line of intersection between the side and bottom of a flat-bottomed or V-bottomed boat. Diagram 12A is a plank-built flat. The chine was formed by the intersection of the side planks with the bottom planks, and the joint was reinforced by a chine stringer and a standing knee brace at each floor frame. Diagram 12-B is a chine-log flat. The chine was created by a carved log hewed from a single tree that formed the entire side of the vessel. A thick board, running the full length of each side of the vessel, was fastened to the top of the chine-log for added freeboard. Standing knees were added for lateral support. The lower portion of the log was rabbeted into which the ends of the bottom planks were nailed.

In the beginning, there were no hard or fast rules for constructing a rice flat. Flats constructed by plantation carpenters would have looked different from flats constructed in a city boatyard or by itinerant workers. As time passed, construction methods became more standardized and functional.

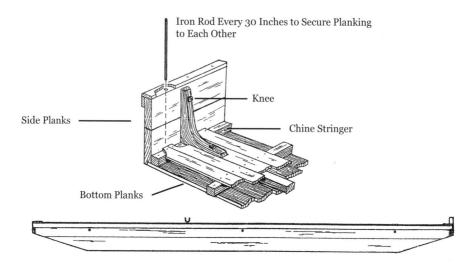

A Plank-built Flat

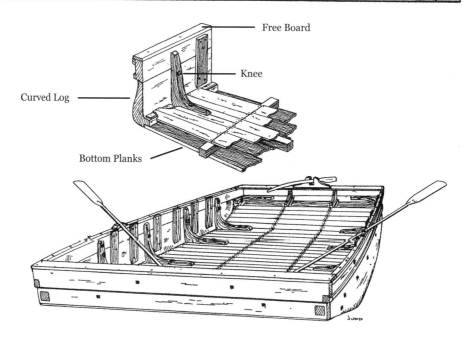

B Chine-log Flat

DIAGRAM 12. Rice flat designs. Illustrations by William Robert Judd.

Canals and Floodgates

The ability to move rice sheaves from field to the barnyard and rough rice to city markets was critical to the success of a plantation. Sheaves left in the field could be damaged by storms and freshets; delays in getting the rice to market meant a decreased demand for the product. Being able to move around the fields on flats for any task related to operation of the rice fields (such as weeding or clearing quarter drains) increased the efficiency of a rice plantation.

The larger tidal rice plantations used a system of canals and floodgates[87] to allow flats to transport rice, workers, and materials within the fields and take rice to the city markets. Louise P. Ford and Marion J. Pelleu commented on the value of canals on a Lowcountry rice plantation: "Fields lying directly along the river were flowed and drained by trunks in the main river bank. But when the fields extended a long ways from the river especially large canals were dug, with flood gates on the river carrying water to the 'backfields.' These large canals, and creeks, were also water therefares [sic]. They were so linked up that a flat boat could be poled within a short distance of any point in any rice field."[88]

The diagram of a typical rice field on the Combahee River shows that several back fields were located where they could not have been flooded with river water or drained into the river (see diagram 13). Since each field had to be supplied with water independently, plantation workers constructed a large canal from the river to the highland and installed a floodgate where the canal entered the river. If the floodgate was used only to flow water into and out of the canal, and not for passage of flats between the river and canal, a *single* gate was used. The floodgate kept a high level of water trapped in the canal to flood the fields.[89] To flood the fields, the gate was opened to the rising tide, and river water flowed into the canal; then water was flowed into the fields from the canal through tide trunks. As the river began to ebb, the gate was closed, trapping as much water in the canal as possible so flooding of the fields could continue and the canal could be used by flats. At the next flood tide, the gate was again opened, filling the canal again so water could flow into be fields from the canal.

To drain the rice fields, the gate was opened as the river ebbed, and the water level in the canal was lowered enough so that water could flow from the fields through the tide trunks into the canal, and then through the open floodgate into the river. Since it may not have been possible to drain the fields completely on one low tide, the gate was closed as the tide changed, preventing water from flowing back into the canal. At the next ebb tide, water was again drained from the fields. Three low tides were probably sufficient to completely drain the fields.

When the growing season was over, closing the gate once the canal was filled maintained a high water level for internal movement of rice flats. Leakage probably necessitated periodically opening the gate at flood tide to replenish the canal water.

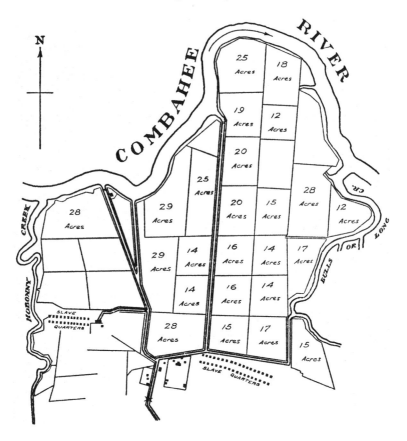

DIAGRAM 13. A typical rice field on the Combahee River. Each field is served by a canal from which a tide trunk (not shown) flowed water into or out of the field. Reproduced from David Doar, *Rice and Rice Planting in the South Carolina Low Country* (1936).

The inside width between the two walls of a canal and the number of gates and their design depended on the intended function of the floodgate. When the floodgate was used only for flowing, one gate was installed. However, on most plantations, the passage of flats through the floodgates was also a regular function during the preparation and planting of the rice fields, harvesting the crop, and in times of pending disaster when freshets and storms caused breaks in the outer banks. Flats loaded with mud to repair the banks would traverse the floodgates day and night until the damaged bank was repaired.

To allow passage of flats, the floodgate was built wide enough for the flats to pass through, and had either two or four gates. In floodgates that had two gates, both shut against a center upright, which could be pushed aside in a slot cut in the crosspiece at the top middle cap and in the shutting sill below. When the floodgates functioned for flowing water into and out of the canal, they functioned similar to the one-gate floodgate. When it was used for flats to pass through, the gates could be opened only when the water level in the canal and river were equal.

David Doar gave a general description of the construction of a floodgate, and field studies of the remains of floodgates in South Carolina and Georgia have verified Doar's description of a four-gate floodgate (see diagram 14).[90] Although diagram 14 is for the diamond gate discussed next, the design was the same except for the gates and the shutting sill. According to Doar, the area where the floodgate was to be placed was banked off; then the water was removed, and rows of plank puncheons (pointed boards), three to four feet apart, were driven crosswise into the canal bed. Large foundation logs, dressed flat on top and bottom, were placed on top of each row of puncheons, and the spaces between the logs packed with earth. Next, plank puncheons were driven into the canal bed, side by side, the length of each end log to prevent venting, the flow of water along the sides of the floodgate, which would have eroded the fill and undermined the structure. Upon the horizontal logs, plank flooring was spiked throughout, running lengthwise. Next a wall was constructed on each side of the plank floor running the length of the floodgate and to a height high enough to retain the embankment. Then the gates were hung.

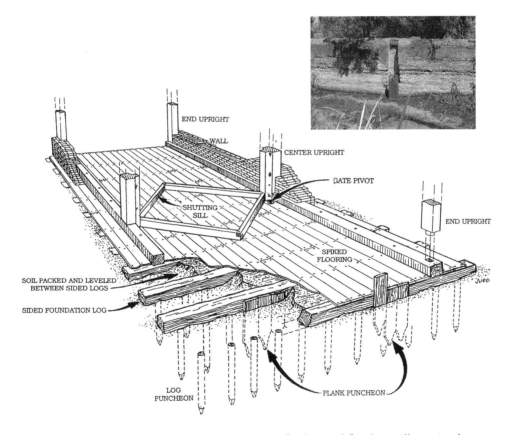

DIAGRAM 14. Construction of the lower part of a diamond floodgate. Illustration by William Robert Judd. *Insert:* ruins of tabby floodgate at Hobonny Plantation on the Combahee River in Colleton County, South Carolina. Photograph by Richard Dwight Porcher, Jr.

Materials used for wall construction varied, brick being more common. Some walls were constructed with vertical posts sided with horizontal planking; a few walls were constructed of tabby, a mixture of sand, lime, and oyster shells (see diagram 14, insert). Tabby formed a cementlike material, which was poured into a wood form about fourteen inches high running the length of the floodgate. When the tabby cured, the form was raised and another section was poured. The process continued until the desired height was achieved. Three tabby-built floodgates have been located, all on the Combahee River in South Carolina.

Floodgates with four gates were called *diamond gates*. The engineering that went into constructing diamond gates was considerable, and their operation was highly sophisticated. Diamond gates had one set of gates opposing the river and the other set opposing the flow from the canals. This design formed a diamond shape—hence the name. Each set

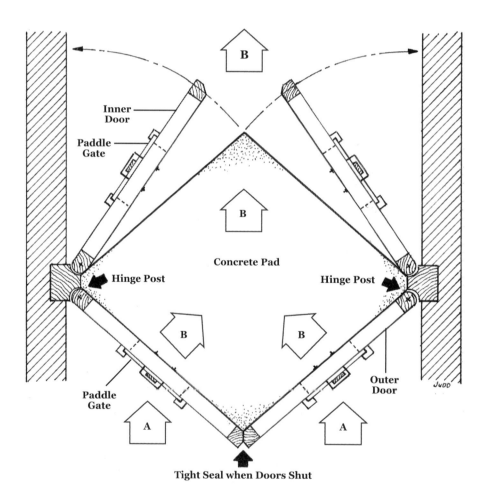

DIAGRAM 15. Plan view of the four gates in a diamond floodgate. Illustration by William Robert Judd.

FIGURE 6. The diamond floodgate at Nieuport Plantation on the Combahee River in Beaufort County, South Carolina. Considerable deterioration of the floodgate has occurred since this photograph was taken in the early 1900s. Nieuport is now part of Nemours Plantation. Courtesy of the Charleston Museum, Charleston, South Carolina.

---

of gates in a diamond gate shut against each other and against a diamond-shaped sill on the floor of the floodgate. The sill in a diamond gate was constructed of squared timber as shown in diagram 14 or a formed concrete abutment pad like the one built at Nieuport Plantation on the Combahee River in Beaufort County (see diagram 15). The diamond gate created a better seal against water pressure than the one- or two-gate floodgate. Water pressure from either side tended to force the gates tight against each other, creating an almost water-tight seal. The diamond gate at Nieuport Plantation is the only diamond gate that has survived reasonably intact (see figure 6).

The gates of the diamond gate receiving the higher level of water (riverside or canal side) were impossible to open because of the water's pressure. If, for example, the water level was higher in the river than in the canal, it would have been impossible to open the outer gates against the river pressure. To overcome this problem, small gates (called paddle gates) were put in the main gates. The paddle gates could be opened against the river's pressure since they were lifted vertically by a lever similar to the operation of a tide trunk. The diamond floodgate had paddle gates on each of its four gates (see diagram 16). The paddle gate on the right is in a closed position; the one at the left is shown in the open position.

The diamond-shaped area that gave the diamond-floodgate system its name was formed by the four gates (see diagram 15). To restrict water from entering the enclosed rice-field canal, paddle gates in the outer doors were let down to their closed position. Pressure from the rising river water (A) forced the outer doors tightly against the shutting sill.

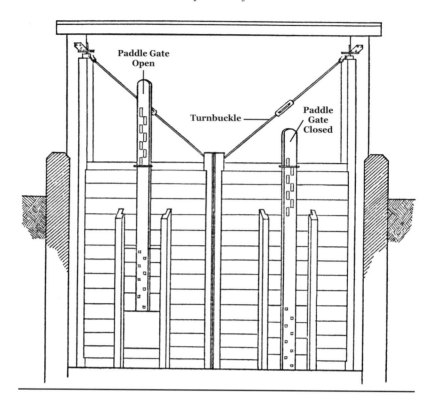

DIAGRAM 16. Two of the four gates in the diamond floodgate at Nieuport Plantation. Reproduced from David Doar, *Rice and Rice Planting in the South Carolina Low Country* (1936).

This force also sealed the outer gates at their hinge posts and at the mitered posts where the gates shut against each other to minimize leakage. To maintain perpendicular miter posts, each gate had a diagonal tie-rod and turnbuckle assembly to adjust any sag out of the gates.

To flood the canal from the river, paddle gates in the outer gates were raised and set open, allowing river water (B) to flow through the diamond-shaped area, pushing open the inner gates, flooding the canal. From the canal, the rice fields were flowed. To retain water in the canal when the river ebbed, the inner gates and paddle gates were closed. Pressure from the water held in the canal sealed the inner gates. To drain the fields on the ebbing tide, the paddle gates in the inner gates were raised and the flowing process reversed.

To allow passage of rice flats into or out of the canal, all four gates were opened when the level of the river water was equal on both sides of the diamond gates. Passage normally could be done once a day, but at certain times of the month the tides would allow passage twice a day, early in the morning and late afternoon.

Hopeton, on the Altamaha River in Georgia, one of the largest plantations in the Rice Kingdom, used diamond gates as part of a lock system. Hopeton was established in 1805 by John Couper and James Hamilton. Their partnership ended in 1826. It was Couper's son, James Hamilton Couper, who had become manager in 1818 (and later part owner and co-manager), who made Hopeton the model plantation of the antebellum South. In 1825 Couper went to Holland, where he studied the system of water control, and he applied what he learned at Hopeton when he returned. In his "Agricultural Notes," Couper described the canal system at Hopeton: "This canal is three miles in length, 15 feet wide at the surface, 10 feet at the bottom, and 4½ deep: the extremities of it terminate at the river. At one of them is a lock-gate 75 feet long, 12 feet wide, with 4 pair of gates, calculated to pass flats 45 feet long, and 11 feet wide at any stage of the tide; except the last ¼ of the ebb and first ¼ of the flood; at the other is a flood-gate 35 feet long, 11 feet wide, with two pair of gates."[91] The three-mile canal meandered in snakelike fashion from the Altamaha past the sugar and rice mills, the cotton gin, and Hopeton House, and then through the fields back to the river (see diagram 17). The canal served as the main irrigation and transportation facility for the plantation. Flats could easily navigate the entrance to the canal and the canal itself. At the head of the canal, near the settlement and mills, was a lock seventy-five feet long with "four pairs of gates," which meant the lock had a diamond gate at each end (diagram 17B). At the opposite end of the canal was a floodgate thirty-five feet long with "two pairs of gates;" in other words, a single diamond gate (diagram 17C).

The Hopeton lock had the convenience of moving flats from the river into the canal or vice versa at any time the water level in the river was above half of ebb tide. Below this level, the river would be at or below the level of the floor of the lock, making it impossible to float a flat. Couper's "Dutch floodgate" was the only one of its kind in the Rice Kingdom.

Deadly Rice Fields

Diseases were always a problem for both slaves and planters. Confined to the plantations, black people suffered from disease in far greater numbers than white people. Death by disease was more prevalent in South Carolina and adjacent rice areas than anywhere else in British North America. Between 1670 and 1775, settlers in South Carolina alone faced fifty-nine major epidemics. Most diseases, and the most deadly, that afflicted settlers and slaves were not indigenous; they were brought by the slaves and white settlers on ships that ported at Charleston, Savannah, or other ports of entry and found their way into the Lowcountry. Yellow fever, malaria, smallpox and typhus were inadvertent passengers traveling with Europeans and Africans, and they quickly spread through the Rice Kingdom. So did vectors of diseases, such as the *Aëdes aegypti* mosquito that became the carrier of yellow fever, and the *Anopheles* mosquitos that carried malaria.

Alexander Hewatt wrote in 1779: "No work can be imagined more pernicious to health, than for men to stand in water mid leg high, and often above it, planting and weeding rice; while the scorching heat of the sun renders the air they breath ten or twenty

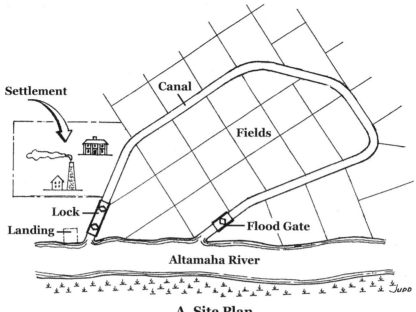

A. Site Plan

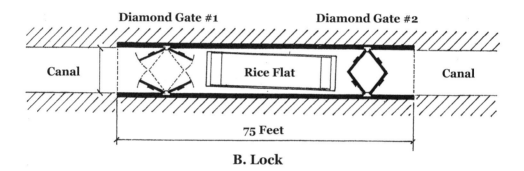

B. Lock

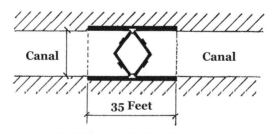

C. Diamond Floodgate

DIAGRAM 17. Site plan for Hopeton Plantation on the Altamaha River in McIntosh County, Georgia. Illustrations by William Robert Judd.

degrees hotter than the human blood, and the putrid and unwholesome effluvia from an oozy bottom and stagnated water poison the atmosphere."[92]

In her *Journal of a Residence on a Georgian Plantation in 1838–1839,* Frances Anne Kemble, wife of Pierce Butler, owner of Butler's Island, a rice plantation on the Altamaha River in Georgia, described the suffering of slave women in graphic detail, as well her husband's unwillingness to tend properly to their needs: "Mr. [Butler] was called out this evening to listen to a complaint of overwork from a gang of pregnant women. Mr. [Butler] seemed positively degraded in my eyes as he stood enforcing upon these women the necessity of their fulfilling their appointed tasks. How honorable he would have appeared to me begrimed with the sweat and soil of the coarsest manual labor, to what he then seemed, setting forth to the wretched, ignorant women, as a duty, their unpaid exacted labor!"[93]

Kemble's description of the slave infirmary at Butler's Island leaves little doubt why, at least on Butler's Island, there was such a high mortality rate. She described the infirmary as

> a large two-story building . . . built of whitewashed wood . . . with half the casements . . . were glazed . . . obscured with dirt . . . the other windowless ones were darkened by the dingy shutters, which the shivering inmates had fastened in order to protect themselves from the cold. In the enormous chimney glimmered the powerless embers of a few sticks of wood, round which . . . as many of the sick women as could approach were cowering . . . most of them on the ground, excluding those who were too ill to rise; and these last poor wretches lay prostrate on the floor, without bed, mattress, or pillow, buried in tattered and filthy blankets . . . hardly space to move upon the floor.—those who, perhaps even yesterday, were urged on to their unpaid task.[94]

On the other hand, historian Julia Floyd Smith wrote in *Slavery and Rice Culture in Low Country Georgia, 1750–1860* (1985), "Slave diet . . . was varied, palatable, and nutritious. . . . Slave houses were of good construction and contained adequate footage to accommodate, comfortably, a family of five . . . , and slave families were recognized and family stability was encouraged. Substantial hospitals were maintained to provide for and treat ill slaves, and pregnant women were cared for during their confinement and delivery of their offspring. Medical care was administered by a physician whose services were retained by the owner."[95]

No matter how dedicated a planter was to the best care of slaves, however, converting free-flowing inland swamps and tidal freshwater swamps to reservoirs and rice fields, combined with the warm climate and long growing season, created a disease-infested, Lowcountry cesspool. The inland reservoirs, which held nonflowing water most of the year, were ideal breeding grounds for mosquitos and other insect vectors of disease. Free-flowing small tributaries were blocked by the banks that surrounded both inland and tidal fields, and the fields themselves became breeding sites for *Anopheles* mosquitos, the vector of malaria. Water-transmitted diseases such as cholera found safe haven in the putrid ditches used as privies near the cabins and in the stagnant waters of the rice field ditches. Both races suffered from yellow fever; however, this disease was a city disease because the *Aëdes aegypti* mosquito preferred to lay its eggs in a water source with an artificial bottom, such as a gourd or barrel, rather than a rice field or ditch. Slaves who were in the city suffered along with white residents.

Around 1790, although they did not necessarily know why, planters and their families began to realize that rice plantations were unhealthy places, especially in May through November. Two Lowcountry sites were found to be freer of illness, especially malaria and yellow fever: the sandy longleaf pinelands and the coastal barrier islands. Science knows today what the planters did not know then: The coastal beaches were free of the *Anopheles* and *Aëdes* mosquitoes, which breed inland in freshwater sites and were kept from the islands by ocean breezes. The soils of the pinelands were sandy and did not hold water for breeding sites, and the pinelands were beyond the flight path of the mosquitoes that breed in the distant swamps. In these pineland and seashore sites, there was adequate clean drinking water, so waterborne diseases were not serious problems.

Having spent most of his life before the Civil War in Philadelphia, Pierce Butler made the mistake of staying overnight on Butler's Island to tend to plantation matters in August 1876. He became ill with malaria and died soon afterward. Field slaves could not voluntarily move to the pineland villages or the seashore—work in the rice fields went on year round. House slaves might travel with their owners to the mountains, pinelands or seashore. Those left behind died horrible deaths at appalling rates even in cooler months.

Year round hard work took a steady toll on the health of the slaves, making them more prone to illness. During the summer weeding required slaves to work for several weeks standing ankle- to mid-leg deep in mud and water in burning sun. In the winter they stood in cold water while clearing ditches and repairing banks and worked outdoors pounding rice in mortars. Black workers were more susceptible to diseases such as pleurisy tuberculosis, influenza, and pneumonia. Year after year they died of pneumonia. Often in the winter an entire plantation might be covered in water from freshets or hurricanes. An old slave song from the Carolina Lowcountry sums up the plight of slaves working the fields in the winter: "In de cole, frosty mornin', it an't so bery nice, Wid de water to de middle to hoe among de rice." And the work had to go on. The banks had to be repaired and the ditches cleaned in preparation for the next year's planting. Broken banks or clogged ditches meant loss of crops.

Asiatic cholera devastated almost entire slave populations on some plantations. Coming from its primordial source in India and gradually penetrating Europe, cholera made its first entry into United States at New York in 1832, after which it spread widely over the country, including South Carolina.[96] Caused by the waterborne microorganism *Vibrio comma*, cholera had long been a scourge of Europe. In 1834, an epidemic struck Gowrie Plantation on Argyle Island: "Fifteen slaves died during a dreadful fortnight in mid-September—most of them 'prime hands' no more than thirty-two years old—and three more perished in the epidemic. Because nine other slaves also died that year from other illnesses, while just two were born, only forty-one slaves were still alive on New Year's Day 1835—only a little more than half of the original number."[97]

Lack of privies and using settlement ditches for that purpose contributed to the cholera epidemics as well as using water from ditches for family use. Overseers and planters who maintained sanitation practices on their plantations could avoid cholera epidemics, but they did not always do so. Sending slaves to pineland camps was also a means to prevent cholera outbreaks. Had Charles Manigault been willing to buy a permanent pineland refuge where the slaves could have been quickly resettled at the first sign of cholera, the epidemics of 1852 and 1854 might have been avoided.

Poor health from inadequate nutrition, clothing, and shelter took a heavy toll in slaves. Boiled rice, cornmeal, and some butter or fat from bacon were a slave's typical meal, supplemented by salt and molasses. Vegetable gardens grown during free time after tasks were completed provided extra nutrition. But it was still a diet high in carbohydrates and low in protein, and it did not provide enough calories for working in the rice fields. Drafty living quarters and inadequate clothing and medicine also contributed to poor health. Poor medical treatment caused slaves to turn to a "slave pharmacology" based on the native and introduced plants around the village, which in some cases served them better. Blood-letting and harsh purgatives provided by "doctors" were in most cases worse than the disease.

How much, if at all, did the advances in market preparation reduce the death toll from diseases in the rice fields? Even an extensive survey of primary records might not answer the question. But the following observations are relevant. Harvesting was done by hand throughout the industry; threshing was manually done until the advent of mechanical threshers in 1830; advances in milling began in the mid-1700s, and by the turn of the century the process was highly mechanized. Both threshing and milling required a small work force. Ditching, clearing canals of weeds, and many other labor-intensive tasks continued throughout the industry no matter what advances were made elsewhere. Advances in market preparation probably did not translate into less manual labor on plantations or into making plantations healthier workplaces. With more efficient machines, more fields were created. The creation of new sites for disease-bearing vectors and disease germs may have offset any reduction of demand on the labor force. And if medical care did not improve, there might not have been any gain at all. Even if significant gains were made, the crop had to be planted, and the plantation was still the site of deadly rice fields.

Obstacles to Overcome

Throughout the entire period Carolina rice was grown, planters had many obstacles to overcome to get their rice to market. How they overcame these obstacles determined the success of the plantation.

FIGURE 7. *Bobolinks on Rice* by Edward von Siebold Dingle. Original painting in the Drayton Hastie Archives, Magnolia Plantation, Charleston, SC.

The *bobolink* (or rice bird) was the most serious pest of the fields.[98] The bobolink (see figure 7), a transient visitor to South Carolina, arrived about April 10 on its spring migration from South America to its nesting sites in southeastern British Columbia and northern Nova Scotia. Huge numbers of flocks visited the fields during spring planting, devouring rice before the fields were flooded. Planters avoided the greatest influx by planting in late March to early April and then again in June. During the fall migration back to South America, the birds fed on the ripening rice. Left alone, a flock could strip a field in a few hours. Planting early ensured that the rice ripened before the fall flocks returned.

During both migrations, all plantation hands were put into the fields to keep the birds moving. The birds' frenzied feeding was so voracious that fear was secondary. When scared from one portion of a field, they moved on to another. Slaves lit fires on the rice-field banks or beat pots and pans in an effort to drive off rice birds. Slaves with muskets stood on the banks and fired from morning until dusk. They also employed white-oak smackers, made by tying together two strips of white oak. When they were smacked together, they caused a loud noise. Post–Civil War estimates placed the loss to bobolinks at 25 percent and upward of the total crop.

Writing in 1833, "An Observer" suggested erecting platforms on the banks and stationing a worker on each one. Four to eight cords or twines would be attached to the platform and extend outward fifty yards in a circle around it. Each cord would be tied to a sapling or hoop pole on a bank, and a cowbell or a hollowed-out gourd filled with scraps of tin could be attached to the sapling or pole, so that a pull of the cord would set the noisemaker in motion. The worker on each platform would be responsible for watching six or eight acres. The writer suggested that bits of cloth should also be tied to the cords, so they too would be set in motion when a cord was pulled. The alarm from the bells and the fluttering of the cloth, like the flying of so many hawks, would scare the birds. He contended that this system could replace the work of many with the work of one and that the platform could be manned by a worker unfit for field duty. [99]

Waterfowl such as ducks were a problem because they could consume enormous amounts of rice. Slaves resorted to many and odd methods to keep waterfowl out of the fields. Some would fire bullets with nicks cut into them, which caused a buzzing noise that kept waterfowl from the fields. Killing an alligator or farm animal and placing it on a bank was another deterrent to waterfowl. The dead carcass attracted buzzards, which waterfowl mistook for hawks.

Red rice, or volunteer rice, was the bane of the rice planter—and still is in the present-day rice-growing regions of the Southwest and California. Red rice is the same species as the cultivated Asian rice (*Oryza sativa* L.) but is a weedy biotype (*Oryza sativa* L. var. *sylvatica*) that shows a wide variability of anatomical, biological, and physiological features apart from the cultivated plant. The seed of one will not produce the other. Red rice comes from sowing seed containing red grains. Once red rice gains a foothold in a field, it increases rapidly from year to year unless eradicated or held in check.

Rice from heavily invested fields was unmarketable. Because red rice had a vigorous growth habit, it was also a strong competitor for field space and nutrients, reducing yields, sometimes as much as 100 percent in a heavy infestation. At seedling stage, red rice plants were difficult to distinguish from the crop. Identification was possible at maturity because of gross morphological differences. Before harvest of the commercial rice, most of the red rice seeds fell off and into the seed bank; the grain that stayed on was harvested along with the regular crop. Because the seeds possessed a remarkable resistance to premature germination, seeds in the seed bank germinated the following year when the next crop was planted. Allowing fields to lie fallow was not a remedy because the seeds of red rice will remain dormant in the soil for five years and germinate when the field is planted years later.

Red rice had a thick, red bran that was hard to remove during milling. When milled, red rice did not have the same pearly, white look as the cultivated rice, thus spoiling the market price. Large amounts of red rice required more milling, resulting in more broken grains.

David Doar listed the ways that planters tried to control red rice:

1. No cattle were allowed to run in the fields to prevent their hoofs from burying the fallen seeds.
2. After harvest, birds were allowed to glean the fields.
3. Fields were flooded to allow ducks to feast on the seeds.
4. Red rice was hoed after the stretch flow was taken off.
5. Planters tried to obtain the cleanest seed possible. From 1 to 3 percent contamination was good; if contamination was 8 percent or more, the planter had to find better seed.
6. Workers removed red ears by hand from stacked rice.[100]

Rats were also a major problem in the fields. In his plantation journal, Charles Manigault recorded in 1844: "That some 16 or 18 years ago soon after Mr. Potter bought the property next below me, the rats were so numerous that they lost the entire planting of seventy acres—the seed being so much eaten by the rats—that Mr. Potter requested Mr. Wallace to go over & look at the devastation, & tell him what he had best do. Mr. Wallace told him that the planting was entirely destroyed, & that the only plan was to plant it over again, & get as many dogs as they could & go to work & hunt them—which was done.[101]

Planters paid overseers or other peoples to catch or kill rats. Manigault wrote that an old man at Mr. McAlpin's killed four thousand rats between harvest and planting and was paid one pound of tobacco for every one hundred rats. At Manigault's old place, a worker called Abram killed about one thousand.

Insects plagued planters and laborers. Water control was the only available method of insect control for most of the industry. The rice grub was the larva of a large beetle (*Chalepus trachypygus* Burm.). The beetle appeared in May when the fields were dry and worked its way into the ground, feeding on the young roots of the rice where it laid its eggs. The larva hatched by June, and for as long as the field was dry, the larva did considerable damage. When the harvest flow commenced, both the larvae and the adult insects drowned because they were not aquatic organisms. In upland rice fields where no water was available for irrigation the grub could do serious damage. Both adult and larva fed only on rice.[102]

The water weevil (*Lissorhoptrus simplex* Say) was one of the main pests in the fields. For many years the planters had been familiar with two insects, the one a minute, white, legless grub, infesting the roots of the rice plant, and known as the "maggot"; the other was a small, gray beetle affecting the leaves of the plant and called the "water weevil." Modern investigations determined the maggot is the larva of the beetle.

The adult appeared in the fields in April and May, feeding on the leaves of the young plants during the stretch flow, but it did no serious damage unless it was present in enormous numbers. It fed during the morning and then migrated down the stalk into the water below to escape the midday heat. The female breeds at this season, laying eggs among the roots of the plant.

The eggs hatched after the harvest flow was applied to the fields, and the maggots fed on the roots of the plant. Since water was necessary for existence of the larva, workers drew off water for a day or two during the harvest flow, which afforded a partial remedy. Often, however, drying the field sufficiently to kill the maggots caused more injury to the crop than the maggots did. Since the maggots fed on native aquatic plants, the field would soon be repopulated from other sources.[103]

Maintaining the hydraulic system was an ongoing problem and required a skilled labor force. Planters and laborers had to be steadfast and vigilant to maintain this vast system of banks, canals, and fields on the rice plantation. Damage to the banks by burrowing animals such as crayfish, rats, muskrats, snakes, and alligators caused the banks to leak and break. Freshets, storms, and hurricanes plagued the plantations on a regular basis, blowing out trunks and breaking banks, both of which required costly repairs, lest the planter be forced to yield his heritage back to the river swamp from whose grasp his father's slaves wrested it a century earlier. Waves from the river eroded banks. Banks had an inherent weakness early planters were not aware of. Vegetable matter in the banks decomposed with age, rendering the banks more susceptible to erosion and collapse. Replacement of a trunk was a herculean task. According to Doar: "If an old trunk was to be replaced, it was ditched all around and either broken up or hauled out and the new one put in."[104]

Fixing a break in a bank also required skill, and repair was accomplished by an unusual method. Often breaks in a bank occurred where the bank was built over quicksand, which made creating a stable foundation almost impossible. According to Doar, workers pursued a "plan of going on the margin outside the original break and running a half-moon bank, then filling in the old break as a supporting margin on the inside."[105] The half-moon was constructed by driving two rows of poles in a semicircle and placing the fill within the poles. One can see remnants of these poles along the rivers today.

Declining *soil fertility* did not affect inland-swamp fields and tidal-swamp fields equally. In Georgia and South Carolina, the two main rice-producing states, soil fertility played out differently. Inland-swamp fields were not flooded with river water laden with nutrients. Instead they were flooded with water from reservoirs that received runoff from the surrounding uplands, which were not nutrient rich, and soil infertility was the result. Played-out fields were one reason planters deserted inland swamps after the Revolution and switched to tidal culture. After years of cultivation, nutrient flow from upland rainwater was not sufficient to sustain the crops, and before the Revolution soil amelioration was not a developed practice. Georgia did not have as serious a problem with soil depletion as South Carolina. Georgia's inland-swamp cultivation began considerably after South Carolina, leaving less time for the soils to become worn out, and by the Revolution inland-swamp fields in Georgia were still producing significant crops.

Tidal rice fields also experienced a decline in soil fertility, but not to the same extent as inland-swamp fields. The flow of river water into the fields through the tide trunks undoubtedly brought nutrients from the river into the fields. Rice fields along major brown-water rivers were especially rich because they received runoff from agricultural fields in the piedmont and watersheds in the mountains, naturally enriching the soil. Fields along the black-water rivers did not receive nutrient-rich waters as did tidal fields; consequently, soil fertility was more of a problem. Planters had other means to enrich the soil: turning under stubble, resting fields for a year, and crop rotation for fields that could be thoroughly drained. Commercial fertilizers were not available before the Civil War, so planters were left with only natural means of soil amelioration. Agricultural reformer Edmund Ruffin traveled for eight months in 1843 through South Carolina touting the value of adding lime to worn out agricultural fields. Although he was primarily concerned with upland fields, he did suggest to rice planters that the same treatment would help their rice fields. Rice planters, however, never took soil amelioration seriously except for crop rotation, turning under stubble, and resting fields, but these means replenished the soils significantly enough that crop loss was never a serious problem for most planters.

Unknown to the planters, clearing the fields of vegetation created a problem later with *soil compaction.* Decomposed organic matter in the soil was not replaced by natural means since there was no leaf fall and rice stubble was often burned off the fields. The soil became compacted and the soil level was lowered. Rice fields at the upper zone of tidal range could no longer drain sufficiently to allow workers in the fields to weed or harvest the crop. Planters had to abandon these fields.

Saltwater intrusion was yet another problem. Fields closer to the salt wedge faced a problem with salt water during a severe drought. With a lessened amount of freshwater to push the salt wedge toward the ocean, salt water moved up the river. If salt water got into rice, it would kill it. In 1828 Roswell King, Jr., overseer on Butler's Island on the Altamaha River, answered a query about what he did when the river turned salty by saying: "Butler's Island is so situated, (up and down the river, about four miles,) that, in dry years, when the river becomes salt, at the lower end of the island, the water is brought in by canals at the upper end, and let into the fields at high tides, when the river is salt."[106]

Planters on other rivers faced the same problem. On the Cooper River, a coastal black-water river that often became salty twenty-five miles from the ocean during a drought, planters found a supply of freshwater in an ingenious way. Inland wetland depressions, acres in extent, were dammed to hold rainwater. A canal was dug from the reservoir to the backside of the river fields. When the river became salty, water was let into the fields from the reservoirs. Sometimes rice was grown in these inland reservoirs. Washo Reserve on today's Santee Coastal Reserve supplied freshwater to Santee Delta fields close to the ocean. Planters without access to inland reservoirs faced crop failures during severe droughts.

The Legacy of the Abandoned Tidal Rice Fields

Scientific studies have documented that abandoned rice fields with breached banks on the Cooper River in Berkeley County are dynamic, not static. Reversion to swamp forest through the process of plant succession is probably the ultimate fate of most abandoned rice fields unless man intervenes. Many fields have already reverted to swamp forest.[107] Whether these secondary swamp forests will have the same composition as the originals is for botanists to answer one day. More than likely, the rate of succession and the climax forest composition will not be the same in each field or river system because of local environmental factors.

Beyond the Cooper River and its tributaries, many fields share the same outcome. For example on Woodbourne Plantation, a long peninsula between the Waccamaw River and Bull Creek, one can paddle down the main canal that fed and drained the fields. Only one familiar with the history of rice culture would realize that Carolina Gold once graced the abandoned fields on either side of the canal. J. Motte Alston's former fields today harbor a mature bottomland swamp forest alive with woodland birdsongs, perhaps the same songs slaves might have heard in the original forest before they turned the swamp land into rice fields.

Fields along the Savannah River share a similar fate. U.S. Highway 17 passes through the Savannah Wildlife Refuge on the South Carolina side. The refuge consists of abandoned rice fields that today support a secondary bottomland hardwood forest. Along the Ogeechee River in Georgia, fields have reverted to secondary hardwood swamp forest.

Presently South Carolina state policy prevents a landowner from repairing breached banks in abandoned fields. The freshwater tides, which still make their twice-daily ebb and flow, carry nutrients and detritus from the fields into the adjacent river and ultimately to the coastal estuaries. The main reason the state does not allow repair of the broken banks is that this detritus forms the base of the aquatic food chain. As long as the banks of the fields remain breached, the cycle of growth, decomposition, and detritus flowing out of the fields is repeated each year as a new crop of annual marsh plants emerges in the spring. If the banks are not restored, the fields will revert to swamp forest. The forest trees are perennial, and most of the sun's energy captured through photosynthesis will remain in the fields as part of the woody vegetation. Banks will prevent the annual exit of detritus from the fields. This counters state policy. Scientists are studying whether breaches can be repaired and significant water control obtained to control woody vegetation, while at the same time allowing flow of detritus through modified tide trunks into the river.

There is also public pressure to keep the banks breached since hunters and fisherman can legally enter the fields from the river as long as the banks are breached. The owners pay taxes on the fields but have little control over their use. Keeping the fields in open water will require mending the banks so water control can prevent woody vegetation from becoming established. When a field has open water, birds of prey such as the bald eagle and osprey feed on fish. Both are common today in the Cooper and Santee Rivers. There is the interesting dilemma: if the fields are allowed to undergo natural succession, swamp forest will replace open water, and waterfowl habitat will be eliminated. So will the public's recreational use.

Outer banks on some former rice plantations have not been allowed to deteriorate. Maj. Pierce Butler's extensive rice plantation Butler's Island is now owned by the Georgia Department of Natural Resources, which maintains the outer banks and therefore can control the water supply to the fields. The abandoned rice fields are maintained for waterfowl. In South Carolina, Bear Island Wildlife Management Area on the Edisto River has many abandoned fields still under bank, the banks having been repaired before the state policy was implemented. Bear Island is a major waterfowl wintering area; where slaves once grew rice, public hunts are now conducted by the South Carolina Department of Natural Resources. Plantations bought by northerners after the rice industry failed have fields with intact banks, which are maintained for private hunting. As long as the banks are intact, the owners can manage the water regime to promote species of plants that attract waterfowl, or they can drain the fields, plant crops such as corn, and reflood the fields, providing feeding habitat for waterfowl. Either way woody vegetation can be kept out of the fields, and open water can be maintained.

Little did planters know the ecological legacy they would leave future generations. Environmentalists, governmental officials, and a concerned public face a daunting challenge to manage these former rice fields.

Harvesting

Until the adaptation of horse-drawn machines in the mid-1800s, American farmers harvested grain crops by hand. This was hard work, and harvesting rice was even more arduous than harvesting other grain crops. Harvesting rice had to be completed quickly because once rice was ripe, it was a race against the weather to get the rice sheaves out of the fields and to the threshing yard for drying in preparation for threshing.

Seven to ten days before harvesting, the harvest-flow water was drawn off the fields to allow sufficient drying for the slaves to work. Harvesting generally began in the end of August or the beginning of September, after the heads were well filled and all but the lower rice hardened. Once the rice was ripe, no time was lost in harvesting it, for if it was allowed to overripen, the rice would shed or scatter from the head. Planters such as Robert Allston recognized the need to harvest on time. In an August 29, 1860, letter to Benjamin Allston he stated: "It would be a great mistake for Pittman or Belflowers to postpone the harvest until 1st Septr. In three weeks thereafter much of their rice would be shelling in the field, and before the close of harvest still more."[1] Some planters harvested before the entire field was fully ripe, leaving green spots to ripen and be harvested later.

The Sickle

The sickle was the most common tool for harvesting small American grain crops until the late 1700s, and it remained the only method of harvesting rice throughout the entire time rice was grown in the Rice Kingdom. The sickle has remained virtually unchanged for nearly six thousand years. Mesopotamian farmers made sickles from clay as early as 3700 B.C.E. Farmers next made them from wood with stone or bone inserts. Iron replaced stone or bone prior to the Industrial Revolution. No matter what type of gain was being harvested, using a sickle was backbreaking drudgery, and at best a laborer could harvest no more than an acre a day.

The general steps in sickle harvesting are the same for rice as for other grain; however, there were minor differences from plantation to plantation and from region to region. In

1848 Robert Allston described the harvesting process that was likely the standard for the industry:

> *Curing.*—The field is to be dried some two or three days before the grains be *fully* ripe, and the Rice forth with cut, laying it of an even thickness on the stubble, the heads being clear of any water.
>
> If the weather is fair, one days sun is sufficient. Accordingly after the dew is off, on the day after rice is cut, it should be tied in sheaves and be borne to the barn-yard, and there stacked before dews falls again, in ricks about 7 feet wide, 20 feet long, and built up as high as a man can pile from a stool two feet high. Here it undergoes a heat which is supposed to mature and harden the grain. The rice will keep well enough in the ricks herein described, until threshed, but it is often transferred to large stacks after the harvest for safe keeping. Stacks from 12 to 6 feet in diameter.[2]

Albert Virgil House gave an excellent description of harvesting with the sickle: "Harvesting of rice in Georgia required five distinct steps. The first was cutting off the 'ear-bearing' upper portion of the stalks with hand sickles. This was a two handed operation in which the field hand grasped a handful of stalks with his left hand and swung the sickle with his right, leaving a generous stubble. As all wielders of the sickle or scythe know this task was easier to perform when done with a high degree of rhythm. With the completion of the rhythmic stroke the hand turned and gently deposited the sheaves of grain on the stubble (see figure 8), there as the second step it was allowed to dry for twenty-four hours at least."[3] The sheaves were then immediately hauled to the barnyard and stacked.

In 1828 Charles Munnerlyn commented: "The task for a laborer must be regulated by the distance he has to carry, and the growth of the Rice. The most common task is, a quarter of an acre to be cut, and the same quantity to be tied and carried into the yard. In a field near the yard, I give more, and sometimes, if the field is a remote part of the plantation yard, and the crop of luxuriant growth, I cannot get so much done."[4] This amount of rice that could be cut by a laborer in a day under the task system was the standard for the industry although it varied somewhat from plantation to plantation and from worker to worker.

Toward the late 1800s, an endless array of reapers, headers, binders, and combines for harvesting grain came on the market. Although they were primarily for grains other than rice, several were modified for use with rice. These machines were used extensively in the rice-growing states of Louisiana, Texas, and Arkansas soon after prairie rice cultivation began in 1886. Lowcountry planters knew of these new machines, and some attempted to use them. Thomas J. Molony, Jr., reported in 1871 on the use of a "Rice Bird" reaper: "It will cut rice, but, like all self-rakers, has not the power to lay the rice evenly in grips, in consequence of the rice being so heavy and green. I think the machine can be improved somewhat."[5] In the same article, he reported on another new machine "tried here upon the plantation of Mr. Smith Barnwell; and the machine can be made a perfect success by

FIGURE 8. "Harvesting the Rice." Reproduced from *Harper's New Monthly Magazine* (November 1859). Insert: cradle scythe. Reproduced from R. Douglas Hurt, *American Farm Tools* (1982).

a few trifling alterations. It cuts a five foot swarth, (or four drills of rice), cuts any height, and lays the rice in swarth on top of stubble, in a manner equal to the best of hand labor. It can be drawn by two good mules or horses and managed by one man accustomed to handling mules or horses. Eight acres of rice can be cut per day when the machine is worked, or an acre per hour if pushed to its full extent."[6]

Rice Kingdom planters made little use of these machines for three main reasons. First, the financial downturn after the Civil War made it difficult for planters to invest in new machines. Second, the soil of the rice fields was not firm enough after draining to support the weight of machines, especially the steam tractors used to pull harvesters. Third, rice fields were generally twenty acres or less in size, and it was difficult for machines to maneuver effectively in small fields, especially machines pulled by steam tractors.

Thus the new machines did not replace the sickle (or "rice hook" as it was often called on the plantations), and it remained the main harvesting implement throughout Lowcountry rice culture.

The Cradle Scythe

By the end of the War of Independence, grain farmers in Virginia, Maryland, and Pennsylvania were using a more effective tool to harvest grain—the cradle scythe. Although a European invention for hay mowing, American farmers redesigned it to cut grain (see the insert in figure 8). As the harvester cut the grain stalks with the scythe, they fell onto the fingers of the cradle. The cradle was then tilted and the stalks fell into a pile where they were raked into a bundle and bound into sheaves. Using the cradle scythe, a harvester could cut nearly three acres of grain a day.

Some rice planters tried the cradle scythe. Charles Cotesworth Pinckney reported in 1837, "The cradle has been tried, and failed, principally because it cuts the straw too low, increases its bulk, and the labor of threshing. Where the threshing-mill is used, this is not an important objection; and though it would increase the labor of bringing the rice out of the field, the longer straw enables the binders to tie larger sheaves, and thus expedite that part of the harvest."[7]

Cutting the plants low meant there was little stubble to turn under as manure for the next year's crop or to lay the sheaves on for drying.[8] Although the cradle scythe enabled the reaper to maintain an upright stance, the 10–12 pound tool required skill and strength to manipulate.

As late as 1845, there was still interest in the cradle scythe. The *Southern Agriculturist* reported on a contest between "an athletic active Hooker" and an "expert, experienced Cradler." The hooker, using the sickle, cut four rows; however, the cradler, using the cradle scythe, cut six rows in a shorter time. There was also less waste with the cradle scythe. When sickle workers tied sheaves, many grains and ears scattered over the ground, but the rice stalks slide from the cradle onto the ground with little waste of grain. Thus it appeared that the cradle scythe was superior. Yet the report also noted the same problems that others noted for drying the rice on stubble cut by the cradle scythe. Also the cradle scythe could not cut plants that had fallen over, and the wooden fingers of the cradle tended to warp in the heavy dew.[9]

On September 2, 1845, the Agricultural Society of South Carolina reported that Dr. Francis Yonge Porcher invited "the society to witness a trial of the Scythe & Cradle in harvesting Rice at his Plantation on Goose Creek." The society appointed a committee of five to examine the operation and report on it at their next meeting. No report of the committee appeared in the subsequent minutes.[10] The cradle scythe never gained use for harvesting rice.

Transporting the Sheaves

Sheaves were carried to the barnyard by various means. Stacking in the fields was avoided, lest a storm or hurricane flood the field and destroy the rice on the sheaves. If the field was adjacent to the barnyard, sheaves were carried there by carts or on the heads of female workers (see figure 9). In distant fields with adjacent canals, the sheaves were loaded onto rice flats and poled to the barnyard and unloaded.

Charles Pinckney described an unusual way to transport the sheaves to the barnyard. On fields where neither flats nor wagons could be used to advantage, "recourse may be had to the Devonshire hook, made of four small branches of young trees, with part of the trunk, so placed on the back of a horse, mule, or ox, that it can be made to carry a large and heavy load. This has been tried with success."[11] No diagram of the Devonshire hook has been found, and this source is the only reference found of its use. Even though Pinckney said it was "tried with success," success apparently did not translate into common use.

In 1835 Thomas Spalding of Sapelo Island decided that transporting rice from the fields by carts or manually was too time consuming. He suggested that a lightweight portable railway be built from the field to the barnyard, on which Jersey wagons would convey the sheaves. The wagons would have their axletrees adjusted to the width of the rails and wooden flanges would be screwed on the outsides of the wheels to keep them on the rails.[12]

Once in the barnyard, sheaves were temporarily stored in ricks for curing. These ricks were slanted on top to better shed water. One planter oriented the ricks on a north and south axis so that each side would receive the benefit of the sun for half a day.[13] When the rice was sufficiently cured and cooled, threshing was done immediately, or the ricks were torn down and the sheaves were placed in long or round stacks, where they remained until they were threshed. Occasionally the sheaves were stacked during the winter.

The curing process was important to prepare rice for threshing and milling. When the rice sheaves were stacked in ricks, heat from natural fermentation of the stalks matured and hardened the rice. This process was monitored to prevent too much heat from

FIGURE 9. Hauling sheaves, early 1900s. Courtesy of Historic American Building Survey, National Park Service.

injuring the grain. Injured rice was said to be "mow-burned." To measure the temperature of a rick a stake was inserted into it at either end. The stakes were examined daily. A worker drew out the stake suddenly, and if he found the inner point of the stake to be too hot to hold in the hand, the rick was pulled down, aired, and built again. As soon as the curing subsided, the rice was ready for threshing.[14]

Harvesting remained a manual process throughout the rice industry in the Lowcountry. Mechanical harvesters never replaced the sickle. But slaves and later freed people, developed an expertise and work ethic so efficient that machines were not needed even if they were available and affordable. Harvesting by sickle was never a major impediment to producing large crops for market.

Threshing

Threshing—removing grain from the stalks of plants—goes back more than ten thousand years, to the dawn of agriculture.[1] Throughout the histories of threshing and milling earlier technologies persist as succeeding ones are employed. For example the beating action of the manual flail is carried through successive threshing machines before being replaced by beaters on the mechanical threshers of 1830 and later. Likewise the fanner basket for winnowing rice starting in 1685 was carried through to the winnowing house and then to the mechanical wind fans employed in the complex post–Civil War threshers. Their basic function was the same: to clean the threshed grain. Often a step in the evolutionary process was drastic and revolutionary. The Emmons thresher of 1829 was a dramatic leap from the flail. Other changes were more incremental. For example the Butts thresher of 1848 improved on the Emmons thresher, basically retaining the main features but refining them.

This chapter is a chronological treatment of the evolution of rice-threshing process, from the flail to the steam-driven threshing machines of the late 1800s and early 1900s. This chronological sequence, however, might not have been followed on a particular plantation. The integration of methods, machinery, labor, and management differed from place to place, plantation to plantation, and individual to individual. One planter may have used flails well into the 1900s, while another planter, who had more resources, might have switched every time a new threshing machine came on the market. Yet another planter might have abandoned a newer machine that broke or wore out because of the cost of repair, and reverted to the flail.

The U.S. Patent Office has on file plans for many machines invented to thresh grain, some specifically for threshing rice. To determine which of these machines were commercially successful in the Rice Kingdom is a difficult task. Some were tried briefly, but no mention is made of them again in archival sources. Some patented machines were never used commercially even though they represented a technological advance. Others were successful and gained widespread use. The destruction wrought by the Civil War forced many planters to revert to the flail. Prewar threshing machines had worn out and were too expensive to repair, or they had been destroyed by former slaves or Union forces. Many planters could not afford the new machines that became available after the war.

There are no existing descriptions of several machines invented for threshing, only mentions of their inventors and functions. These machines were evidently never patented. David Ramsay mentioned that a Mr. DeNeale invented an improved method of threshing. No description or diagram of this machine exists, and it would be unknown were it not for Ramsay's note.[2]

Another problem with detailing the threshing process in the Rice Kingdom was created by the loss of threshing patents in the 1836 U.S. Patent Office fire, including patents by citizens of South Carolina. The following are South Carolina citizens' rice-threshing patents lost in the fire:

Benjamin S. Hort, Georgetown, patent dated February 21, 1812
Elias B. Hort, Charleston, patent dated February 18, 1828
Jehiel Butts, Georgetown, patent dated May 10, 1830
William Mathewes, Charleston, patent dated August 27, 1835[3]

The Flail

The flail (see figure 10) was the first instrument used for threshing rice in the Rice Kingdom, and it remained the primary threshing implement until the advent of the Emmons thresher in 1830. Despite improved machines, however, threshing with flails continued on some plantations until the end of the industry. Planters who could not afford mechanical threshing machines continued to rely on the flail. Historian Lewis Gray mentioned that

FIGURE 10. "Threshing." Reproduced from *Frank Leslie's Popular Monthly* (February 1879).

treading was also employed; however, no evidence was found that this process of trampling the sheaves with human feet or animal hooves to dislodge the seeds was a significant threshing method in the Rice Kingdom.[4]

Threshing with the flail was an onerous and slow process, and could be done only in good weather. On average, men could thresh ten bushels a day, and women could thresh eight. Threshing with the flail initially involved four steps. The first step was beating the sheaves with the flail to dislodge the rice. Second, the stalks were shaken to remove embedded rice. Third, the stalks were hauled away. Fourth, threshed rice was winnowed to remove the chaff. A fifth step, using a rolling screen, was added in the mid-1700s. The rolling screen was a screen with different-sized openings to separate smaller defective rice and larger trash not removed during winnowing. These five steps were incorporated into all the mechanical threshers that came on the market starting in 1829.

The flail has three parts: a long handle called the "staff" is attached by a piece of rawhide, the "swivel," to a short, clublike piece of wood called a "swingle" or "bob." The long-handled staff allowed the worker to remain upright during threshing. A worker held the staff and whirled the swingle in a circular motion over his or her heads, bringing it down with force on the sheaves and dislodging the rice from them.

The flail was the best threshing implement in its time. The output in bushels from threshing with a flail was determined by the size of the sheaves. Charles Manigault, in his plantation journal, commented on the average number of bushels that should be obtained in one day and the problems associated with maintaining this average:

> I have heard much discussion among Planters respecting the Quantity averaged in threshing with the flail stick. 12 Bushels are said to be the task while some planters say that with another they do not get more than 5 or 6 Bushels of cleaned winnowed rice. I have this day made a calculation with Mr. Bagshaw & measured carefully. I have say 25 full hands on the floor 12 men who do 600 sheaves each 13 women & weak hands 500 each. Mr. Bagshaw says the usual sized sheaves should always give 2 bushels every hundred sometimes 2½ bushels which latter [*sic*] would turn out 15 Bushels the 600 sheaves. But Negroes are cunning enough to remember that what they are harvesting they will have to thresh, & will tie as small sheaves as they can. B[agshaw] says at this last harvest he told them that he would give them 110 for every 100 if they went on tieing [*sic*] such small sheaves. Well, the last weeks threshing on an average each man & woman's threshing when winnowed out is ten Bushels. I told Mr. B[agshaw] that some planters went entirely by measure, & had a tub on the threshing floor which when filled of the threshed rice just from under the flail mixed with threshed particles of straw & tailings so as to come at a due estimation of their being what will produce when winnowed 12 Bushels. "Yes" said he but the negro will then knock away so as to break up & intermix the greatest possible quantity of cut straw &c. with his days task & thus defeat justice by rendering things uncertain & unequal.[5]

FIGURE 11. The threshing yard at Middleburg Plantation on the Eastern Branch of the Cooper River, Berkeley County, South Carolina, early 1900s. Note the wooden frame in the right foreground, which was used to thresh seed rice. © image, reproduced by permission of the Gibbes Museum of Art/Carolina Art Association, Charleston, South Carolina.

FIGURE 12. *Shaking the Rice from the Straw after Threshing* by Alice Ravenel Huger Smith. © image, reproduced by permission of the Gibbes Museum of Art/Carolina Art Association, Charleston, South Carolina.

FIGURE 13. Fanning rice. Photograph courtesy of the Charleston Museum, Charleston, South Carolina.

Virtually all plantations had a yard similar to Manigault's where sheaves were brought for threshing (see figure 11). The sheaves were laid on a bed of packed clay, a bed of mixed sand and tar, or a wooden floor (figure 10) which made it easier to dislodge the rice with the flail. Duncan Clinch Heyward described the process:

To thresh the rice, the planters had for many years resorted to flail sticks. The bundles of rice were placed in rows on the ground, with the heads joining each other. Workers walked "between the rows, swinging the flail-sticks above their heads, and bringing them down on the heads of rice, this beating off the grain."[6]

On some plantations the sheaves were taken to a building that housed a threshing floor. A convenient size for a threshing floor was 110 feet by 60 feet. According to James Bagwell, "The sheaves were spread out on the floor and beaten by slaves wielding flat sticks. The separated rice fell through slits in the threshing floor onto a sub-floor made of two inclined planes which collected and funneled the rice into tubs of one-bushel capacity."[7]

After threshing, the straw was shaken by hand to separate embedded rice (see figure 12), then raked and hauled away. Some planters had the driver examine the straw to ensure that all rice was shaken out before it was discarded. The rice and chaff were swept into a pile or collected in a bin and stored for winnowing. Threshed rice had to be cleaned before it was milled because it picked up unwanted trash, called "tailing" or "chaff." Initially cleaning was done with a fanner basket (see figure 13), using wind to

FIGURE 14. A winnowing platform. Reproduced from Edward King, *The Great South* (1879).

blow away the lighter chaff. A fanner basket was a circular, shallow straw basket about two feet in diameter with a raised lip. The contents of the basket were rotated so that the lighter chaff moved to the outside, where it was repeatedly tossed in the air. The wind blew the lighter chaff away, and the heavier rice fell back into the basket.

Winnowing platforms were also employed. The winnowing platform (see figure 14) was a raised floor with a grated opening in the center. When the wind was sufficiently strong, the mixture of rice and chaff was hauled to the top the platform and dropped through the grated opening. The wind blew away the lighter chaff; the heavier rice dropped down into a collecting basket or other collecting container or onto a hard clay surface under the platform. Workers with brooms swept remaining trash from the rice.

Enclosing a winnowing platform created a winnowing house (see figure 15). The winnowing house was a room ten feet square raised about fifteen feet above the ground, with a grated hole in the floor and an outside stairway leading to the room (see the insert in figure 15). Workers stored rice in the winnowing house until the wind was right for winnowing, so they did not waste wind time while hauling rice to the winnowing house.

Placement of a winnowing house was important for capturing the best breeze. In an 1848 letter to his overseer Charles Manigault stated: "The next job will be the new Winnowing House, (which has been long wanted) the old one as you say was badly placed, the mill during one of the most prevalent winds keeping the breeze from it. You will select the best spot for the new one, which I presume will be in front of the Barn Yard gate."[8]

FIGURE 15. The winnowing house at Mansfield Plantation on the Black River in Georgetown County, South Carolina. In the background are the threshing barn, steam engine, and chimney. The barn burned after this photograph was taken by Charles N. Bayless in 1977. Courtesy of the Historic American Building Survey, National Park Service. *Insert:* the operation of a winnowing house. Illustration by William Robert Judd.

Seed Rice

To insure its viability, seed to be used for planting was threshed differently from rice used for food. David Doar pointed out that in hand threshing, when the rice sheaves were spread too thinly on the threshing floor, the rice seeds would be broken and be "not so good" for planting.[9] Planters also found that mechanical threshing split the hull, and when the seeds were planted, water oozed through the split, rotting the seeds.

Many methods were devised to thresh seed rice. Whipping the sheaves against a barrel, log, or plank produced unbroken seed. On Middleburg Plantation on the Eastern Branch of the Cooper River in Berkeley County, an elevated, square wooden frame was employed (see figure 11). An inclined trough in the center of the wooden frame caught the rice after the sheaves were beat against the frame.

On Argyle Island in the Savannah River, the location of Gowrie and East Hermitage plantations, owned by Charles Manigault and his son Louis, seed rice was never threshed by mechanical threshers or flails. Charles Manigault advised Louis on January 10, 1856, to "beat it *Carefully* over the edge of the Barrel until all the good size ripe grain fall *therein.* The straw with the rest of the grains, on it must be passed thro' the thresher as soon as convenient. You will thus have first Rate Seed."[10] With two hands working on each barrel and as many as six barrels being filled at a time, it was possible to thresh all the seed rice needed for the next season in a week or two.

Before it was planted, seed rice was winnowed, early on with a fanner basket and later in a winnowing house. After the wind fan became available, J. Bryan stated in 1832: "Immediately before planting, the seed-rice is passed through the wind-fan, and a sieve, to take out any light grains or grass-seeds that may be in it."[11]

Some planters sold seed rice, or had an intermediary sell it. Erasmus Audley of 14 Elliott Street, Charleston advertised that he had "FOR SALE, *Four Thousand Bushels of excellent,* SEED RICE, As free from red as any in the State."[12] Planters who could supply quality seed rice could supplement their income from planting.

Seed rice was carefully guarded from pilferage by workers. In at least one surviving structure, seed rice was kept under lock. At Middleburg Plantation, on the Eastern Branch of the Cooper River in Berkeley County, stands a double-celled structure where seed rice was secured (see color plate 6).[13]

The Kogar Wind Fan

The wind fan was the next advance in threshing and replaced the winnowing barn for cleaning rice. Wind fans were employed in connection with mechanical threshers throughout the remainder of the industry, albeit modified and improved with time. The *Journal of the Commons House of Assembly* records a 1754 petition for a wind fan invented by Joseph Kogar of Colleton District: "And the Petition of Joseph Kogar was read setting forth, That the Petitioner, a German Protestant, had been about twenty years in this Providence, a great part of which time he had spent in contriving and inventing Machines for a more easy and expeditious way of manufacturing Rice, first a Wind fan, after which many trials and much time spent upon it the Petitioner at last brought to perfection and to be of great Advantage to the Public."[14]

There is no surviving patent description or diagram of Kogar's wind fan. Charles Drayton recorded the use of Kogar's wind fan in 1807: "Went to Freers mill. It has a ground floor, first floor & loft. The threshed rice, from the loft passes thro a tube into a fanning Screen below; which seperates it from Straws and dust: from thence, it passes by a hopper, into the Stone grinding mill, situated a little below the first floor. From there it passes into a wind-fan like Cogers [Kogar's] to drive off the flower [flour] dust & chaff. From thence it is carried by hands into the pounding mortars. From thence it is passed through a wind fan like Cogers. I saw no sifting of flower. The mill is to be furnished with different Screens or sives [sieves], mounted on a cylinder frame, about 12 feet long & [no number] inches at the greatest diameter."[15]

1. A small-stream floodplain on Huger Creek in Berkeley County, South Carolina. Photograph by Richard Dwight Porcher, Jr., 2011.

2. Penny Dam, an inland-swamp reservoir at Fairlawn Plantation in Charleston County, South Carolina. Photograph by Richard Dwight Porcher, Jr.

3. Former rice fields at Drayton Hall on the Ashley River in Charleston County, South Carolina. The remains of the three banks that separated individual fields are labeled "Outer Bank," "2," and "3." *T* marks the location of the lift-gate trunk. Aerial photograph by Richard Dwight Porcher, Jr., 2007.

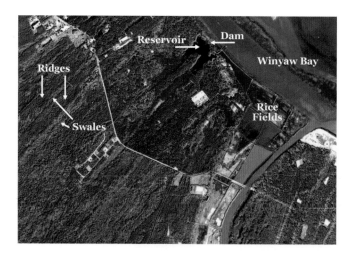

4. Estherville Plantation reservoir and rice fields on Winyah Bay in Georgetown County, South Carolina. North Inlet–Winyah Bay poster of a National Aerial Photographic Program near-infrared image, Estuarine Reserve Division of the National Oceanic and Atmospheric Administration, U.S. Department of Commerce.

5. A tidal freshwater swamp on the Waccamaw River in Georgetown County, South Carolina. Photograph by Richard Dwight Porcher, Jr.

6. The seed barn at Middleburg Plantation. The doors to the two adjacent rooms where the seeds were stored are missing. Photograph by Richard Dwight Porcher, Jr., 2009.

7. Rolling screen at Chicora Wood Plantation.
Photograph by Richard Dwight Porcher, Jr., 2010.

8. Rolling screen at Kinloch Plantation. The chute fed rice into the screen. Hopper 1 collected clean rice and sent it to a chute, which delivered the rice into a vehicle. Hopper 2 collected trash and sent it to a chute through which it was removed from the barn. Photograph courtesy of Norman Sinkler Walsh, 2010.

9. Remains of the Butts thresher at Mansfield Plantation. Photograph by Richard Dwight Porcher, Jr.

10. The rake enclosure in the Chicora Wood threshing barn.
Photograph by Richard Dwight Porcher, Jr., 2009.

11. The threshing barn at Chicora Wood Plantation.
Photograph by Richard Dwight Porcher, Jr., 2011.

12. The storage barn at Chicora Wood Plantation. Visible on the river side of the storage barn are the two portals through which rice was loaded into a waiting vessel. The insert shows the portals close up. The lower portal had a sliding door on the inside so laborers could close it when it was not in use. A chute was inserted into the opening, and rice flowed from the barn into a waiting vessel. Photograph by Richard Dwight Porcher, Jr., 2010.

13. The Invincible thresher, patented circa 1884, at Kinloch Plantation. Photograph by Richard Dwight Porcher, Jr.

Historian R. Douglas Hurt described a fanning mill imported from England that was used in the 1770s. This mill was the same as a wind fan: "The fanning mill consisted of a series of wooden paddles approximately 18 to 24 inches long which were attached to a rod geared to a crank. The paddles or fans were enclosed in a box-like frame which also housed several screens. As the grain and chaff was poured into the container at the top, the farmer turned the crank which caused the fans to take in air through apertures in the sides and blow it across the screens. The grain fell onto the screen and sifted to the bottom as the forced air blew the chaff away. The cleaned grain poured out into a basket below."[16]

These early wind fans were the forerunner of more complex wind fans, or "chaff fans" as they were often called, used to clean virtually all grain crops, including rice. They operated on the same principle, with modifications to the riddles depending on the grain being cleaned. A "riddle" is a course sieve that is used to clean large material, such as grain, from chaff and straw. Several riddles were grouped in a "shoe," which was activated with a side-to-side motion, causing all the riddles to shake in unison.

Diagram 18 shows the basic fanning mill of the 1800s. Grain was fed into the hopper. A slide regulated the amount of grain fed from the hopper to the top riddle of the upper shoe. A crank connected to the drive shaft made the shoe shake. As grain was fed onto the top riddle, the shaking spread out the grain. A blast of air from the fan entered at the bottom of the shoe and moved upward, carrying the lighter chaff away and forcing any substances like straw to tail over. The process was repeated as the grain fell through the

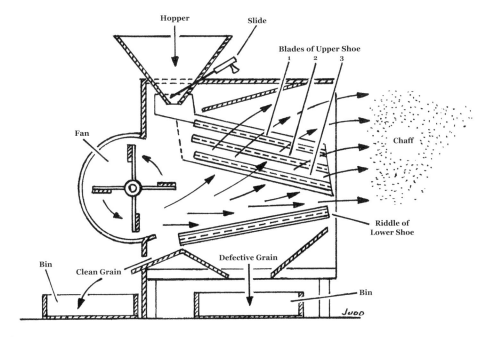

DIAGRAM 18. Operation of a typical 1800s wind fan. Illustration William Robert Judd, adapted from a diagram in B. W. Dedrick, *Practical Milling* (1924)

pores of the first riddle to the second riddle below, and then to the third riddle. As the grain fell onto the lowest riddle, the wind blast from the fan had a full sweep and blew out any remaining chaff.

The grain next dropped onto a riddle of the lower shoe. This riddle had finer mesh and sifted out any defective or light shriveled grain, which dropped through the hopper below into a catch bin. The good, clean grain passed over the end of the riddle directly under the fan into another catch bin.

Replacing winnowing houses, wind fans could clean rice immediately after it was threshed because the use of wind fans was not dependent on the vagaries of the wind. One 1775 source stated that winnow rice "was formerly a very tedious operation, but now much accelerated by the use of a wind-fan."[17]

Wind fans, either based on Kogar's invention or the English model, were commonly used in mechanized threshing barns and mills. As water or steam power was employed, wind fans were driven by mechanical power.

The Andrew Meikle Thresher

Threshing grain with a flail involved long hours of debilitating work. Consequently there was an impetus to develop a better method of threshing. The first successful mechanical thresher for grain—the type of modern threshers—was invented by Scotsman Andrew Meikle in 1786 (see diagram 19).

Animal power ran Meikle's thresher. Straw was passed from the feeder board into two fluted feeder rollers, which directed the straw to the beater cylinder. The beater loosened

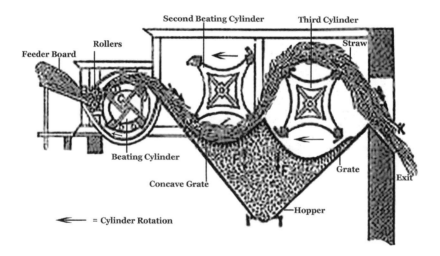

DIAGRAM 19. Interior view of Meikle's Threshing Machine, 1786. Reproduced from Theodore Dwight Woosley, *The First Century of the Republic* (1876).

the grain before it passed the straw to a second beating cylinder, which operated over a concave grate. The grate did not allow the straw to pass through, only the grain (and small bits of chaff). Loose grain fell from the straw through the grating into the hopper below. A third cylinder raised and loosened the straw, shaking out the remaining loose grain, which passed through a second concave grate into the hopper below. The straw was ejected through an opening in the wall of the building. Around 1800 a wind fan was added to separate the chaff from the grain, and the machine became a true separator, threshing, cleaning and delivering the cleaned grain in one operation. There is no evidence that Meikle's thresher was adapted to thresh rice, but the design was evidently incorporated in later rice threshers.

Thomas Jefferson may have been responsible for introducing the Meikle thresher into the United States. Jefferson saw a model of Meikle's thresher in Arthur Young's *Annals of Agriculture,* and asked Thomas Pinckney, who was in London, to send him a model of the thresher. In 1793 Jefferson wrote in his memorandum book: "Gave order on bank US for 62.8 to John Vaughan for his bill for 13–13 sterl. on Byrd, Savage & Byrd paiable [sic] to T. Pinckney, and inclosed it to T. Pinckney to pay for threshing model."[18] Jefferson then wrote to James Madison: "I expect every day to receive from Mr. Pinckney the model of the Scotch threshing machine. . . . Mr. P. writes me word that the machine from which my model is taken threshes 8. quarters (64. bushels) of oats *an hour,* with 4 horses and 4 men. I hope to get it in time to have one erected at Monticello to clean out the present crop."[19] The model, a treble-geared machine run by animal power, arrived in Virginia at the end of 1793. Jefferson used his thresher for wheat.

Thomas Spalding (1774–1851) mentioned a Scottish threshing machine: "Accordingly, threshing by machines in Scotland was considered more complete than the use of the flail. Not so, the American imitations of the Scotch thresher, which have heretofore been inferior to the flail, and required its aid in threshing out the bands and other portions of the sheaf." Spalding said the main reason threshers in the Rice Kingdom failed was "from men desiring to design something new, that they may patent, rather than something old, which other men, and long use, have perfected."[20] Inventors apparently followed Spalding's advice. Future threshers were based in part on previous proven ones, though they were modified by their inventor.

Rolling Screens

Planters wanted to send threshed rice as clean as possible to the mills, so they installed rolling screens to remove heavier particles, such as defective rice and sand, that were not removed by wind fans. Presumably they received a higher price for their clean rice. The first documented date for using rolling screens to clean threshed rice is around 1782–83. An 1802 letter from N. and D. Sellers to Henry Laurens gives their version of the history of rolling screens. Though it is not certain that they are referring to rolling screens used in threshing and not milling (discussed later), their reference to cleaning rice and the date of their letter makes it more likely they are referring to threshing. They wrote: "In answer to thy question, relative to the rolling screens for cleaning wheat, We beg leave to inform that what our father John Sellers was the first person who took up the Wire work or screen making in America about 44 years ago, at which time he also built a Grist and Merchant Mill about six miles from Philad: and we believe the first rolling screens he made, & the first made in America of wire was put up in his Mill upwards of 40 years ago, which answering a very valuable purpose & becoming known brought him many applications for them."[21]

The rolling screen was subsequently adapted for rice. The Sellers' 1802 letter also contains a history of wire mesh and screens for rice from their company in Philadelphia, which is pertinent to rolling screens for rice:

> We are favored with thy letter of the 9th [illegible]: in answer to which We shall beg leave to trouble thee with a short history of the progress of the business of making wire works for cleaning rice generally. So long ago as the year 1782 or 83. Joseph Smith of Charleston, then with his family in Philad. seeing us in the wire Business advised us to turn our attention to a wire work suitable for dressing rice. We told him it was our professional Business and nothing more was necessary than to give us samples of the subject or grains to be separated: that we could readily ascertain the Mesh & were in the constant habit of doing it for every purpose to which that kind of work could he applied. He furnished us with samples of Rice from which We discovered the requisite Mesh & then in the year 1783 made for his Son James Smith a number of sieves for dressing Rice, and [illegible] for him & other persons in Charleston a considerable Number.[22]

Three rolling screens survive and may be found at Chicora Wood (see color plate 7), Cockfield (see figure 16), and Kinloch (see color plate 8) Plantations. The screen at Chicora Wood is twelve feet long with intact wire mesh. Each of the six sides is twelve inches wide. The mesh opening at the upper end is square and 0.8 centimeters; the mesh opening at the lower end is rectangular, 2 centimeters long and 1.25 wide. Sand grains and small trash could pass through the square mesh but the whole rice could not. It passed through the mesh at the lower end. The Cockfield screen contained a section of wire mesh and was in its original mount, which was removed before measurements were made. The Kinloch screen is still in its original mount in the cupola of the threshing barn (see color plate 14), but it has no wire mesh and is 95 inches long with eight sides, each side thirteen inches wide. The location in the cupola documents the location of the screen in the threshing system. The gears that drove it from the power source are intact as are remnants of the two hoppers below the screen. One hopper carried away the chaff and the other sent clean rice to a wind fan for further cleaning.

To construct a rolling screen, wooden ribs were fixed to a metal spine; spokes and wire mesh were fixed to the outside of the frame. Rolling screens were either round, hexagonal, or octagonal. The screen at Chicora Wood is hexagonal; the one at Kinloch

FIGURE 16. The rolling screen at Cockfield Plantation on the Combahee River in Colleton County, South Carolina. This rolling screen was later removed and discarded when the barn was converted into a residence. Photograph by Richard Dwight Porcher, Jr., 1986.

---

is octagonal; and the one at Cockfield is round. No records of rolling screens used in the threshing process indicate any operational advantage of one shape over another. The choice of shape may have been based on whichever design was easier and cheaper to construct, or easier to clean. Charles Manigault wrote to overseer James Haynes from Paris on March 1, 1847: "Mr. Daniel Heyward got the idea of this leather slapper from me, & he fixed a piece of Band leather close to his rolling screen, so that when revolving each angle of it caused this leather to flap against the wire, & clean it. And this he called 'a perpetual flapper' as it operated all the time the rolling screen was in motion."[23]

Diagram 20 depicts a rolling screen used for rye or buckwheat. No diagrams of rolling screens used for rice were found in archival sources, but all rolling screens worked in a similar manner. To operate the screen, threshed grain was poured into the inlet and passed down a chute into the rolling screen. The screen turned at about fifteen to twenty revolutions per minute, and because it was on an incline, its contents gradually worked their way down to the far end. The upper end of the screen had the smallest mesh. If rice was the grain being cleaned, there were only two sizes of meshes. (See figure 16 and color plates 7 and 8, where physical evidence indicates only two size meshes.) Small, broken pieces of rice and trash fell through the small mesh and passed out through a hopper. At the far end, with the larger mesh, whole rice passed through the mesh into a hopper, where it was collected and placed in a barrel for shipment. The products that came

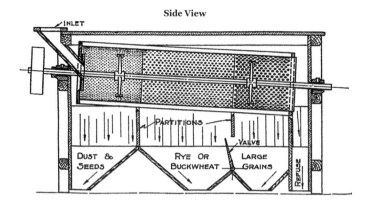 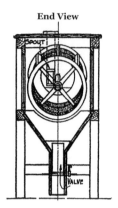

DIAGRAM 20. Operation of a rolling screen.
Reproduced from B. W. Dedrick, *Practical Milling* (1924).

---

through each size of mesh fell into a hopper below the screen and were removed. The diagram shows that at the lower end large refuse passed out the end. Threshed rice did not contain large refuse, so a refuse bin would not have been necessary.

The Bernard Thresher

In 1808 Aaron Dyer announced that his son in Wilmington, North Carolina, had built a threshing machine using a plan invented by Benjamin B. Bernard. Worked by oxen, it threshed between three and four hundred bushels of rice a day.[24] Bernard's threshing machine was primarily built for wheat and oats, but Dyer's son modified it for rice (diagram 21).

Bernard's thresher incorporated the basic element of Meikle's: a beater cylinder that loosen grain from stalks. An ox (or other animal) attached to an arm turned a cog wheel, which drove a band wheel. The band from the wheel turned a shaft to which wooden beaters were attached inside a drum. The rice stalks (or other grain stalks) were placed on a feeder board that fed the stalks into the drum. The revolving beaters removed the rice from the stalks.

While Bernard's thresher was not built for rice and did not gain widespread use, it may have been the template for other threshing machines. All rice threshers that became commercially viable relied on revolving beaters to thresh rice free of the stalks.

During the 1820s, as crops increased in size and the flail was no longer adequate on large plantations, the need for a mechanical thresher for rice became acute. In 1827 the Agricultural Society of South Carolina, based in Charleston, offered a reward of one hundred dollars "for the best constructed Machine for threshing rice, which has been in operation for two Months adapted to general use, by threshing perfectly fifteen Bushels per hour with the force of three horses, or four Mules, or three Yoke of Oxen."[25]

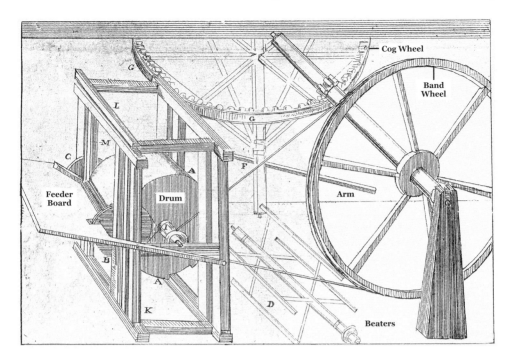

DIAGRAM 21. Patent diagram for Benjamin Bernard's thresher, 1808. This diagram was published with an advertisement in an unidentified newspaper. From the Pinckney Lowndes Papers of the Ravenel Collection, South Carolina Historical Society (folder 11/332A/22).

As late as 1829, a practical thresher for rice still had not emerged. In that year "A Friend to Improvement" pointed out that the time needed to thresh the crop was so tedious that the time for planting the next year's crop arrived before the threshing was done: "Our threshing, which all must allow, when the crop is a heavy one, is a most tedious and laborious operation, is almost universally performed by the simple flail. No machine, therefore, can, at the present time, be more desirable that the Threshing Mill on our plantations, upon a cheap and effective plan, which will relieve our laborers from that operation, and enable planters to appropriate the whole of the winter months to the preparation of their lands, for the succeeding crop."[26]

The Calvin Emmons Threshing Machine

New Yorker Calvin Emmons came to the aid of the rice planters. On July 27, 1829, Emmons received patent no. 5,584 for an "Improvement in the Threshing Machine." Fortunately Emmons's diagram (see diagram 22) and description were restored after the fire in the U.S. Patent Office and a new number (5584X) was applied. Emmons may have based his machine on the Scotch thresher or Bernard's thresher. In 1834 a writer identified only as "W" said that "the iron teeth of the beaters [of the Emmons thresher], and a joint in

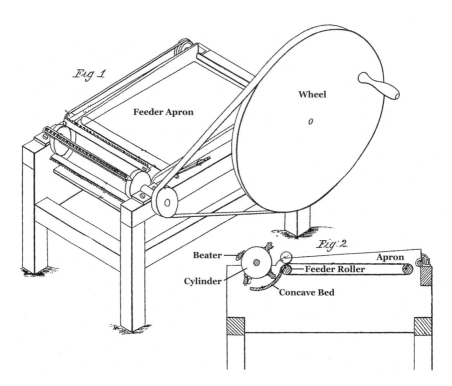

DIAGRAM 22. Patent diagram for Calvin Emmons's improved threshing machine, 1829. United States Patent Office.

the iron which attached them to the cylinder, and gave centrifugal force, constituted its only variation from the old Scotch thresher."[27]

In the February 19, 1831, issue of the *Georgian,* several planters attested to the quality of Emmons's thresher. J. Hamilton, Jr., stated that "Messrs Emmons & Chambers have been for the last two months on my place, instituting a series of experiments with their Machine, which I think have resulted in a satisfactory discovery of that long waited desideratum in our Southern Agriculture, a cheap [illegible] mechanical [illegible] for the threshing of Rice." Daniel Elliott Huger and W. M. Heyward "were gratified by the manner in which the Rice was threshed, whilst the grain was sufficiently separated from the foot stalk, it was quite as little broken as with the common flail." Thomas Young of Savannah said, "I conclude in opinion with the above subscribers, with respect to the merit of Mr. Emmon's Threshing Machine; and think the Rice is less broken than with the common fail." Peter Evans of Rice Hope said that "it threshed in four hours, ninety two bushels and three pecks, making over twenty three bushels per hour. I am satisfied that a machine five feet in width and six feet in length, with the power of two mules or horses, with ten hands, would thresh from four to five hundred bushels per day."[28] The planters were clearly impressed with the demonstration.

The original patent shows a hand-cranked machine (fig. 1 in diagram 22), two and a half feet wide; this is probably the machine Emmons demonstrated for General Hamilton. It is unlikely that many were operated by hand on plantations once his thresher proved successful. By 1831 steam engines were coming into common use, and although animal and water power could drive the thresher, most planters probably used steam power.

The four hinged iron beaters with teeth attached to the cylinder were the main feature of his thresher. Rice sheaves were untied, and the stalks were placed on the feeder apron, which carried the stalks to the feeder rollers. The feeder rollers fed the stalks into the space between beaters and the wooden concave bed below. The two feeder rollers and the shaft were powered by different bands from a large wheel. As the iron shaft rotated, centrifugal force caused the hinged beaters to extend as far from the shaft as the joints by which they are secured allowed, and formed a cylinder when in motion. As the stalks passed between the beaters and bed, rice was combed from the stalks. The stalks were ejected from the back of the thresher. The threshed rice more than likely fell through slots in the concave bed along with the chaff and was winnowed. By 1830 mechanical wind fans were available, and it is likely that one was employed in conjunction with the Emmons thresher. The Emmons thresher did not employ a device to remove embedded rice. Workers shook stalks to remove loose rice that was not separated in the threshing machine.

Emmons did not construct his thresher specifically for rice. In his patent he said it was "for Thrashing all kinds of small grain Viz. wheat, rye, oats, Rice etc." Fortunately it was so versatile that it was easily adapted for rice. Emmons continued to modify his thresher. He substituted steel plates for the original wooden bed, replaced the original apron with an inclined feeder board, replaced the iron beaters with steel beaters, and replaced the iron teeth of the beaters with cast-steel teeth. (The iron teeth were readily worn by impact with the rice.)

Emmons's machine revolutionized rice threshing. Not only was it a success, but it may well have been the prototype for later threshing machines that were more reliable and efficient. In a May 1843, letter to Robert Allston,[29] Emmons stated that in the winter of 1830–31:

> I was first induced to introduce the machine in the rice-planting districts, by the urgent solicitation of General James Hamilton, who had seen a model or hand machine an agent of mine had in Savannah, Georgia, and which General Hamilton had taken over to his plantation near the city, and with the operation of which he was so well pleased that he at once sent me a bale of rice in the sheaf, that I might further experiment with it here; and wrote me, offering every facility for trying the experiment by animal power on his plantation, if I would bring out a machine and driving geer [sic] the following fall or winter, to which I acceded; and after the erection of which machine, General Hamilton invited the neighboring planters to witness the experiment, the result of which proved satisfactory.[30]

Robert Allston used the Emmons thresher at Chicora Wood on the Great Pee Dee River and stated in his "Rice": "The invention, which is now in very general use, yielding, when worked by animal power, from two to three hundred bushels per day, and when propelled by steam, 450 to 700 bushels each, is due to the ingenuity and mechanism of Calvin Emmons, of N. York."[31]

Emmons's thresher was in general use throughout the tidal rice-growing region, especially by the more affluent planters such as James Hamilton Couper and Pierce Butler on the Altamaha River and Charles and Louis Manigault on the Savannah River.

William Emmons's Improvement

On February 7, 1831, William Emmons of New York, was granted a patent no. 6,366X for an "Improvement in the Thrashing Machine" (given no. 6,366X after the fire). His machine was similar to that of Calvin Emmons but featured an improved threshing cylinder and concave bed (see diagram 23). The teeth of the threshing cylinder were fixed, not hinged, and bent backwards. The patent does not state why backward-directed teeth functioned better. The concave bed housed a series of notched iron strips, which the

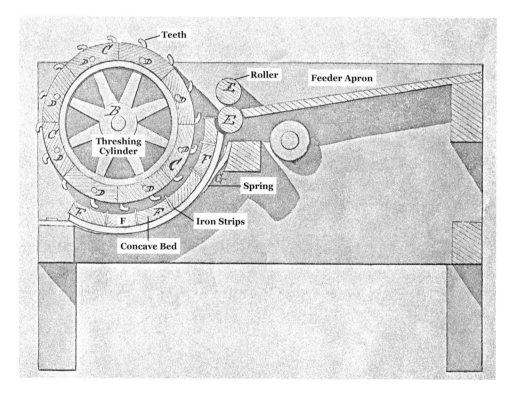

DIAGRAM 23. Patent diagram for William Emmons's improved threshing machine, 1831. United States Patent Office.

teeth of the cylinder operated against to strip off the rice. A feeder apron fed sheaves to the feeder rollers, which in turn fed the sheaves into the concave bed. The straw and rice exited at the back of the concave bed. Motive power could have been water or steam.

The concave bed of William Emmons's thresher was supported by springs. The function of the springs is not stated in patent. They likely allowed the space between the cylinder and bed to adjust depending on the volume of straw. Threshers introduced after the Civil War featured a method to adjust the space between the cylinder and bed. They may have been based on William Emmons's thresher.

Ludlow's Thresher

Writing in 1834, "W" described another thresher sent to Charleston at about the same time the two Emmonses' threshers were invented. This was Ludlow's thresher, built on the same general plan: "This thresher consists of a number of steel blades, two and a half inches long, of a particular shape, firmly driven on a spiral line into a wooden cylinder, this cylinder revolves in a concave bed, also filled with the same kind of knives, among which those of the cylinder pass, not allowing space for the rice to escape, which is thus torn off."[32]

"W" said Ludlow's thresher "threshed an equal quantity of rice with Emmans' [*sic*], and was propelled by steam." It worked on the same principle as the Emmonses' threshers, using steel blades as the beaters, but the beater blades passed through another set of steel blades (knives) embedded in the concave bed, which combed the rice from the stalks. Although he said the Ludlow thresher was more durable and efficient than the Emmonses' threshers, no evidence was found in archival sources that it was actually used on a plantation. Straw was shaken to remove loose rice and raked away by hand. Then the rice was winnowed to remove the chaff. Ludlow's patent was one of the many lost in the Patent Office fire.

Mechanical Improvements

With the advent of mechanical threshers, wind fans, separators, elevators, and other labor-saving machinery came into common use. By 1832 threshing was mechanized using new machines. Although no diagrams of the mechanical workings of any threshing barns have been found, descriptions of the process are known. An 1832 bill of scantling written by Daniel Blake lists the materials needed to construct a threshing machine on the Ogeechee River in Georgia:[33]

1832				
March	23	To a small threshing machine & horse power for Ogeeche	$100.00
July	28	White pine & poplar for fans and rakes	10.06
		8 gudgeons & 8 bands for straw rakes	3.75

		boxes for rakes and fans	2.00
	15	2 grudeons & 4 bands for drum shaft	2.50
		shafts, flanges & arms for two fans	16.50
		screens, bolts, boxes & washers	9.69
		turning pullies	2.00
		turning gudgeons & shafts	2.50
		2 pullies for draft animals	4.30
		hands from 22 May to 31 August—88 days @ $1.25 per day	110.00
		leather bands for Ogeechee	28.00
		altering 2 machines	60.00
		small machine & horse power for Combahee	180.00
		2 sets of castings for horse	174.62
October	1	10 days at Ogeechee for putting up	40.00
December 13		6 days at Combahee & expenses	28.00
		a fan for Combahee	80.00
		1 [illegible] of cast iron wheel	3.00
			$937.42
		[illegible]	496.97
			$440.45
		By Cash at Sundry times	$470.00
		14 days work of [illegible]	14.00
		Bill of Rice at Steam Mill	12.97
			$496.97

As "W" stated in 1834, "The rice after passing through the thresher, falls by gravity into a hopper, which conducts to a fan, from which it descends thoroughly winnowed to the floor. The straw is cast from the thresher, and caught by two revolving rakes, is thoroughly shaken and thrown to an inclined slatting which conducts to a door, whence it is taken away."[34]

A threshing barn was two-story building with a loft, and the thresher was on the second floor. Attached to the side of the building was a lean-to structure, sometimes open to the elements, but normally enclosed, which housed the steam engine and boilers. The chimney for the boilers was located outside and adjacent to the building.

Archival research has produced only one drawing of a steam-powered threshing barn.[35] Based on archival photographs and inspection of the threshing barn at Chicora Wood, the drawing has been redrafted for clarity as diagram 24. The chimney location is based on a note attached to the diagram that stated: "You did not say which end the chimney was to go I suppose it is on this end."

The two most important features in the original drawing are the placement of the thresher on the second floor and the two vertical studs that represents the location of

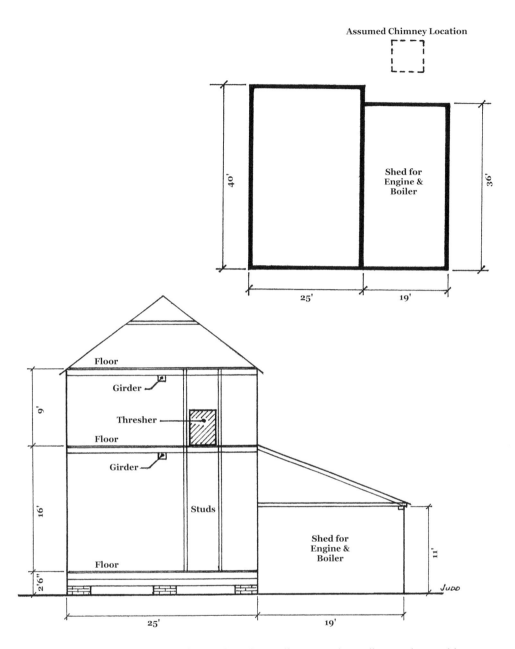

DIAGRAM 24. Diagram of a threshing barn. Illustration by William Robert Judd, redrawn from a period drawing in the South Caroliniana Library, University of South Carolina, Columbia.

the exterior chute for removing the threshing refuse. The location of the thresher on the second floor allowed rice to fall into a large hopper feeding the chaff fans (wind fans) located on the first floor.[36]

Machinery replaced all the steps that were done manually. Beaters operating over a concave bed removed the rice from the straw, replacing the flail. A wind fan blew away (winnowed) chaff after the rice was threshed, replacing the winnowing house, and revolving rakes shook the straw to remove any rice left by the thresher and cast the straw outside the barn. By this time there may have been a conveyor belt to carry the sheaves to the thresher on the second floor. The threshing barn was either a separate building with its own power source or was coupled to the power system of a rice mill.

Revolving rakes were a new addition to the thresher; they solved the problem of rice being passed out with the straw that was ejected from the thresher. This was a common problem dating back to the flail; laborers had to shake the straw after flailing it to remove embedded rice. Blake's 1832 bill of scantling and the 1834 article by "W" both indicate that rakes were used during the time of the Emmonses' and Ludlow's threshers, and it seems likely that some type of rakes were installed in conjunction with these threshers. While no drawings exist, the barn at Chicora Wood contains remnants of four rakes.

Steam power became the power of choice for planters who could afford a steam engine. By the 1850s a steam-powered threshing mill cost $8,000 (about $227,000 in 2009 money),[37] confining its use to only the wealthiest planters. Steam power turned the cylinders fast enough with an even rotation, successfully combing rice from the stalks. Water power was still used, however, and it was not uncommon for animal power to be used if the threshing machine was small, if no adequate water source was available, or if the plantation was too small to afford a steam engine.

Planters made good use of the rice straw. As Calvin Emmons explained in 1840, "A portion of the straw is fed to the cattle, the remainder is put on the high sandy grounds, where corn and sweet potatoes are planted. On those plantations which have no high grounds, it is either burned or thrown into the river. More recently some use the straw as fuel for the steam engines that drive the threshing machines."[38]

Mathewes's Machine for Threshing Rice

On August 27, 1835, William Mathewes of Charleston District was granted a patent for a "Machine for Threshing Rice," given number no. 9,059X after his original patent was lost in the 1836 fire (see diagram 25, figs. 1 and 2).[39]

Mathewes's machine had the same basic construction as the threshers patented by the Emmonses and Ludlow. Mathewes described the operation of his thresher: "The straw containing the grain is spread upon the feeding table and by the motion communicated from the beating cylinder B to the roller next the feeders. The cloth is set revolving over its two rollers and carries the grain and straw forward between the feeding rollers J and R where it is caught by the teeth of the beaters and driven over the concave bed D, which

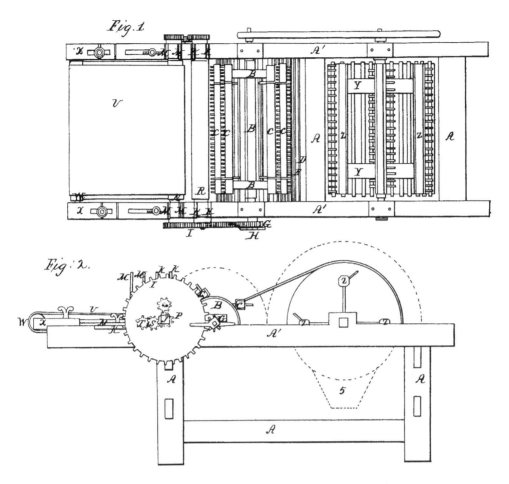

DIAGRAM 25. Patent diagram for William Mathewes's rice thresher, 1835. United States Patent Office.

breaks out the grain and throws all over to the bed of bars under the comb Z. Here the grains fall through, and the straw is driven to the back over the frame being separated from the grain."

Mathewes claimed several improvements over the two previous threshers. First, he substituted a small cog wheel to drive a large cog wheel, which was fastened to the feeder rollers. (There are two feeder rollers, an upper and lower; the lower roller is hidden below the upper roller.) He claimed that the small cog wheel would not slip like a pulley or belt and would give a continuous motion to the feeder rollers. Second, the forward-feeding table roller could be adjusted either nearer to or farther from the feeding rollers by means of a thumbscrew and groove. This allowed the feeder cloth to be kept stretched and allowed for a different length apron depending on the length or strength of the rice stalks. Third, he employed a beater with stationary bars that had either square or oblong teeth. Fourth, the method of driving the comb was different. The comb removed embedded rice from the straw and discharged the straw from the machine. Mathewes added a wind fan that could be installed in the hopper to winnow the chaff from the rice as it fell into the hopper, but he said that the comb generally carried off all the chaff. Mathewes's thresher was apparently the first to include a mechanical device to shake the embedded rice from the straw. He called his shaker a "comb"; later the comb was called a "rake." The rake was separate from the thresher until after the Civil War, when advanced machines combined devices in the thresher to remove embedded rice.

Jehiel Butts's Improved Threshing Machine

On May 23, 1848, Jehiel Butts of Charleston received patent no. 5,600 for a threshing machine called an "Improvement in Machines for Threshing Rice." Butts's thresher became the most popular threshing machine in the decade before the Civil War. The general design appeared to have been based on the threshing machines of the Emmonses, Ludlow, and Mathewes; however, Butts's thresher had several improvements that made it a far superior machine.

Butts claimed his thresher (see diagram 26, figs. 1 and 2) was superior to the previous machines for several reasons. Made entirely of iron and steel, it was more durable than others. It required half the speed of revolutions of the beaters (350 revolutions per minute), so it ran with less friction. It required fewer attendants and could thresh the sheaves while they were still tied, saving the labor of untying them. Butts's machine was compact and threshed the rice free of rice straw. It could be attached to other machines to share a power source. Finally it threshed nearly three times as much rice as previous machines.

The concave bed (B) was a key to Butts's thresher.[40] The bed consisted of a series of curved cast-iron segmented plates. Each iron segment had a rib on its inner edge, which formed an abrupt edge (H) and met the rice as it progressed through the concave bed. Each segment had a small concave surface (h), which extended from the upper part of the edge of each rib (H) to the lower point between each rib. The small concave surface turned the rice grains partly around before they reached the next abrupt edge.

When the Butts thresher was in operation, the beating cylinder turned at 350 revolutions per minute. The sheaves were placed on the inclined feeder table, which moved the stalks toward the two feeder rollers, fluted to better grasp the stalks. The rollers forced the stalks to contact the beaters (E') forcing the stalks through the concave bed (B) between the ribs (H) on the concave surface and the beaters. The rice was forced to contact the edges of the ribs (H and I), which combed the rice from the stalks. The threshed rice was discharged at the back part of the machine, and the straw passed to revolving rakes, which removed embedded rice.

Charles Manigault wrote from Paris to James Haynes on January 1, 1847: "For amongst other things I even feared the ridicule in case I should have added myself to the long list of failures in attempts at threshing machines—particularly in those propelled

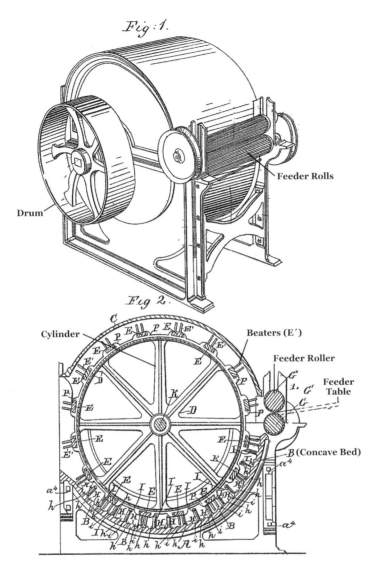

DIAGRAM 26. Patent diagram for Jehiel Butts's improved rice thresher, 1848. United States Patent Office.

---

by water power like mine—nearly all of which have proved failures. This iron Beater of mine being the first of its great dimensions which has ever been made—my being the first to order such a one, & its triumphal performance, is really a source of pride as well as a pleasure to me."[41]

Both steam and water power could drive the Butts thresher. A band from the power source ran to the drum that turned the cylinder armed with teeth (the beaters). Another band from the power source was attached to the shaft of the drum and turned two feeder rollers.

A feature of the Butts thresher that even Butts did not recognize, but that Charles Manigault did, was that it acted as a flywheel. He wrote to Robert Habersham & Son from Paris on February 1, 1847: "One of the most important principles in *this* iron Beater is, that from its size & weight & its velocity, it acts as a flywheel, by regulating its own movement & and the rest of the machinery also. It is upwards of 3 feet in diametre—& about 22 inches wide, with 16 Bars each having 2 rows of teeth."[42]

Although most planters placed their Butts thresher in its own threshing barn, Manigault coupled the thresher at Gowrie to the outer end of the pestle shaft of the water-powered mill, and both could be worked simultaneously. Manigault wrote to Robert Habersham & Son about how pleased he was with his Butts thresher: "And such have already been the advantage to my plantation in having it, that, after the short period had elapsed in getting it in full and successful operation, it then threshed out with the day tide alone nearly double each day what my whole force previously effected with the flail stick."[43]

One Butts thresher survives in relatively good condition. The threshing barn at Mansfield Plantation burned, but the thresher survived the fire and is preserved by the owner (see color plate 9).

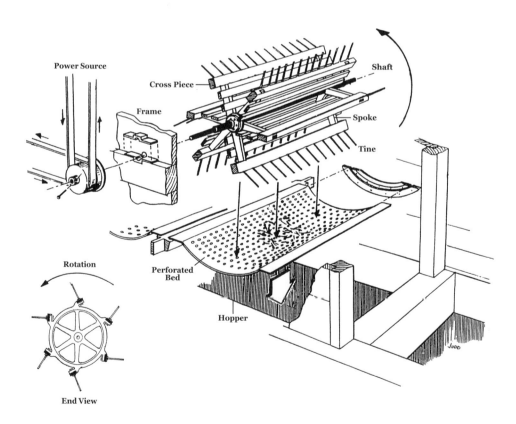

DIAGRAM 27. The rake at Chicora Wood Plantation. Illustration by William Robert Judd.

Wind fans had become standard equipment by 1846. Charles Manigault wrote to his overseer: "Mr. Frederick Rutledge told me that such is the Quantity of Rice which this newly invented Iron thresher beats out, that one of them which Mr. Butts put up for him on Santee required two wind fans to clean the rough rice which it threshed out—while one wind fan has hitherto answered for all Beaters."[44] The Butts thresher was so efficient that it required two wind fans.

Rakes

With the advent of the Butts thresher, rakes and wind fans separate from the actual thresher became standard equipment. Revolving rakes removed embedded rice in the straw and cast the straw from the barn. On April 22, 1847, overseer James Haynes wrote to Charles Manigault: "He [Butts] has however sudgested [sic] other alterations, which is to put in two more rakes. This is to be done by slipping the beater back from its present place which would bring it nearer the door and put the two additional rakes in front of it connecting them by a frame to the present rakes. This is to prevent the rice being thrown out with the straw. This work can be done by the carpenters at home rakes and all, excepting the two Iron shafts for them & boxes & there will be no other expense attending it and it will save considerable rice going out with the straw which has to [be] shaken out by the hands before taking away the straw & which is not always perfectly done.[45]

The rakes at Chicora Wood Plantation were constructed of wood with iron parts attached (see diagram 27) while the rakes at Mansfield Plantation were constructed entirely of cast iron (diagram 28). Both were cylindrical in shape, approximately thirty-five inches long and twenty-eight inches in diameter.

An individual rake at Chicora Wood consisted of two iron hubs, each with six wooden spokes fastened to six crosspieces. Each crosspiece had a series of iron tines six inches long. The iron hubs were fixed to each end of an iron shaft, which fit into each side of a frame that housed the rakes (see color plate 10). The head rake was turned by a belt-and-pulley system from the main power source. Each additional rake received power from the pulley of the preceding rake. A perforated bed under the rakes allowed the rice to fall into a hopper below; however, the chaff was too large to pass through the bed and was removed.

The Mansfield threshing barn burned in the 1980s, when a prescribed fire in the adjacent marsh got out of control. Cast-iron parts retrieved from the barn were used to diagram an individual rake. A rake had a one-piece cast-iron wheel at each end. Each wheel had six spokes radiating from its hub. Six horizontal flat bars, each with multiple tines, were bolted to six angled castings around the circumference of each wheel, forming the cylindrical shape. The hub of each wheel was keyed to an axle. Attached to the axle was a gear drive which was rotated by the power source which turned the rake.

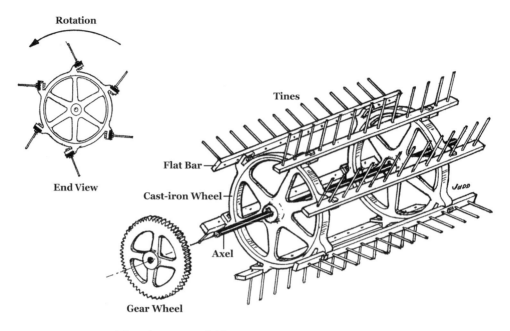

DIAGRAM 28. The rake at Mansfield Plantation. Illustration by William Robert Judd.

The Chicora Wood Threshing Barn

Chicora Wood and the other plantations owned by Robert Allston and his family were major rice producers in Georgetown County. The threshing barn at Chicora Wood still stands (see color plate 11), but, a yard shed, a winnowing barn, and a shed attached to the threshing barn are gone (see figure 17).[46] Absent any archival records or extant physical structures and equipment that would suggest rice was milled at Chicora Wood, one must conclude that Allston likely sent his rice for milling either to Waverly Rice Mill on the Waccamaw River or to Charleston.

What model thresher Allston used at Chicora Wood prior to the Civil War is uncertain. He may have installed a Calvin Emmons thresher, which he described in his essay on rice.[47] During the decade before the Civil War, most planters shifted to the Butts thresher because of its greater output, durability, and efficiency. Allston was in the vanguard of planters who sought to raise rice production to its highest level, so he probably had a Butts thresher installed at Chicora Wood.

The company of Thurston & Johns of Charleston, wrote to Charles Petigru Allston of Georgetown, on August 6, 1869: "We . . . have seen Mr. Eason about the repairs to the Beater of the Threshing Mill at Chicora, he tells us that he has no idea of the amount of work necessary and that unless you can give him this information, or he can see the beater, he cannot make an estimate which would approach correctness. If you will send down the Beater we will have the repairs done and return it as soon as possible, or if you

FIGURE 17. The barnyard at Chicora Wood on the Great Pee Dee River in Georgetown County, South Carolina, circa 1900. Reproduced by permission of the South Carolina Historical Society, Charleston, Allston Family Papers.

will describe exactly the quantity & kind of work to be done to it we will obtain & forward an estimate for you."[48]

Only part of the threshing machinery at Chicora Wood has survived. Today the barn houses two chaff fans, a rolling screen, and sections of the rakes and wooden frame that housed the rakes. Remnants of the steam system that once powered the threshing machinery still exist. An intact brick chimney and underground duct lead to the partial remains of the brick furnace that supported the steam boilers. Next to the furnace is the brick mount for the steam engine. The missing seam engine and boilers were housed in a shed attached to the threshing barn.

Archival records, remnants of the steam system, and remaining artifacts in the threshing barn provided sufficient information for the authors to reconstruct in diagrams the threshing system that existed at Chicora Wood (see diagrams 29, 30, and 31).

Diagram 29 depicts the different floor levels within the threshing barn and the steam system that ran the machinery. On the first floor, a large pulley wheel was connected to the engine's flywheel axle. A belt ran from the pulley wheel to an overhead pulley-and-axle assembly, which powered all the machinery in the barn. On the first floor were two side-by-side chaff fans (see diagram 29, insert), each fed from overhead by a large hopper placed directly below the thresher and the rake enclosure located on the second floor. An opening in the second floor rear wall, in line with the rake enclosure, led to an external vertical chute for disposal of straw from the rakes.[49]

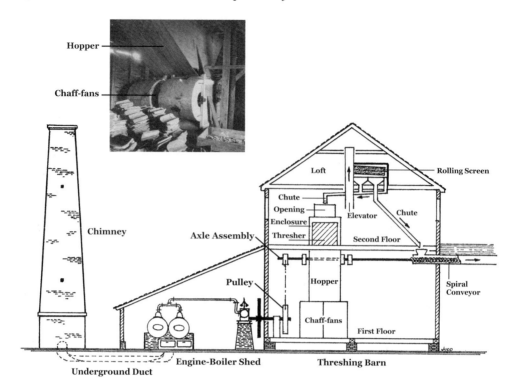

DIAGRAM 29. The threshing-barn complex at Chicora Wood Plantation. Illustration by William Robert Judd. *Insert:* Interior of the Mansfield Plantation threshing barn (now destroyed) showing two chaff fans (wind fans) below the hopper that fed threshed rice from thresher and rakes to the chaff fans; photograph by Charles N. Bayless, 1977. Courtesy of the Historic American Building Survey, National Park Service. The same system was in place in the Chicora Wood threshing barn.

Rice from the first-floor chaff fans was scooped up by small buckets attached to an enclosed elevator belt and raised to a rolling screen on the loft floor. The rolling screen separated sand and small debris from the whole, market rice. The refuse passed through a chute and was discharged from the barn. The market rice passed through a similar chute and down into a hopper in the first floor loft. That hopper fed an enclosed, horizontal, elevated spiral conveyor. The conveyor exited the threshing barn and entered the rafter area of the storage barn on the river, a distance of approximately 250 feet. The covered, elevated conveyor assembly ran toward the storage barn at a slightly downward angle, which allowed the rice to move easily. Power for the conveyor was supplied by the rotating overhead pulley-and-axle assembly on the first floor. The first spiral conveyor section, with its universal joint, is still in place in the threshing barn as are the last few sections in the rafted area of the storage barn.

Threshing 141

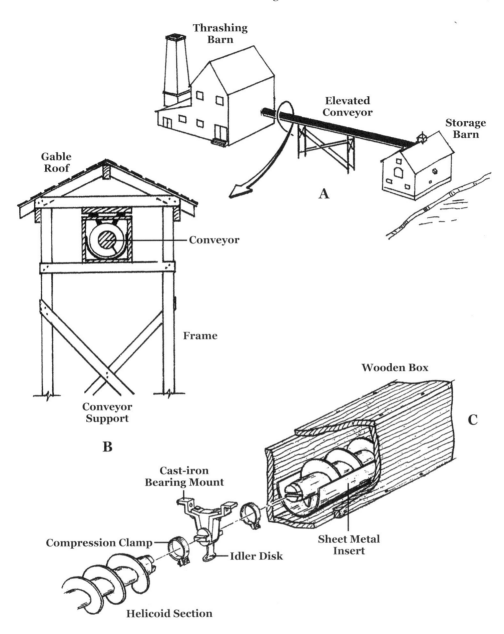

DIAGRAM 30. Elevated spiral conveyor. Illustration by William Robert Judd.

Diagram 30 illustrates the enclosed spiral conveyor complex. Diagram 30A represents the elevated conveyor between the threshing barn and storage barn. Diagram 30B is a section view of a conjectural framework supporting the elevated conveyor. The framework was topped with a narrow gable roof to protect the conveyor from the elements.

The spiral conveyor (diagram 30C) was an assembly of eight-foot-long helicoid sections, each resembling an auger. The conveyor was housed in an enclosed horizontal wooden box with a concave sheet-metal insert below conforming to the circumference of the spiral sections. The sheet-metal insert held rice as the blades of the conveyor pushed the rice toward the storage barn. The conveyor's assembled length was restricted only by its intended destination. The sections were assembled by a cast male-and-female coupling and suspended overhead by multiple cast-iron bearing mounts. Each helicoid section's end was slotted to receive a raised matching key cast on both sides of an idler disk that freewheeled within the bearing mount. The coupling of each section was secured by a compression clamp. Rice was let in at one end, and the revolving spiral blades drove it to the other end.

After drying, sheaves were brought to the feed house. Writing around 1906, Elizabeth Pringle commented, "The feed house is packed up to the very roof with the rice from P. D. Wragg, and I want to get it threshed out to allow Vareen to be brought out of the flat and stowed in the feed room."[50] The missing feed house at Chicora Wood was likely the leaning shed attached to the threshing barn (see figure 17). Some planters built a shed to store the sheaves to prevent them from getting wet after they had been dried in stacks. Governor Aiken had a similar arrangement at Jehossee Island: "The sheaves are brought from the stacks in the great smooth yard to a large shed where all the sheltered grain can be saved, and are there opened and laid on carriers . . . which carries them up to these machines [threshers] in the second story, where the grain is separated from the straw."[51]

Sheaves were carried to the second floor manually or by a conveyor belt (see diagram 31). Next laborers fed the sheaves into the feeder rollers of the thresher, which passed the stalks to the beater.[52] (The Butts thresher is used in the diagram of the threshing system.) Rice was combed off as the stalks passed between the beaters and the concave bed, falling through holes in the concave bed and passing down an incline into the hopper below a framed enclosure that housed the rakes. From the hopper the rice and chaff passed into the two chaff fans (only one shown) on the first floor.

Even though rice was physically removed from the stalks by the beaters, much remained embedded and did not fall from beneath the beater. Four rotating rakes were used to remove the embedded rice. Straw from the thresher passed to the first rake, which shook the straw and released the loose rice. The straw passed in turn to the second, third and fourth rakes, each rake repeating the shaking process. Straw from the last rake was ejected through an opening in the barn wall down a vertical chute to a waiting horse-drawn wagon, where it was hauled away each day to prevent chance of fire. The straw was used for animal feed, green manure, or fuel for the steam boiler.

Anne Simons Deas recounted her visit to the threshing barn at Buck Hall on the Cooper River in Berkeley County, South Carolina, just prior to the Civil War: "So we made our way along by the boxes [that housed the rakes] to this 'dormer-window', where the straw comes rolling out, and through it is pushed by a woman, with a stout stick."[53]

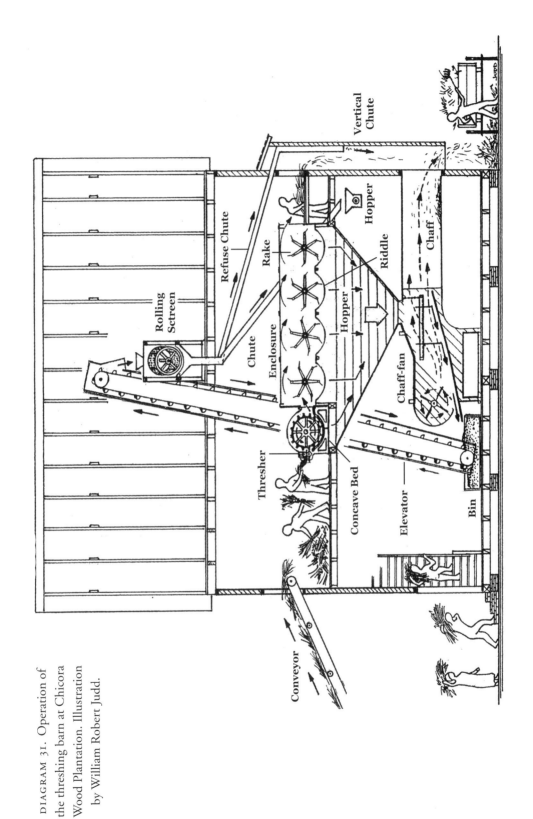

DIAGRAM 31. Operation of the threshing barn at Chicora Wood Plantation. Illustration by William Robert Judd.

The rice and chaff from each rake fell onto a perforated bed or riddle. Although the perforated beds or riddles are missing in the Chicora Wood barn, physical evidence to their location is illustrated in color plate 10. Attached to each side of the enclosure beneath where each rake was connected is a mounting support made with two pieces of wood fashioned in a semicircle and spaced one above the other to receive the end of the riddle. Four rakes were mounted in the enclosure; the locations of the first three rakes are shown in color plate 10.

Loose rice and chaff fell through the riddle's perforations into the same hopper that received the rice from the thresher. The riddle prevented the straw from passing into the hopper below while allowing the rice and chaff to pass through. Rice and chaff then passed from the hopper into the chaff fans on the first floor. Robert Allston used chaff fans to blow away the chaff.[54] A current of air produced by the fans blew away the lighter chaff, which was discharged out of the barn and hauled away. The cleaned rice passed down an incline into a holding bin, where it was scooped up by buckets attached to an enclosed elevator belt and lifted to the barn loft, where a rolling screen separated trash from the market rice. An intact rolling screen is present in the barn (see color plate 7), but it was removed from its support—hence its placement in the loft is uncertain.

The rolling screen at Chicora Wood was slightly elevated at the end where the rice entered. It was constructed with two sizes of meshes. The mesh at the elevated end was the smaller and allowed trash and dust to fall through to a refuse chute. The larger mesh at the lower end discharged the market rice through a chute down to a hopper in the overhead on the first floor, which fed an enclosed conveyor that carried the rice to the storage barn at the river's edge to await shipment to the city mill.

The Chicora Wood Storage Barn

On plantations where milling was done off-site, rough rice was shipped in bulk or in large barrels to the city mills. Plantations close to the city mills sent rice by flats while plantations distant from the city used sloops (see diagram 32) or deep-draft schooners to carry rough rice. The vessels were either plantation owned or hired.[55]

On plantations that did not have storage barns adjacent to the river, rough rice was hand carried in baskets from the threshing mill to the wharf and dumped into the hold of the vessel. Plantations without deep-water wharves were at a disadvantage when transporting rice on a deep-draft schooner, forcing it to lie some distance from the shore. A temporary wharf could solve the problem. Louis Manigault wrote to his father, Charles Manigault, from Gowrie on February 27, 1853: "I now ship by Schooner Moore 4170 Bushels Rough Rice & 17 Bbls. on deck. I write Middleton & Co. & please read the letter I write to them. The Moore loaded first rate. There being little straw near the mouth of Canal we got the Moore in close shore, she grounded at low tide & the Capt. rigged a gang way from one flat to the other & so on to the Schooner. It worked Charmingly and

DIAGRAM 32. A plantation sloop, based on a wreckage discovered in the Ashley River (length: 35 feet; beam: 12 feet; depth: 5 feet; displacement: 23 tons). Illustration by William Robert Judd.

---

the Hands walked from the Brick Thresher & did not stop until they pitched the Rice in the Schooner's hold."[56]

Most large plantations had storage barns, generally adjacent to a river. Duncan Clinch Heyward described the storage barn, or "rice house," at a Combahee River plantation:

> As the grains of rice were threshed from the straw, they were passed through a fan on the first floor of the mill, and then in elevator cups were carried again to the

FIGURE 18. A threshing barn and a storage barn on a plantation on the Ogeechee River in Georgia. Reproduced from *Harper's Weekly,* January 5, 1867.

second floor and run through a large screen. Again falling to the first floor, they ran through a 'market fan,' after which elevator cups carried them back to the second floor and emptied them into wooded tubs, each usually holding fifty bushels.

A Negro boy sat on one or the other of these tubs, and as soon as one tub was full he would turn the stream of rice into the other tub, and let the rice in the full tub run down a chute into the rice house.[57]

Storage barns were usually separate from threshing barns, but on some plantations the storage barn was attached to the threshing barn. Storage barns did not have power sources. Rice was probably stored in some type of large, open bin. Temporary storage was necessary since there was not always a sloop or schooner available to on load the rice, and it was not feasible to shut down the thresher to wait for a vessel because it entailed shutting down the steam engine. Charles Manigault wrote to his brother: "As soon as you please you can spout out 3 or 4000 bushels into the Flat & send it by canal down to the mill to get it out of your way, & if the Yemassee can return for a 3d load I will send you another vessel."[58]

One remarkable sketch of a threshing barn and storage barn survives. It was published in an 1867 issue *Harper's Weekly* and shows a scene from a plantation on the Ogeechee River in Georgia, dated 1867 (see figure 18). An elevated conveyor passed rice from the storage barn into the rice flat for shipment to a city mill.

Chicora Wood had a similar threshing and storage operation (see color plate 12). Enough artifacts remain in the Chicora Wood storage barn to allow the authors to describe the plantation's storage system (see diagram 33).[59] The spiral conveyor from the

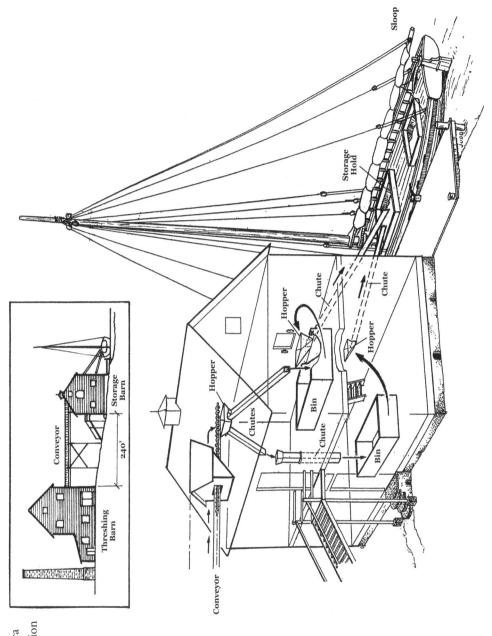

DIAGRAM 33. The Chicora Wood storage barn. Illustration by William Robert Judd.

threshing barn passed the rice into a hopper in the loft of the storage barn. A double chute from the hopper directed the flow of rice either to a storage bin on the second floor through a vertical chute or to a storage bin on the first floor. A sliding tin shunt in each chute could be manually raised or lowered to control the flow of rice to the desired bin.

When a vessel arrived, rice from the bin on the second floor was scooped into a hopper at floor level, which fed into a wooden chute that passed rice to the vessel's storage hold. Rice from the first-floor bin was poured into a hopper, which fed a similar chute that passed the rice to the vessel's hold. The Chicora Wood system was complex, but similar ones were likely in place on other large plantations.

Antebellum Threshing

Archival material and observations in the threshing barns at Chicora Wood and Kinloch allow for a description of the status of threshing prior to the Civil War. A contract between Anthony Weston and Henry Augustus Middleton in 1851 included rakes, fans, elevators, screens, and carriers, all of which were used in the threshing system at Chicora Wood and by threshers throughout the Rice Kingdom:

> Memorandum of an agreement between A. Weston & Mr. H. A. Middleton
> I will agree to make & part for Mr. H. A. Middleton one thrashing machine 1, 40 in iron beater & rakes, 2 fans, 1 set of Elevators, 1 screen, 1 carrier & all of the impelling powers for the sum of fourteen hundred & fifty dollars.
> I will also agree to furnish the bands for two hundred & fifty dollars   250.00
> The cloth I will furnish for one hundred & fifty dollars   150.00
> $1850.00
> I will deliver the mill to Mr. Middleton by 25th of Nov. 1851
> Mr. Middleton to bear the expense of the machinery up & my hands may up & down & furnish my hands with provisions on the plantation.
> All of the lumber to be found on the plantation.
>
> Anthony Weston
> Henry A. Middleton[60]

Archibald Hamilton Seabrook's "Reminiscences" include an account of an 1862 trip to Eldorado Plantation on the South Santee River in Charleston County, South Carolina. His description of the threshing operation includes screens or sieves and corroborates the information in the Weston and Middleton contract revealing that screens were used in the threshing process. As Seabrook wrote, "The last of the rice crop was then being harvested and we loved to see the mill operation. The grain was threshed off the straw on the first floor. On the second floor the grain passed through screens or sieves, separating it from small litter and short pieces of straw. On the third floor it was passed through more fans, thoroughly taking off all the litter. All this work was done by large cups attached to belts around pulleys. The grain was then ready for shipment.[61]

Threshing became highly mechanized in the decade prior to the Civil War. The threshing barn was a two-or-three story building. Often the first floor was built of brick while the second and third floors were constructed with wood, but others were built with wood throughout. The threshing barn usually stood on the edge of the river or a creek so that vessels could easily be unloaded or loaded. Rice sheaves were brought from the stacks and laid on the carriers, which conveyed them up to and through a window in the second floor to the feeding table of the thresher (or multiple threshers in some barns). The beaters separated the rice from the straw, and the rice fell down through a hopper into the winnowing machine below, where it was cleaned of chaff. Rakes shook the straw and removed embedded rice which was also conveyed to the chaff fans. The straw was removed and either burned as fuel for the boilers that ran the steam engine or used for green manure (or a combination of the two). Elevators carried the rough rice from the chaff fans to second-floor loft (or a third-floor loft if one was present), where it passed through a rolling screen to remove additional debris and defective rice. The rough rice was stored in a separate barn to await shipment to local mills. If the storage facility was detached from the threshing barn, it was conveyed either manually or mechanically to the storage facility. This system was in place until after the war when advanced mechanical threshers became available.

Most thresher, elevators, screen, and conveyors were driven by steam engines, but some smaller threshing barns used water power as late as the 1850s.

Machine Maintenance and Fire Prevention

With the shift from the manual flail to mechanical threshing machines, equipment repair became an everpresent problem. On April 22, 1847, Charles Manigault's overseer on Gowrie Plantation, James Haynes, wrote to Manigault concerning the Butts thresher: "You wish to know what conclusions I have come to respecting the improvements to the present thresher. I would inform you that Mr. Butts has been on to remidy [sic] the evil I complained of viz. the bottom feeding roller stopping. And he says it will be quite useless to make any alterations in the roller chords or pulleys, that they will run perfectly easy hereafter. He did not succeed."[62]

Fire was a concern even before steam engines with chimneys were used to drive the threshing mills. Failure to move the straw away from the barns was always a concern for the planter. Charles Manigault wrote his son Louis on January 11, 1859, "He [probably the overseer] told me yesterday that Robbs Rice Mill had also caught fire. And there is so much burning of Mills now, that you had best have an eye to no fire being left near, or about the Thresher, or straw left near it, &c. You should now keep all the rest of the Straw in a heap at a safe distance from the Thresher, for the Mules, Cows &c."[63] Plowden Weston warned his overseer: "The Proprietor considers an Overseer who leaves *any* straw or tailing during the night within 300 yards of the mill, as unfit to be trusted with the care of valuable property."[64]

On September 29, 1860, B. T. Sellers, overseer at Nieuport, on the Combahee River, wrote to Williams Middleton concerning the thresher: "When you were up last winter, you told me that you had given orders, for the Blacksmith at Hobonny to put the mill at Nieuport in order, but that order was not carried out, all that was done to the mill, was to put in new teeth in the Beater, & to get the mill so that I could start, I had to depend upon the goodness of my neighbors, in allowing their Blacksmiths to work for me in their own time. I do not know who you will think is in fault in this matter, but I do know that I cannot get even a Set of Gate Hinges made at Hobonny. The mill at Nieuport is now broken down & I do not know when I will be able to get it to work."[65]

Some plantations had experienced slave laborers who could make major or minor repairs. Other plantations had to depend on mechanics from the cities or other plantations. But the machines had to be fixed. Downtime was a severe drain on the plantation economy, especially during the market-preparation season.

DeBow's Review reported that rules on the rice estate of Plowden Weston of South Carolina included steps to ensure that "the mill is to be closed in time to allow the whole [threshing] yard to be cleaned up by sunset."[66] Planters coated the roof with fireproof paint or applied slate shingles or zinc (tin) roofing for protection from chimney ashes.

Postwar Threshers

After the Civil War a series of new mechanical threshers came on the market. According to David Doar: "After the war a smaller thresher, containing all appliances necessary for threshing and cleaning the rice, came on the market, and many planters had these installed in their old mills as they required less power, less space, and fewer hands to operate, but passed out the rice also through old fans, screens, etc., thereby getting a better sample of rough rice."[67]

These postwar threshers contained all the steps in one machine. They were built primarily for threshing on the burgeoning grain farms of the Midwest, but several manufactories made them for rice because of the development of rice growing in Louisiana, Mississippi, and Arkansas. The postwar threshers completely automated the threshing process, with all the functions contained within one steam-driven machine. Beaters replaced the flail; wind fans replaced the winnowing barn and fanner basket; shaker pans with riddles that separated the rice from the straw replaced manual shaking; and conveyor belts and stackers replaced laborers who carried the rice to the thresher and removed the straw.

Typical of these new threshers was the 1876 Invincible (see color plate 13) housed in the threshing barn at Kinloch on the North Santee River (see color plate 14).[68] The stalks were brought to the threshing barn and passed through an opening in the wall to the feeder belt of the thresher (see diagram 34).

The feeder belt carried the stalks to the beating cylinder (A), where they passed between the cylinder and the concave bed, where the rice was loosened. The concave bed

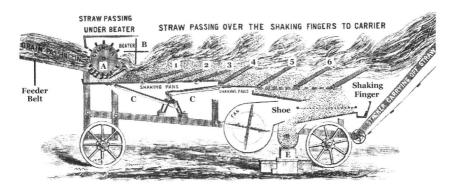

DIAGRAM 34. Skeleton view of the Invincible thresher. Collection of Richard Dwight Porcher, Jr.

---

was perforated and allowed rice to fall to the shaking pans below. A straw beater (B) just back of the beating cylinder prevented the straw from winding around the beating cylinder. The straw beater passed the straw to the first series of shaking fingers. Seven rows of shaking fingers, each row consisting of six fingers, passed the straw toward the back of the thresher. Rows 1, 3, 5, and 7 were raised; then these rows were lowered and rows 2, 4, and 6 were raised. The shaking motion from the alternating raising and lowering of the rows of fingers separated the rice and carried the straw backward. Rice fell into the shaking pans below. By the time the straw had passed down the rows of shaking fingers and reached the discharge end, all rice had been shaken out. The shaking fingers passed the straw to the stacker, which carried off the straw.

The shaking pans were operated by double connecting rods (C) from the double crank shafts, which gave the pans an alternate motion that moved the rice forward to the discharge end. At the discharge end, rice fell from the shaking pans into a shoe that housed a series of stacked riddles. The rice fell through the riddles into the blast of the wind fan, which blew away the chaff, which exited with the straw. The rice fell into a collecting device (E), clean of chaff and straw.

Portable threshers driven by portable steam engines were available to rice planters after the Civil War. They were expensive, and it was easier to haul the rice to a threshing barn equipped with a stationary machine than to purchase a portable thresher. However, a planter who had several adjacent plantations could have benefited from being able to move the thresher, along with a portable steam engine, from one plantation to another.

No portable threshers have survived, but one was successfully used on a Cooper River plantation. Sanford William Barker of St. John's, Berkeley, a plantation on the Cooper River, said: "With one of Sinclair's Portable Thresher, thirty-inch cylinder, driven by a portable engine of eight-horse power, (six is sufficient) using one-third of a cord of wood per day, I have threshed fifty bushels per hour, when the machine has been fed properly.

But as it is difficult to get the majority of negroes to do this as it should be done, the average work of the machine is not to be rated as high.[69]

In spite of modern improvements in threshing, the flail was used on some plantations until the end of the industry. On some plantations the threshing yard of 1860 or 1900 was the same as one of the 1700s. These plantations either could not afford mechanical threshers or were unable to repair the prewar machines. A photograph of Middleburg Plantation (see figure 11) on the Cooper River in Berkeley County in the early 1900s shows an extensive threshing yard. Many other photographs of threshing yards in the early 1900s exist in the archives.

6

Milling

Milling was the final step in market preparation of Carolina rice. Milling removed and separated the inedible, siliceous hull, bran, and germ from the edible endosperm. Brushing was the final step of milling and produced clean, polished rice for the market. Although the mortar and pestle did not produce rice as clean and white as later methods did, it was still the goal of the early rice planters to mill rice as clean and with as few broken grains as possible.

Tidal irrigation presented the planters with a production dilemma. The larger crops that resulted could no longer be processed with the mortar and pestle before the next planting season began. Planters were forced to replace preindustrial milling methods with more efficient mills. Craftsmen, planters, and inventors all attempted to design new rice milling machinery, and many succeeded. By the late 1700s, mills had been developed to handle the increased harvest output and to produce a clean product for market.

No matter what type of mill was used, milling of rice set the schedule for the season's plantation calendar. Milling occurred between the August-September harvest and the resumption of planting in March-April. Milling took place at the same time as other plantation tasks such as processing indigo or cotton if these crops were also grown, harvesting provision crops for the labor force and planter families, repairing trunks and gates, clearing the ditches and canals of weeds, burning the stubble or turning it under, destroying volunteer red rice, shipping processed rice to market, and other typical rice-plantation tasks. Since rice contained in its hull was nonperishable, it could be stored after threshing for the duration of the milling season.

Descriptions and/or diagrams of many rice-milling machines devised over the years were never recorded or preserved. Many machines were produced on plantations and are known today only by passing references in plantation journals. For most of these machines the description is so vague that there is no way to reproduce them. Others are mentioned with the names of their inventors in the *Statutes at Large of South Carolina.* In some such cases no diagrams or descriptions are included while others include brief descriptions but not enough information for the scholar to understand their construction. Examples of these entries in the *Statutes at Large of South Carolina* include the following. In 1743, no. 698, George Timmons "hath found out a new method of cleaning rice. . . ," and the General Assembly granted him the "sole privilege and advantage of making and framing the said new machine or engine, for the cleaning of rice." In 1756, no. 853, Adam Pedington "discovered a new machine for cleaning rice," and the General Assembly granted him sole right to his invention. In 1788, no. 1400, Samuel Knight was granted by the General Assembly sole and exclusive right to a machine he invented for "beating out rice."[1] No descriptions or diagrams of these machines were recorded.

How many other machines were invented, proposed, or tried, but never described or diagramed, is lost from history, but undoubtedly there were many. Henry Laurens was already aware of this situation in 1773, when he stated: "How many plausible Models are now laying useless about the Garret and Dark Rooms of the House of Assembly in Charles Town, from which no Benefit could ever be derived, not withstanding all the pretences [*sic*] of the Projectors."[2]

Fixing the date when a particular machine was first used for milling is also difficult. A planter or traveler in the Rice Kingdom might mention a machine used for milling rice in a letter or other document, but these mentions establish only that the machine was being used at the time the letter was written. They do not state the date on which the machine was first used, which might have been years before the visitor saw it.

Many milling patents were registered but lost in the 1836 fire at the U. S. Patent Office. Most were not restored, and these are lost forever. The fire was especially damaging for South Carolina inventors. The following is a list of South Carolinians' rice-milling patents that were lost in the fire.

"Winnowing Screen Pendulum," Lewis Dupre, April 1, 1807
"Hulling and Cleaning Rice," Jonathan Lucas, Jr., July 12, 1808
"Hulling and Pounding Husks," Jacob Read, June 9, 1809
"Hulling Rice and Polishing," Jonathan Lucas, Jr., November 6, 1819
"Hulling Rice by Steam," John L. Norton, December 16, 1823
"Hulling Rice," John Ravenel, May 7, 1828
"Hulling Rice," Asa Nourse, July 19, 1828
"Hulling Rice," Asa Nourse, April, 1829

Throughout the rice industry there was a continual quest to improve milling. As late as 1827, years after Jonathan Lucas I had applied steam to mills, the Agricultural Society of South Carolina appointed a committee to draw up a resolution "for the purpose of solicitation the Rice Planters, and any others, to subscribe any sum or sums, as an inducement, for the Invention of any plan or Machine, whereby Rice can be pounded or prepared for market, in a cheaper or better manner than now in use."[3] Whether any machines were invented solely because of the resolution is unknown.

Morphology and Terminology of the Rice Fruit

The literature on milling presents a confusing array of terms for the rice fruit, many for the same morphological structure. For example, "bract," "chaff," "shell," "pellicle," "husk," and "hull" have all been used to refer to the hulls that surround the rice fruit, in technical terms the "lemma" and "palea" (hulls). Other sources used "husk" and "chaff" to refer to the bran that remained after the lemma and palea were removed. Rice seed with the bran left intact is called "brown rice," a nontechnical term that leads to confusion. This book will consistently use the same technical term throughout for a particular structure and correlate the technical term with common usage. These structures and their technical terms are seen in diagram 35.

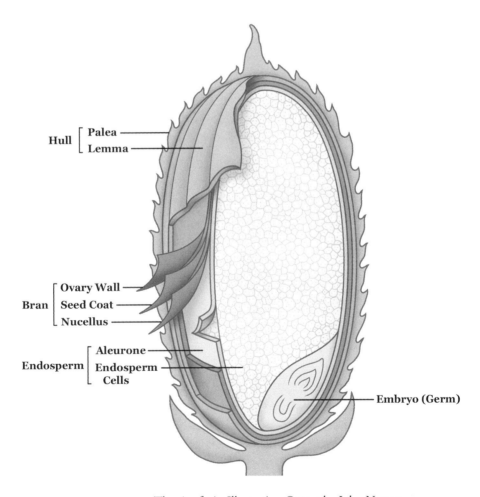

DIAGRAM 35. The rice fruit. Illustration © 2013 by John Norton.

Rice belongs to the grass family. Most grass species contain both male and female reproductive parts within the same flower. (Corn and wild rice, which have unisexual flowers borne on different parts of the plant, are notable exceptions.) The reproductive parts are the female pistil (comprising the ovary, style, and stigma) and the male stamens (comprising the anthers and filaments). The ovary comprises the ovule(s) that become(s) the seed(s) after fertilization. The seed comprises the seed coat, nucellus, aleurone, endosperm, and the embryo. The mature ovary wall (technically called the "pericarp") and the seed(s) contained within are the *fruit* of a plant. In rice, as in all grasses, the ovary wall is fused to the underlying seed coat of the contained seed. (Members of the grass family have only one seed per fruit.) The fruit is technically classified as a grain, or caryopsis. In grasses the fruit is subtended by two accessory, indurate structures, the lemma and palea, which are not technically part of the fruit. The lemma and palea will be called the "hulls," a milling term. The structure removed from the rice plant during threshing is the fruit along with the hulls.

In milling terminology the combination of ovary wall, seed coat, and nucellus is called the "bran" and the embryo is called the "germ." The aleurone, technically the outer layer of the endosperm, is rich in proteins, fats, mineral salts, and vitamins. The aleurone secretes enzymes that break down the endosperm starch into sugars that are absorbed by the germinating embryo as its food source until it produces its first photosynthetic leaves. The starch-containing endosperm provides the human food source.

Milling rice removed the hulls, bran, aleurone, and germ.[4] The first step in milling removed the hulls, which are indigestible by humans. The hulls were used as fuel for steam engines and for soil amelioration. Rice with intact hulls is called "paddy," or "rough" rice.[5] With the hulls intact, rice will keep fresh in storage much longer. For plantation use, the hulls were removed with the mortar and pestle just prior to cooking. The next step in milling removed the outer layers of the bran and the germ. The mixture of bran and germ is a powdery, brown product of slightly sweetish taste, rich in fat, minerals, and vitamin B; it was used during plantation days as domestic food and livestock feed. It is used today as human food and for animal feed.

After the outer layers of the bran were removed, rice was called "undermilled" and was ready for the next and last step of milling, brushing or polishing. Polishing removed the inner bran layers not removed by the mortar and pestle, the aleurone, and some endosperm; the combination of the three was called the "rice polish." The polish is a powder or flour of a definite sweet flavor, and it contains a relatively larger proportion of carbohydrates and less protein, fat, and crude fiber. It was also used as food on the plantation.

Unhulled rice consists of the hull (18–28 percent) and the fruit (71–83 percent). After the hull is removed, the fruit consists of the bran (5–8 percent), the germ (2–3 percent), and the edible endosperm (89–94 percent). When the bran and germ are removed, *white rice* is the result. White rice is 90–94 percent starch, and 6–10 percent protein. White rice is more attractive in appearance, requires less cooking time, and is easier to digest than brown rice. Consequently white rice is more desirable as a food, so most commercial rice today is highly milled white rice (with thiamine added). White rice lasts longer in storage than brown rice. Fats contained within the bran become rancid in hot, humid climates. Removal of the bran was necessary during export of Carolina rice because rice shipped with the bran intact often became rancid before it reached markets overseas.

Removal of the bran and most of the aleurone resulted in a loss of the vitamin thiamine, a dietary lack of which causes beriberi. Beriberi is apt to occur among people whose major food is machine-milled rice unless thiamine is obtained from another source. Whenever rice is undermilled, beriberi is not a problem. Slaves did not suffer from beriberi as long as they had a sufficient diet of plantation rice. Rice for home use was milled in a mortar and pestle, which left parts of the bran and aleurone intact. Today, with modern methods to store or ship rice, the bran can be left intact without becoming rancid. More and more people are realizing the value of the bran and are opting to eat rice as "brown rice" in spite of the minor inconvenience of longer cooking time.

The milling machines that evolved in the Rice Kingdom were powered by manual, animal, water, or steam power. There is no straight-line evolution from one power source to the next. Some machines in the early 1800s could be powered by either water or steam depending on circumstances at the individual plantation, and conveyor belts and brushes could be driven by water or steam power. Some machines designed for water power were later converted to steam power.

Manual-Powered Mills

Mortar and Pestle

Mortars and pestles have been used by civilizations throughout recorded times. Along the Eastern Seaboard of North America, Native Americans used the mortar and pestle long before the Spanish and English established settlements in the New World. John Lawson, appointed by the Lords Proprietors to survey the interior of Carolina in 1700–1701 reported: "The Savage Men never beat their Corn to make bread; but that is the Womens Work, especially the Girls, of whom you shall see four beating with long great Pestils in a narrow wooden Mortar; and every one keeps her Stroke so exactly, that 'tis worthy of Admiration."[6]

Although colonists were undoubtedly familiar with the Indian-type mortar and pestle, the African-type mortar and pestle in the hands of African slaves was the initial method of milling (see figure 19). The mortar and pestle was the means by which rice was prepared for food in Africa, and it dates back to the dawn of rice cultivation there. Some only first or second generation removed from the homeland, African slaves brought to the Lowcountry by the Goose Creek Men around 1670–72 would have had knowledge of the mortar and pestle. Since the mortar and pestle could be made of native materials, it was not necessary that they be imported with slaves during the Middle Passage. Once in the Lowcountry, African slaves made mortars and pestles on the plantations from native pine and cypress.

The mortar and pestle removed the hull and bran. The process is often called "pounding," but this term is somewhat misleading because the goal was not to grind or pound rice into meal (as with corn or wheat), but simply to remove the hull and bran, leaving the rice whole. Slaves put a few pecks of rough rice into a mortar and pounded the rice with the pestle, a motion that caused the rice grains to rub against each other, creating friction that removed the hull and the outer layers of bran. Later the rice was winnowed with the fanner basket to remove the hulls and bran. At some point screens were employed to remove broken rice, which was kept for consumption on the plantation. The mortar and pestle did not entirely remove the bran and germ, leaving much of the rice's nutritional value but decreasing its market value. The mortar and pestle did not separate the inner bran and aleurone from the outer bran. The process of brushing to remove the inner bran layers and aleurone came later.

The operation of the mortar and pestle by slaves may have been more complicated and sophisticated than is often understood. The Reverend William Ellis, in his *History of Madagascar* (1838) pointed out that three successive poundings of a single quantity of rice were employed:

> The rice is prepared with great care, and involves considerable labour: when first brought from the granary, it is put into large stone or wooden mortar, about eighteen inches or two feet deep, and twelve or eighteen inches wide. Here it is carefully beaten in a peculiar manner, with a large wooden pestle, about five feet in length, so as to break and remove the outer husk [hull] without breaking the grain. The rice is then taken out, and separated from the husk [hull] by winnowing; it is then beaten in the mortar a second time, for the purpose of taking off the inner skin [bran], which is also removed without breaking the grain, after it is again submitted to the winnowing-fan, and the pieces of earth or small stones carefully picked out. The rice is then a third time submitted to the operation of the pestle, to remove any remaining portion of the inner covering [bran] of the grain.[7]

In *Slave Songs of the Georgia Sea Islands* (1942), Lydia Parrish made the following observation on use of the mortar and pestle in Darien, Georgia, during the early 1900s: "The mortars were generally hollowed out of cypress or hard pine logs and used upright, although on Sapelo I have seen some made of immense pieces of squared timber laid flat so that the beating took place against the grain. The pestles had two ends, one sharp for bruising the husks [hulls] and removing them, the other flat for whitening."[8]

At some point in the Rice Kingdom a two-step milling process was employed. The first step, beating the rice with the pointed end of the pestle, removed the hull; the second step, beating the rice with the round or blunt end, removed the bran. Evidence that this practice was widespread is seen in the mortars and pestles in the archives. Most of

FIGURE 19. Milling rice with a mortar and pestle on Sapelo Island, Georgia, circa 1915. The pestle held by the worker on the left is pointed or angled on one end rounded on the other. Reproduced by permission of Georgia Department of Archives and History, Athens.

the pestles in museums or private collections have a sharp, pointed end, and a blunt or round end, and virtually all photographs in which both ends of pestles are visible show pointed and blunt ends. The lower end of the pestle is *pointed* or angled while the upper end is *blunt* or rounded (see figure 19). The use of the two-step process was undoubtedly based on the characteristic of the rice. The hull is more easily removed than the bran, which tightly adheres to the underlying layer. How the pointed end was better suited to remove the hulls and the blunt end the bran is unknown.

Except for these two references, the two-step method is not mentioned in the archival sources. Did the two-step process originate in the Rice Kingdom, and if so, when? Or was the process introduced from Africa with the beginning of the slave trade? Why are there only two extant records of the two-step process? The surviving pestles are evidence that the two-step process was used, but not when. No surviving mortars and pestles in historical collections date to the beginning of rice culture. This does not mean that the two-step process was not used then, only that it has not been documented. It may well have been the standard method of milling from the beginning.

Milling rice with the mortar and pestle was an arduous task for the slave work force. In years of large crops, processing the entire crop might take well into spring, and the toll on the slave force was magnified. Milling with the mortar and pestle was often haphazard because the process varied from worker to worker according to skill, strength, and enthusiasm for the task. Because much of the rice was broken in the mortar or was only partially hulled, the rice sent to market was often deemed substandard by buyers. Some planters suspected that the workers deliberately produced poor market rice because the broken rice was used for domestic food.

Planters were also concerned about the debilitating effect preindustrial methods had on the health of slaves. Hand milling with mortar and pestle fatigued and debilitated workers, especially in years of heavy crops. More than one planter was troubled by this problem. Peter Manigault instructed his overseer in 1794, "If the Rice made at Goose-Creek is not yet beat out, I wd. wish to have it sold in the rough, to save labour of the Negroes."[9] All planters did not necessarily share Manigault's concern for slaves' health; those who did were rewarded with a healthier work force and more profit in the long term.

Understandably planters were not satisfied with the mortar and pestle. From the time rice became a commercial crop, better methods of milling were a necessity. A series of new milling machines came one after another, and by the mid-1800s milling was fully mechanized, highly efficient, and steam driven. All these machines, however, still retained some type of mortar-and-pestle component.

The Guerard Pendulum Engine

The French Huguenot Peter Jacob Guerard is credited with developing the first "engine" to mill rice within the Rice Kingdom, but it is doubtful that his invention was an "engine" in the sense the term is understood today.[10] On September 26, 1691, the General Assembly of South Carolina gave Guerard the right to bring legal action against anyone who reproduced his Pendulum Engine without his consent: "Whereas, Mr. Peter Jacob Guerard, hath at his proper cost and expense of time, lately invented and brought to perfection, a Pendulum Engine, which doth much better, and in lesse time and labor, huske rice, than any other heretofore hath been used within the Province."[11]

How Guerard's "engine" functioned is lost to history. Neither a diagram nor a description of it exists, and it is unlikely that Guerard ever created a diagram of his Pendulum Engine. One source suggests that it was nothing more than a pestle attached to a limber pole supported between two uprights that helped to raise and lower the pestle. Another source suggests that the pestle was attached to a pole that acted on a fulcrum. The worker stepped on the end of the pole, lifting the pestle. When the worker stepped off the pole, the pestle fell down into the mortar.

Yet neither of the two implements described operated like a pendulum, which is an object attached at one end to a fixed support while the other end swings back and forth under the influence of gravity. Guerard was a goldsmith by trade, so the action of

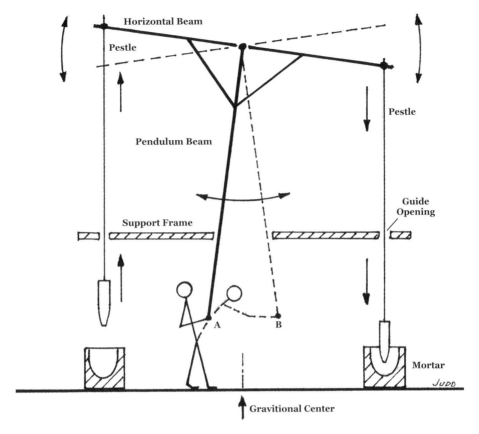

DIAGRAM 36. Possible design of Peter Jacob Guerard's "Pendulum Engine," 1691. Illustration by William Robert Judd.

pendulums in clocks may have influenced his design. Assuming Guerard's mill acted on a pendulum motion, diagram 36 shows how it might have operated.

A pivoting T-shaped assembly consisting of a horizontal beam fixed to the top of a pendulum beam represented the pendulum engine. (The structural framework supporting the engine has been omitted from the diagram to show the inner workings more clearly.) At each end of the horizontal beam a pivoting pestle operated through a guide opening aligned with the mortar below. An opening or framed encasement contained the swing of the pendulum, thereby controlling the distance of a pestle's descent into its respective mortar. The operator pulled the pendulum beam to position A, and the end of the horizontal beam moved downward, forcing the pestle into the mortar and raising the opposite pestle. The operator then pushed the pendulum beam to position B, raising the end of the horizontal beam and lifting the pestle out of the mortar, while at the same time forcing the opposite pestle into its respective mortar. The operator repeatedly moved the pendulum beam back and forth, causing the ends of the horizontal beam to rock up and down in a seesaw manner, raising and lowering each pestle. Only one worker would have been needed to operate this "engine."

The action of the pendulum was similar to that of a flywheel. Once the operator initiated the motion of the pendulum beam, its back-and-forth momentum reduced the labor applied by the operator. When the pendulum moved away from dead center, gravity, acting on its weight, brought the pendulum back toward and past dead center, where it was met by its opposite force, sending it back in the other direction. To stop the mill and fill or empty the mortars, the operator simply stopped the pendulum's movement.

Whether Guerard's invention was ever used, or how successful it was, is not recorded. It is unlikely it was used extensively because all records after his "patent" in 1691 refer to slaves manually using the mortar and pestle. Seven years after his invention, the General Assembly was still seeking a productive rice mill.

The Spring Mill

John Lucas, a grandson of millwright Jonathan Lucas I (1754–1821), wrote the most important document about the early history of rice milling. Lucas described four kinds of early mills: *spring, pecker, cog,* and *water*.[12] Since he never witnessed the use of the first three mills, it is uncertain on what material he based his account.

Lucas's crude sketch of a spring mill (see diagram 37, top) has never been mentioned in previously written accounts of rice milling. Judging by Lucas's description of its operation, the spring mill obviously worked on the principle of resiliency: "A spring pole was used to do hand work and worked through two uprights 1 & 2. A spring pole was fastened to the Pestle which after the man had driven into the mortar by its spring was raised out of the mortar ready for another blow." Upright 1 in Lucas's sketch is the fulcrum post, and upright 2 is the anchor post. Lucas scribbled marks at the top of each upright to denote the spring pole's attachment points.

Diagram 37 (bottom) illustrates the operation of the spring mill. Rice was placed in a mortar. The worker pulled the pestle downward, deflecting the spring pole downward, forcing the pestle into the mortar. The worker next released his grip on the pestle. The spring pole rebounded upward, lifting the pestle out of the mortar. The worker repeated the process again and again until the rice was milled. The worker's labor was reduced because the spring pole lifted the pestle and he had only to drive the pestle downward.

How extensively the spring mill was used in the Rice Kingdom is unrecorded. Multiple units of spring mills would have been a marked improvement in plantation milling, lessening the labor per mortar. Both the Pendulum Engine and spring mill share one similarity: simplicity.

The Italian Pestle Mill

Figure 20 depicts a manually operated mortar and pestle mill invented in Italy during the 1500s. Although there are no records of this machine being used in the Rice Kingdom, it is evident that in some parts of the world during the 1500s, mortar and pestle

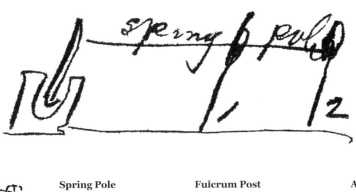

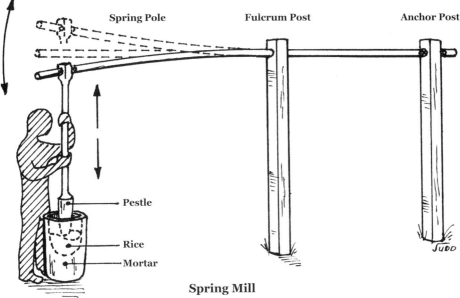

DIAGRAM 37. John Lucas's sketch of a spring mill (top). South Carolina Historical Society, Lucas Family Papers 1792–1936, folder 11/270. Diagram of how the spring mill functioned (bottom). Illustration by William Robert Judd based on Lucas's sketch and description.

---

technology was far more advanced than many historians of rice culture have perceived. This Italian machine also has similarities to the animal-, water-, or steam-powered mortar-and-pestle machines that became standard and continued to be used until the end of the industry. This Italian machine may have been the model of the Deans's rice-pounding machine used in the Rice Kingdom during the eighteenth century.

The print depicts four sets of mortars and pestles housed within a wooden framework. The pestle shafts slid through guide openings in the framework in an up-and-down direction, allowing each pestle to move into and out of its designated mortar. A lever arm hinged on each pestle shaft pivoted on a fulcrum post. Mounted on a separate framework at a ninety-degree angle to the lever arms was a horizontal axle shaft rotated by a hand crank at each end. The worker on the left end turned his crank counterclockwise; the

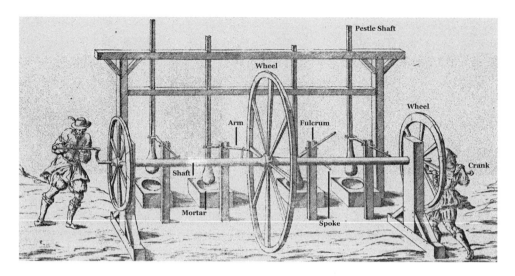

FIGURE 20. An Italian pestle machine, 1578, by Jacques Besson. Reproduced from the Dibner Library of the History of Science and Technology, Smithsonian Institution Libraries, Digital Edition, 1999.

worker on the right end turned his clockwise. Spokes, one for each pestle, were fixed to the axle shaft at staggered intervals around and along its length and set at an angle. When the shaft revolved, each spoke rotated in succession over the top of the lever arm, forcing it downward and lifting the pestle attached to the opposite end. As each spoke slid past the end of the lever arm, its pestle fell by gravity into the mortar below. The pestle was again lifted by the next rotation of the spoke. Each pestle was lifted and dropped, one after the other.

The three wheels acted as flywheels and induced a more uniform rotational speed to the axle shaft, while at the same time reducing the amount of labor exerted at the hand cranks. Flywheels became standard equipment on steam engines for the same purpose. Evidently the machine's designer was well versed in the physics of centrifugal force when he added the flywheels.

Although the existence of this machine was probably not widely known, surely it would have been known to the English, who had been trading with Italy for rice as early as 1600. Furthermore, since wealthy Lowcountry planters sojourned in Europe, a planter or someone with an interest in planting rice probably would have become aware of this machine, and it could have reached the Rice Kingdom at some time in some form. Although the English consumed almost no rice, they exported it to parts of Europe as a trade commodity. The Lords Proprietors, however, were evidently unaware of the machine, or any other machine for milling rice, as the General Assembly requested that the Proprietors "procure and send . . . by Ye: first oppertunity a moddell of a rice mill." No records indicate the Proprietors supplied the colonists with a machine to mill rice.

Whether the Italian design at some later point was the inspiration for the first mortar and pestle machines having an axle with spokes is unknown. Mechanical similarities are evident in both the Stephens and the Veitch pestle mills of the eighteenth century.

Deans's Rice-Pounding Machine

In 1772 Henry Laurens referred to Robert Deans's labor-saving machine, which may have been similar to the Italian pestle machine, which Deans could have learned about in England. Deans, a surveyor and builder who had been in Carolina for many years, perceived the hardships of slaves milling rice, and he contrived to invent a labor-saving machine. At some point he returned to England and developed a model in which four men could do the work of twelve and much easier. He made a large machine and tried it in the presence of several gentlemen and planters in London, and said it met with their approval.

No diagram of Deans's machine has found. There are, however, accounts by Laurens and Deans of their attempt to have a model sent to Mepkin, the Laurens plantation on the Cooper River.

On March 2, 1772, Henry Laurens wrote to Gabriel Manigault from Westminster in England about the Deans machine: "I am now causing at my own Expence a Machine for pounding Rice, to be built. I'll trust no longer to Models for such machines, but will have ocular Demonstration. I have procur'd a Barrel of Rough Rice which I intend to have pounded when the machine is ready. If the work is perform'd to the Satisfaction of Governor Wright, and some other great Rice planters who are at present in England, then I will send all the parts of the Machine over to Carolina."[13]

Laurens wrote again to Manigault on March 20, 1772: "The Pounding Machine which I lately spoke of is nearly finis'd, and in its present Appearance, promises to be useful. One Slender hand it is said will beat 30 Strokes in a Minute, and tend 4 Mortars at once with easy Labour."[14]

On April 21, 1773, Laurens wrote from Dover to J. A. Eckhardt: "Mr. Deans had Invented & set up a Rice Machine which performed tolerably well, but you, having seen & considered its Construction & operation proposed to make such Improvements upon it, as would render it quite perfect, which certainly will be the Case if you can lift the pestles of the same Weight, in the same time (not in proportion of time) with greater Ease or less Power or labour, Eight Inches higher, & all at the same expense, or but a trifle if any thing more.[15]

Evidently Eckhardt said he would suggest improvements that Deans could implement to make the machine better; Eckhardt later asked Laurens for money to make the improvements, which Laurens declined. Laurens wrote Deans from Geneva on May 28, 1773, that he was sorry Eckhardt had been so ambiguous on his intentions about improvements to the machine and said: "I wish you may succeed in your new Attempt. If you do, nothing shall be wanting on my Part, to obtain some proper Consideration for you from the Public in Carolina, and at all Events I will take the Machine which you are to make off your hands. I wish you would think of means for turning your Machine by a Horse or by Water."[16]

Laurens, again dissatisfied with the progress Deans was making on the machine, wrote from Westminster in February, 1774, to John Lewis Gervais in South Carolina saying he would send Gervais a bill of scantling to have Mr. McCullogh build the machine according to Deans's model. The following April Laurens wrote to Gervais that "our friend Deans immediately altered his design & intends to set up a complete Machine here and Pound some bushels of Rough Rice." Deans planned to accompany models of his machine to America, but he became ill and could not sail. Deans sent several models by ship with a Captain Urquhart.[17]

Nothing else appears in the *Laurens Papers* about the Deans pounding machine, and his machine is mentioned nowhere else in archival sources. Whether a functional machine was built according to the model is unknown. No diagram of the Deans machine has been located or is known to exist, but the description that does survive suggests it worked on a method similar to the Italian pestle machine.[18]

Wooden Mill (Wooden Rotary Quern)

Naturalist Mark Catesby first described a wooden mill that gained widespread use in the Rice Kingdom in *The Natural History of Carolina, Florida, and the Bahama Islands*, first published in 1731–43. In a section titled "Of the Agriculture of Carolina" is the entry "Oryza: RICE," in which Catesby explains, "About the middle of September it is cut down and housed, or made into stacks till it is thresh'd, with flails, or trod out by horses or cattle, then to get off the outer coat or husk [hull], they use a hand-mill, yet there remains an inner film [bran] which clouds the rightness of the grain, to get off which it is beat in large wooden mortars, and pestles of the same, by Negro slaves, which is very laborious and tedious.[19]

The hand mill is likely the wooden mill described here. Catesby's note establishes the earliest date on which the use of this mill was recorded the Rice Kingdom. This wooden mill was another advance in the milling process. It removed the hulls, and then a mortar and pestle was used to remove the more tenacious bran. This two-step milling process may well have been based on the earlier use of the two-ended pestle.

Governor James Glen gave another account of the wooden mill in 1761: "The next part of the process is grinding which is done in small mills made of wood, of about two feet in diameter; it is then winnowed again and afterwards put into a Mortar made of Wood, sufficient to contain from Half a Bushel, when it is beat with a pestle of a Size suitable to the Mortar and to the strength of the Person who is to pound it; this is done to free the Rice from a Thick Skin [bran], and it is the most laborious Part of the Work. It is then sifted from the flour [bran] and Dust, made by pounding; and afterwards by a Wire-sieve called a Market-sieve, it is separated from the broken and small Rice, which fits it for the Barrells, in which it is carried to Market."[20]

14. The threshing barn at Kinloch Plantation. The rolling screen was housed in the cupola. Photograph by Richard Dwight Porcher, Jr., 2010.

15. Ruins of the water-powered mill and millrace floor at Rose Hill Plantation on the Combahee River in Beaufort County, South Carolina. Photograph by Richard Dwight Porcher, Jr., 2010.

16. Ruins of the millrace at Stoke Rice Mill on the Western Branch of the Cooper River in Berkeley County, South Carolina. *Insert:* Curved slot for the gate in the millrace wall. There was a slot in each millrace wall. Photographs by Richard Dwight Porcher, Jr.

17. A cast-iron waterwheel hub. The only known hub of any kind, it was cast in one piece and recovered from a rice mill built by Jonathan Lucas I in 1792 on Wambaw Creek. The hub is located at the Village Museum, McClellanville, South Carolina. Photograph by Richard Dwight Porcher, Jr., 2010.

18. Millstones. The millstone came from Waverly Rice Mill and is displayed at the intersection of Waverly Road and Rice Mill Road in Georgetown County, South Carolina. This millstone is six feet in diameter and eight inches thick. Insert A is a close-up of the surface of the millstone; insert B is its sweep; insert C is a runner millstone with grooves on the working surface, located at the Rice Museum in Georgetown, South Carolina; insert D is an abandoned millstone with dressing channels at Rochelle Plantation on the North Santee River in Georgetown County, South Carolina. Photographs by Richard Dwight Porcher, Jr.

19. A cast-iron mortar bottom, which was located at Middleburg Plantation in Berkeley County but subsequently lost. Photograph by Richard Dwight Porcher, Jr., 1976.

20. Steam engine at Mansfield Plantation. Photograph by Richard Dwight Porcher, Jr., 2010.

21. (*above*) Steam engine at Cedar Hill Plantation on the Eastern Branch of the Cooper River in Berkeley County, South Carolina. Photograph by Richard Dwight Porcher, Jr.

22. (*right*) Steam engine at Middleburg Plantation. Photograph by Richard Dwight Porcher, Jr., 2010.

23. Smith & Porter Steam Engine, 1858–60, at Roundhouse Museum, Savannah, Georgia. Photograph by Richard Dwight Porcher, Jr., 2010.

Henry Laurens described the wooden mill in 1772: "In answer to your opinion of the Mill for cleaning rice, I would advise you to be cautious of going to any expence lest it should prove a useless one, for you may depend upon it, no Grinding Mill will answer the purpose. Rice is ground first for the Mortar by a Wooden Mill & the softest kind of Pine is chosen for that service. The husk [hull] is ground off very clean, but nothing less than the Pestle will take off the Inside Coat [bran], & shew the neat whiteness of the Grain."[21] Thus Laurens's letter substantiates what Glen stated nine years earlier: The wooden mill removed the hulls. The hulls were winnowed away, and then a mortar and pestle were used to remove the bran. Then the rice was sifted through a sieve to remove the flour [bran] and dust and a wire screen was used to remove broken rice.

In *A Tour of the United States of America* (1784) John Ferdinand Dalziel Smyth also described a wooden mill. Smyth's book is based on his travels during 1769–75, a period which coincides with Laurens's and Ford's writings about this mill. Smyth, however, described the mill in more detail:

> This mill is constructed of two large flat wooden cylinders, formed like small mill-stones, with channels or furrows cut therein, diverging in an oblique direction from the center to the circumference, made of an heavy and exceedingly hard timber called lightwood, which is the knots of the pitch pine. This is turned with the hand like the common hand mills, for they have not as yet arrived as such a state of improvement and perfection in this business as to make use of horse mills, which might certainly be rendered much more advantageous and useful.
>
> After the rice is thus cleared of the husks [hulls], it is again winnowed, when it is fit for exportation.[22]

Smyth evidently erred when he said "After the rice is cleared of the husks, it is again winnowed, when it is fit for exportation." Rice for export generally had the bran removed, which would have followed removal of the hulls by the mortar and pestle.

Smyth did not give a size for the wooden mill he saw; however, three other references do give measurements from which the general dimensions of the wooden mill may be established. Glen stated that the mills were "of almost two feet in diameter."[23] In 1847 Robert Allston stated: "The rough rice was sometimes ground by being passed between wooden blocks twenty inches in diameter by six inches thick worked by hand."[24]

In November 28, 1785, Timothy Ford recorded the wooden mill in his diary: "To separate them [hulls] is another distinct process; and it is done by friction between two blocks which are thus prepared. They are cut from live oak, about 2 feet through, the under one 2½ high the upper one 12 to 16 inches. These are cut from their centers to their edges into threads or nuts much like a millstone and in every respect work like them (tho by hand) the grain being fed at the center & thrown out at the circumference together with its disengaged chaff [hulls]."[25]

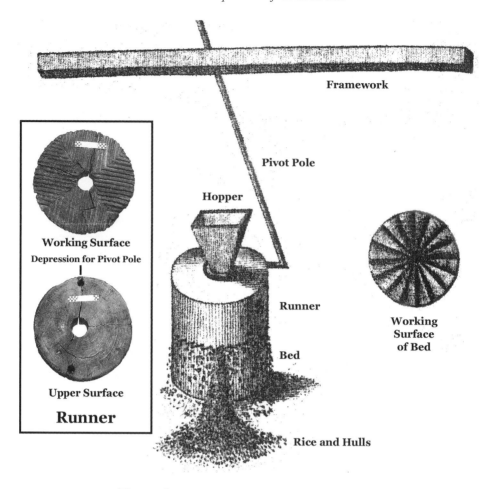

DIAGRAM 38. The wooden rotary quern Luigi Castiglioni saw during his 1785–87 travels in the United States. Reproduced from Luigi Castiglioni, *Viaggio* (first published, 1983). *Insert:* the wooden mill runner found at Snee Farm, Charles Pinckney National Historic Site, Mount Pleasant, South Carolina. Photograph courtesy of National Park Service.

The hand mill Catesby mentioned was likely similar to the hand-operated wooden mill described by Glen, Laurens, Ford, and Smyth since no mention appears in any archival source of any other type of hand mill that used wooden blocks.

The design of the wooden mill was unknown until a chance discovery was made at Snee Farm, the Charles Pinckney National Historic Site in Mount Pleasant, South Carolina, where workers discovered the runner of a wooden mill (see diagram 38, insert) in an abandoned well. (In this discussion of the wooden mill, the upper wooden block is called the runner, and the lower, stationary block is called the bed.) The runner was made of pine, 24 inches in diameter and 6 inches thick, with lands and grooves on the working surface one inch apart and the grooves ½–¾ inch deep. On the upper surface were two depressions in which to insert a short stick or pivot pole to turn the runner manually. The bed was not found. How the spacing between the runner and bed was maintained is unclear because whatever structure kept them apart was lost. As the runner turned, pressure was exerted on the rice, causing friction between the grains, which ground off the hulls.

The two depressions in the runner classify the wooden mill as the ancient type called a "rotary quern," which was used in the millennium before Christ. The primitive rotary quern consisted of two flat circular stones; the top (runner stone) was turned by hand against the stationary bottom one (bed stone). The runner stone was made with a hole in the center, through which was extended a spindle, on which the runner stone was held at a slight distance above the bed stone by means of a bearing. The spindle kept the runner stone from coming in contact with the bed stone and allowed the operator to position the runner stone at the proper distance over the bed stone to process the desired grain. The grain was hand fed through the hole in the runner stone. A stick or pivot pole was inserted into a depression to one side or on the top so the miller could turn the runner stone easily. The stick used to turn the runner stone evolved into the pivot pole. The upper end of the pivot pole was inserted into a hole in a framework directly above the runner stone to act as a pivot. The pole's lower end was inserted into a depression in the top edge of the runner stone. The operator grabbed the lower end of the pole and rotated the stone.

This rotary quern spread throughout Europe and most of the rest of the world, wherever grains were processed for food. When the colonists set out from England to settle the New World, they brought a variety of millstones with them, including rotary querns, mainly to process corn and wheat. When rice became a cash crop in the Rice Kingdom, wood was substituted for stone, and the mill technically became a *wooden rotary quern* employed to grind off the hull and leave the remaining part of the rice intact.

A diagram of this type of wooden rotary quern appears in *Travels in the United States of North America, 1785–87,* by Luigi Castiglioni (diagram 38).[26] Where Castiglioni saw this wooden rotary quern is unknown; but his diagram appears to be an accurate depiction of the wooden mill described by Glen and Allston. Castiglioni's travels in the United States came about a decade after Smyth's ended. Castiglioni's diagram shows a hopper that funneled rice into the hole of the runner. A manually operated pivot pole rotated the runner. Rice and hulls exited from between the runner and bed.

How common the wooden rotary quern was in the Rice Kingdom is uncertain. Judging from the writings of Allston, Glen, and Smyth, it must have gained significant use, at least for a short period. One planter used the wooden rotary quern as late as 1793. Charles Drayton recorded in his diaries that on January 12, 1793, "Doctor Warings hand mill for grinding rice, grinds one bushel in five minutes."[27]

The wooden rotary quern reached Louisiana, where rice was grown in the Mississippi River valley. R. A. Wilkinson described the typical Louisiana rice grower in 1848 as a small farmer "not wealthy, and hitherto almost entirely uneducated, and unable, from the smallness of their means, to vary their crops from the general routine."[28] They did not have the capital to acquire advanced milling and threshing machinery, so they relied on far more laborious and cruder methods that required less capital than those used in the Carolinas. When a steam-powered mill was introduced, "the prejudice against it was so strong . . . that the mill could at last neither buy, nor get rice upon toll, and was abandoned." They reverted to simpler mills, one of which was the wooden rotary quern. Wilkinson's description of the operation of a wooden mill used by the Louisiana rice grower fits the description of the wooden rotary quern used in the Rice Coast, from where it was probably obtained: "It is then turned in a small hand-mill of wood, like a common corn-mill, and partially hulled; then placed in a mortar, or four mortars in a row, where the like number of pestles pound it till the balance of the hull, and the skin that has a yellow appearance, is taken off."[29] Wilkinson did not say what type of mortar and pestle was used to clean the rice after the hulls were removed.

Animal-Powered Mills

Animal-powered mills were common throughout the Rice Kingdom during the 1700s. The earliest record of an animal-powered mill dates to 1710, when Thomas Nairne wrote: "Rice is clean'd by Mills, turned with Oxen or Horses."[30] Nothing is known of the construction and operation of this early mill, or who introduced it.

In a 1768 issue of the *Georgia Gazette* Basil Cowper advertised a Mr. John Smith's rice plantation for sale on the Carolina side of the Savannah River that had "a spacious barn 110 feet long, including a walk of 40 feet square for the cattle that work the machine, which is of the best kind and quite new, with a wharf before the door."[31] Around the time of the Revolution, Elkanah Watson said that Carolina planters milled rice in mortars "ten or twelve in a row, each containing about half a peck," their pestles attached to a shaft powered by horses.[32] Joseph Habersham wrote to John Habersham in October 1789, from Augusta: "If the Oxen will not work in the Machine, I must request the favor of our friend Gibbons or yourself to purchase three Machine Horses, as the crop [rice] must be beat as early as possible."[33] Advertisements about animal-driven mills were common in the newspapers of Charleston and Savannah.

Animal-powered mills operated only mortars and pestles. Unlike manually operated pestles with two different working ends, the machine pestles had only one working end, which was rounded and used to remove both hulls and bran. Since the pestles were fixed onto a machine, it was probably too expensive or too complicated to have interchangeable pestles, one with a blunt end and the other with a beveled or pointed end. Because they had more pounding pressure, the animal-powered machines probably produced enough friction between the rice to remove both hulls and bran.

In *A View of South-Carolina* (1802), John Drayton (1767–1822) described three early milling machines developed for use during the 1700s: pecker, cog, and water mills.[34] The pecker and cog mills were in use before Drayton's time, so he must have described them

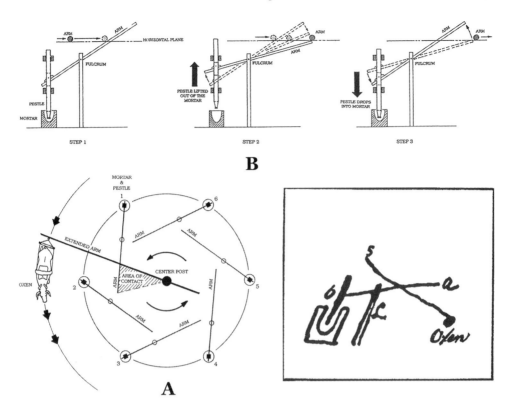

DIAGRAM 39. A pecker mill. Illustration by William Robert Judd, based on the John Lucas diagram shown in the insert. South Carolina Historical Society, Lucas Family Papers 1792–1936, folder 11/270.

based on historical sources or from observations of machines still in use. Unlike John Lucas later, Drayton did not include diagrams of the pecker and cog mills, but other sources allow us to know how they were constructed and operated.

The Pecker Mill

John Drayton described the "pecker mill" as "the most simple; and, probably, that which was first in use. It is so called, from the pestle's striking somewhat in the manner of a wood pecker, when pecking a tree." It was probably the same type of pecker mill described and diagrammed by John Lucas (see diagram 39, insert). According to Lucas: "There were three Kinds of Mills at First the Pecker Thus. This Kind were placed [the mortars] round in a circle. The Arm A was fastened to the Pestle had as a [illegible] on fulcrum Post C. The center Post had an Arm S which moved by oxen round the [illegible] and over the arm a fastened onto pestles which were thus made to raise the pestle until the arm reached the End when the Pestle dropped into the Mortar."[35]

The pecker mill described by John Lucas was probably the same mill advertised in the *South Carolina Gazette* by Peter Villeponteaux in 1733, making it the first documented animal-powered rice mill in the Rice Kingdom:

> Whereas it hath been represented to His Excellency, and to His Honorable Council and Assembly, by Peter Villeponteaux, that the Pounding of Rice by Negroes, hath been of very great Damage to the Planters of this Province, by the excessive hard Labor that is required to Pound the said Rice, which has killed a large Number of Negroes.
>
> Whereas it has pleased His Excellency, and the Hon. Council and Assembly, to grant to the said Peter Villeponteaux a Patent to make and erect an Engine or Machine to clean Rice, which Machine has been tried, and by several Planters approved to be of great Use to the Publick: Therefore the said Peter Villeponteaux hath taken, into Co-partnership one Samuel Holmes, who has great Experience in making those Machines, and We are ready to treat with any Planter on reasonable Rates to erect the same.
>
> Peter Villeponteaux,
> Samuel Holmes[36]

Subsequent to the above advertisement, Villeponteaux published in the *South Carolina Gazette* several times in 1734, a bill of scantling for a rice mill, which listed the lumber needed to construct his machine:

> Oak plank, 100 feet, 5 inches thick
> 4 pieces pine, 12 feet long, 6 inches square
> 12 pieces pine 7 feet long, 22 × 18 inches
> 2 pieces pine 30 feet long, 7 × 5 inches
> Middle post 8 feet long and 18 inches in diameter[37]

This list of materials seems sufficient to construct the pecker mill described by John Lucas, further evidence that Lucas described Villeponteaux's machine. The long plank could have been used for the long arm moved by an ox, and the middle post could have been used as the fulcrum for the long arm. The remaining pieces could have been used to construct the pestles and pestle arms.

The diagram of the pecker mill (see diagram 39) is based on Lucas's diagram, Villeponteaux's description, and the bill of scantling. Although the crude Lucas sketch shows only one mortar and pestle, from his description it is clear that several mortars were arranged in a circle within a framework. (The framework has been omitted from our diagram.) An extended arm was attached to an ox. As the ox moved in a circle around the mortars and pestles, the extended arm passed in succession over arms that were attached to the pestles. The movement of the extended arm (crosshatched in diagram 39A) depressed the end of the pestle arm and raised the pestle. When the extended arm slid past the end of the pestle arm, the pestle fell back into the mortar by gravity. Each pestle was activated in turn as the ox moved around the mortars. The action of the pestle in the mortar of the pecker mill was similar to that of the hand-operated mortar and pestle.

Villeponteaux stated that a gentleman interested in having a mill constructed would have to furnish seasoned timbers and four horses and that he would supply the iron work and the skill. When the pecker mill was driven with four horses, Villeponteaux said, it could clean two thousand pounds of rice daily. The cost of the mill would be "65 pounds Currency of the Province," and Villeponteaux invited any gentleman to observe his machine in operation on James Island. Still, it was hard to sell, but Villeponteaux was a persistent advertiser. At certain seasons he offered it at half price.[38] Even at half price, £30 was not a small sum for a planter to pay. By September 1734 the original machine had been so greatly improved that, when it was driven with only two horses, it could clean five thousand pounds of rice with ease during a working day; if four horses were employed, the machine would turn out one thousand pounds an hour without breaking the rice.[39]

It is difficult to determine how widespread the pecker mill was. They were advertised for sale as late as 1801. In the March 7, 1797, issue of the *Charleston City Gazette and Daily Advertiser* appeared this notice of a tract of land for sale in St. Luke's Parish, South Carolina, which noted that "the Pounding Machine, which is a Pecker, will clean more Rice, and with less labour, than any of the kind in the state." On August 29, 1799, the *City Gazette* had an advertisement for "Two PECKER MACHINES, That pound upwards of six Barrels of Rice a day each." And in the January 1, 1801, issue of the *Carolina Gazette,* Solomon Cohen of Georgetown, South Carolina, advertised for sale a tract of land on the Pee Dee River including "a Pecker Machine, a new Barn, and some Negro Houses." Whether these later pecker machines were the same as those invented by Villeponteaux is unknown. No other pecker machines, however, appear in archival sources relating to rice culture, so it is likely they are the same as, or modifications of, Villeponteaux's pecker mill. Since they are all called "pecker" mills, they probably at least worked like Villeponteaux's machine. Even though more efficient animal-powered and water-powered mills were in use by the turn of the nineteenth century, it was not uncommon for a planter to keep using an earlier machine that met his needs instead of absorbing the cost of a newer one.

The Drum Mill ("Cog Mill")

John Lucas also described a mill he called a "cog mill": "The Cog was a large drum having on its sides inclined wings to which the pestles were made to fit—so that the inclined wing will cause them to rise out of the Mortars until the end of the inclined wing is reached when they drop. The Mortars & Pestles were placed in a circle. The drum being inside, the drum was turned by oxen fastened to an Extended arm. There were a number of mortars around in both planes."[40] His sketch of this mill is shown in an insert in diagram 40.

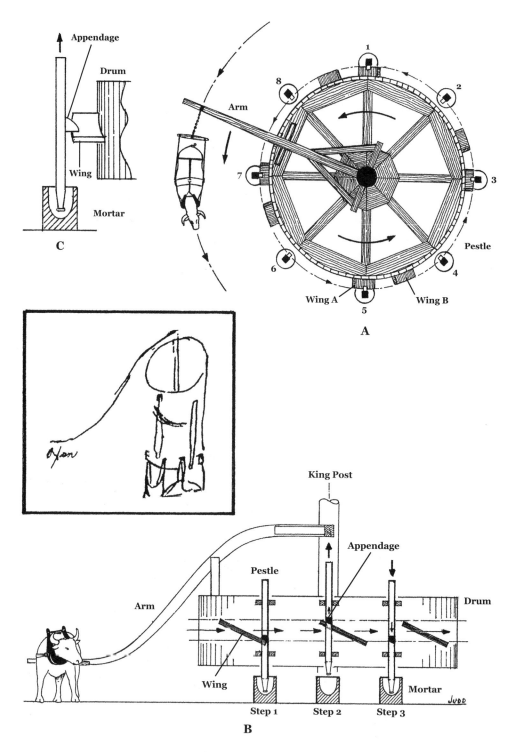

DIAGRAM 40. A drum mill (or "cog mill"). Illustration by William Robert Judd based on the John Lucas diagram shown in the insert. South Carolina Historical Society, Lucas Family Papers, 1792–1936, folder 11/270.

The mill John Lucas described, however, did not employ a cog in the sense that the term is usually defined. In the most common usage, "cogs" are teeth on the rim of a wheel that intermesh with teeth on another, similar wheel to create motive force. Yet an older use of "cog" in carpentry defines it as a "tongue" that fits into a slot. Lucas was apparently thinking of this definition when he wrote about "a large drum having on its sides inclined wings to which the pestles were made to fit." According to this description, however, the "cog mill" should more appropriately be called a "drum mill." Diagram 40 is an interpretation of the mill Lucas described and diagrammed.

Diagram 40A is an overhead view depicting a circle of eight mortars surrounding a large vertical drum with multiple wings projecting from its perimeter. Animal power rotated the drum. (The framework of the mill has been omitted.) Only four pestles are lifted at one time, easing the labor placed on the one animal. Each pestle weighted approximately 230 pounds, so four together totaled 920 pounds. As the drum rotated, the first set of inclined wings (A) lifted and dropped pestles 1, 3, 5, and 7 into their respective mortars. The next set of wings (B) moved forward at the same time to begin the lift and drop of pestles 2, 4, 6 and 8. This process was repeated with each revolution of the drum, which lifted and dropped the pestles sixteen times. If eight additional wings were added, the pestles would be lifted and dropped thirty-two times, probably requiring additional animal power.

Diagram 40B is a conjectural elevation view. An ox walked in a circle pulled an extended arm attached to a king post, which rotated a large vertical drum. Steps 1, 2, and 3 depict the operation of a single pestle as an inclined wing moved along its horizontal plane. The pestles worked vertically through guide openings in the support framework. In step 1 the lower end of the inclined wing struck the angled underside of the pestle's appendage. As the wing moved along its horizontal path in step 2, the appendage rode up the incline of the wing, lifting the pestle out of the mortar. As the top of the wing moved past the appendage, the pestle fell by gravity into the mortar below. This cycle, steps 1, 2 and 3, was repeated as each wing came in contact with the appendage of each pestle.

The mill and animal track were both on ground level and may have been enclosed only by a roof, its sides exposed to the elements. When this mill was first employed in unknown, but its simple construction suggests it was used during the early to mid-1700s. Yet, seventy-two years after the first recorded water-powered pestle mill made its appearance in 1744, Charles Drayton saw an animal-powered mill like the drum mill on a trip he made to the estate of Mr. Boyle on May 26, 1816: "His rice machine of inclined planes on a circular rim of 32. feet Diameter pounds 12. bushels of milled rice in 1. & ½ hour. the planes elevate 2. f. 6 inches. 4. oxen work with ease. It has six mortars, but there is space for 8."[41] With its references to oxen power and "inclined planes on a circular rim," Drayton's description makes Boyle's mill seem like the same type of mill described by Lucas. No description of any other kind of animal-powered rice mill with inclined planes has been found in archival sources.[42]

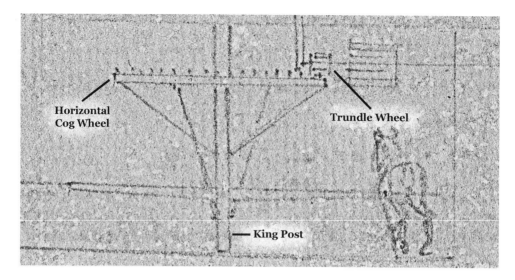

DIAGRAM 41. A cog mill used for a cotton gin, March 30, 1808, by Charles Drayton. From Charles Drayton Diaries, Drayton Papers Collection, reproduced with permission of Drayton Hall, A Historic Site of the National Trust for Historic Preservation.

It was common for a planter to retain a mill he had used for a long time if he kept a sufficient labor force and was not wealthy. Newer mills were more expensive and took time to construct. By 1833, the average water mill cost between seven thousand and eight thousand dollars (equal to approximately two hundred thousand in 2009 dollars). For a small plantation this would have been too great a cost for a minimal increase in the output of finished rice. Consequently planters often retained old mills that worked well and satisfied plantation needs long after newer mills were in use. Charles Drayton, for example, went to see Morris's water-powered mill and recorded that he was not impressed: "It did not turn out that much more rice than [John Drayton's animal-powered] machine."[43]

Some planters may not have had the choice of switching from animal power to water power because their plantations had no water source to drive a mill. Mr. Boyle may have been one of these planters.

The Cog Mill

While John Lucas described a "cog mill" that was really a drum mill, John Drayton described a true cog mill, which consisted "of a large horizontal cog wheel, turning a trundle wheel; working upright pestles, nearly on the same principles as a madder mill."[44] Drayton's use of "cog" refers to one of a series of teeth projecting upward around the rim of a horizontal wheel.

Diagram 41 is Charles Drayton's 1808 sketch diagramming an animal-powered cotton gin that employed a similar horizontal cog wheel to turn a trundle wheel. The trundle wheel rotated a barrel drum. A band (not shown) motivated by the drum drove the mechanical components of a series of cotton gins located on the loft floor. By placing the cog wheel and trundle wheel on the second floor, the transition from a cotton gin to a mortar and pestle rice mill would have been simple. The motive power and gearing would have been the same.

Diagram 42, a cog mill for rice, is based on John Drayton's description and Charles Drayton's sketch of a cog mill for ginning cotton. Both were animal-powered. The cog mill was housed within a barn having an earth floor and a floored second level containing the mill's machinery. To start the milling process, an attendant on the second level placed rice into the mortars below each pestle. The mill was set in motion by animals walking a circle track on the ground level and pulling the ends of a horizontal beam that rotated a vertical king post. The king post extended upward through an opening in the floor of the second level. Attached to the king post on the second level was a horizontal cog wheel that rotated a trundle wheel fixed to the end of the spoked shaft of the pestle machine. As the spoked shaft revolved, each spoke rotated through a slot in its respective pestle, lifting the pestle upward in sequence. As each spoke slipped out of the slot in the pestle, the pestle fell by gravity into the mortars. It took approximately two hours of pounding to clean the rice in each mortar.

Only two kinds of lifters for pestles are mentioned in historical sources from the time this type of mill operated: cams and spokes. Spokes were used more often in the mills described in period sources; accordingly, spokes are used in diagram 42. The difference between a "cam" and "spoke" used to lift pestles is shown in diagram 43. The earliest recorded "machine" that used spokes to lift pestles dates back to the Italian pestle mill of 1578 (figure 20).

John Drayton did not give a date for the cog mill he described, so it is uncertain where it fits into chronological sequence of animal-powered mills. This type of rice mill probably gained widespread use around the middle to late 1700s, and it would have more appropriately been called a "pestle mill."

Drayton did not mention who invented this mill, and no other archival source mentions an inventor's or maker's name. It may well have been the creation of someone in the Drayton family.

Sieves, Screens, and Rolling Screens in Milling

Trying to sort out the differences among sieves, screens, and rolling screens used in milling is difficult. The terms "sieve" and "screen" have been used interchangeably at times, at least early in the history of rice culture. In broad terms, a *sieve* is a wire mesh with small holes used for sifting small material, such as rice bran and polish. *Screens* had larger mesh than sieves and were used to remove larger material such as broken rice.

When and who first introduced milling sieves to the Rice Kingdom is uncertain. James Glen, writing in 1761, was the first to mention sieves and screens: "[the rice] is then sifted from the flour [bran] and dust, made by pounding; and afterwards by a Wire-sieve

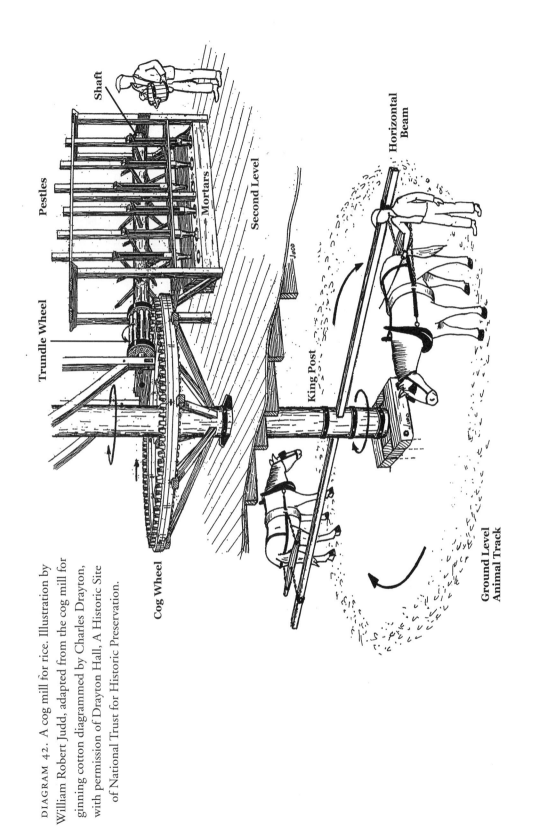

DIAGRAM 42. A cog mill for rice. Illustration by William Robert Judd, adapted from the cog mill for ginning cotton diagrammed by Charles Drayton, with permission of Drayton Hall, A Historic Site of National Trust for Historic Preservation.

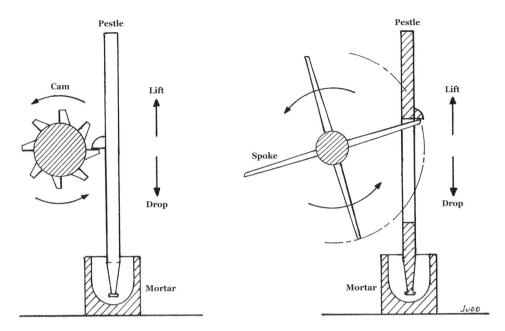

DIAGRAM 43. Cam and spoke pestle machines. Multiple cams produced shorter lift and rapid action but with less dropping force. Four long spokes produced higher lift with greater dropping force. Illustration by William Robert Judd.

---

called a Market-sieve, it is separated from the broken and small Rice, which fits it for the barrels, in which it is carried to Market."[45] The contents of the mortar were first poured into a sieve that was shaken (initially by hand, later by some sort of mechanical source) and the bran passed through the sieve. Glen's "wire-sieve" or "market-sieve" was not a sieve but a screen with openings larger than those of a sieve. After sifting out the bran, the rice was poured onto a screen and shaken. The openings were small enough for broken rice and small rice to pass through, but the whole rice could not.

In 1775 a writer identified only as "An American" explained, "In order to free [rice] from the flour [bran] and dust made by this pounding, it is sifted; and again through another sieve, which separates the broken and small rice, after which it is put up in barrels, and is ready for market."[46] The "another sieve" was undoubtedly the same as Glen's large-mesh "wire-sieve" that was actually a screen.

Four years later, in 1779, Alexander Hewatt wrote virtually the same description of a sieve and screen: "To free it from the dust and flour [bran] occasioned by pounding, it is sifted first through one sieve, and then, to separate the small and broken rice from the large, through another. Last of all, it is put into large barrels of enormous weight, and carried to the market."[47] This "another" was a screen, not a sieve.

Diagram 44 illustrates a conjectural hand-operated sieve and screen separator. The smaller wire mesh (A) separated the bran and flour from the rice. The rice was then

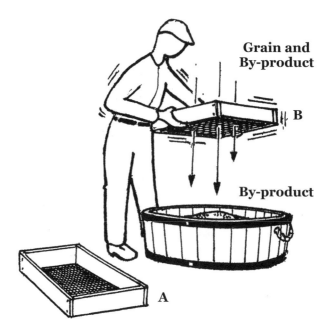

DIAGRAM 44. Hand-operated sieve and screen.
Illustration by William Robert Judd.

placed in a second separator with a larger wire mesh (B), which separated the broken and small rice from the whole, market rice.

Both the sieve and screen were eventually incorporated into one rolling screen with four different sized openings (see diagram 20) to separate four different products from the milled rice. According to Robert Allston, "The grain thus pounded is again elevated to the upper floor, to be passed through a long horizontal rolling-screen, slightly depressed at one end, where by means of a system of wire-sieves grading coarser and coarser towards the lower end, are separated, first the flour [bran], second the 'small rice' . . . third the 'middling rice,' . . . fourth, and last, the 'prime rice,' the larger and chiefly unbroken grains, which fall through the largest wire."[48]

Rolling screens were not without operational problems. Charles Manigault wrote from Paris to his overseer, James Haynes, in 1847: "I fear too the Rolling Screen was neglected by Stephen as usual. It is surprising how quickly the wires get choked up with grains of rice—when all the broken rice, & black particles have to move forward & go into the whole rice barrels. I found at last that the only plan for keeping the Rolling Screen clean was to have a Boy there *all the time* & with a piece of Band Leather with a wooden handle to it he should be gently slapping the wires while the screen is in motion, so that I found it not all necessary to stop the screen to clean it."[49]

Manigault also mentioned in the same letter that Daniel Heyward fixed a piece of band leather close to his screen, so that when it revolved, each angle of the screen caused the leather to flap against the wire when the rolling screen was in motion, thus cleaning the screen. He went on to say that it was not better than the manual method. Manigault's use of "angle" suggests that his screen was hexagonal in shape.

The Veitch Pestle Mill

The South Carolina Commons House of Assembly appointed a committee to inspect a machine invented by George Veitch. The committee reported on January 7, 1768, that they had seen the machine pound two separate rounds with twelve mortars. During the first round, lasting two hours and twelve minutes, Veitch's machine pounded 580 pounds of rice.[50] During the second round, lasting one hour and fifty minutes, the machine pounded even more rice. The committee reasoned that if six horses were used, the machine could beat out 600 barrels of rice in a season. Six days later, the Commons House of Assembly awarded Veitch a sum of £3,500, provided he would publish a bill of scantling for construction of the machine. On January 30, 1768, the House "Ordered, That the following Bill of Scantling, of a new Machine for pounding Rice, invented by George Veitch, be published in *Mr. Timothy's Gazette,* for the Information of the Public."[51] The bill of scantling was published in the February 1–8, 1768 issue of the *South Carolina Gazette.*[52]

Veitch's bill of scantling included a list of materials to build the barn and the milling machine. The barn was "36 Feet by 70" and was covered with seventeen thousand shingles. Here is his list of materials for the milling machinery with twelve mortars and pestles:

For the Inside Work of the Machine:

	Length in Feet	Inches	Thickness in Inches	
One King-Post	14		19 × 19	
One Axis	10		14 × 14	
One ditto	18		18 × 18	
Three [illegible] for Block Mortars	13		24 × 24	To be cut in shorter Lengths for twelve mortars
Eight Posts	13		8 × 6	
Four Pieces	25		7 × 7	To be of Good stock, for a frame to put under the Mortars
One ditto	12		7 × 7	
Two Pieces Pine	31		9 × 3½	For Shafts to draw by.
Four Spokes	13		10 × 3½	To be of good Pine.
One Pine Plank	25		12 × 4½	
One ditto	25		11 × 3½	
Eight Planks	14		16 × 3½	To be of Ash or good pine, for the Horizontal Wheel.

Three Spokes	7	6	8 × 3½	Of Pine, for the Spur Wheel.
Three Planks	9		12 × 2½	To be of Ash, for the Spur Wheel
Three ditto	9		12 × 4	
180 Pieces white oak	1	1	3½ × 2½	For Cogs, for the Horizontal and Spur Wheels.
Fifty Pieces ditto	2	9	4½ × 3½	For the Arms.
Eighteen Pieces ditto	1	6	5 × 4½	
Twelve Pieces good pine	13	6	6 × 6	For the Pestles.
Twelve Pieces Lightwood	2	2	6 × 6	For the Bottoms of ditto.
Twelve Pieces		3	2½ × 2¾	To be of the Heart of Hickory for Rolls of the two Lantern Wheels

Veitch's terminology in this list of components misleading. The two lantern wheels should have been called "trundle wheels." The term "lantern wheel" (or "lantern pinion") is applied when the wheel is used in a vertical position. When it is used in a horizontal position as in this mill, the term "trundle wheel" is used.

The horizontal wheel and king post listed in the bill of scantling indicate the Veitch mill was animal powered and worked on the same basic plan as the cog mill described by Drayton (see diagram 42). Based on the materials listed in the bill of scantling, however, we can conclude that the Veitch mill was more complex. Veitch added an extra trundle wheel. His bill of scantling does not list the diameters of the spur-wheel components or their numbers of cogs or rounds. It lists only the total number of cogs used. The additional gearing from the second trundle wheel improved the performance of operating twelve pestles at one time.

The Veitch mill operated twelve mortars and pestles, which were located on the second floor of the mill above the animal track (see diagram 45, insert). The bill of scantling called for four pieces of good lightwood twenty-five feet in length and seven by seven inches thick to build a frame under the mortars for structural support of the second floor. The bill also called for one king post fourteen feet long. A king post was used between the animal power and the elevated horizontal wheel. A king post fourteen feet tall would end on the second floor of a mill building.

Although others familiar with mill construction might read the list of materials differently and reach another conclusion as to how the one spur wheel and two trundle wheels were arranged, we have concluded that the spur wheel and one trundle wheel were probably fixed on the same axle and housed in a framework that supported a second trundle wheel fixed to the end of the spoked shaft. Animals walking a circle track on the ground floor rotated a king post and the horizontal cog wheel on the second floor in a counterclockwise direction. The cog wheel turned the first trundle wheel and the attached spur wheel; this spur wheel, in turn, rotated the second trundle wheel. The second trundle wheel turned the shaft, causing its spokes[53] to rotate upward, lifting the pestles out of the

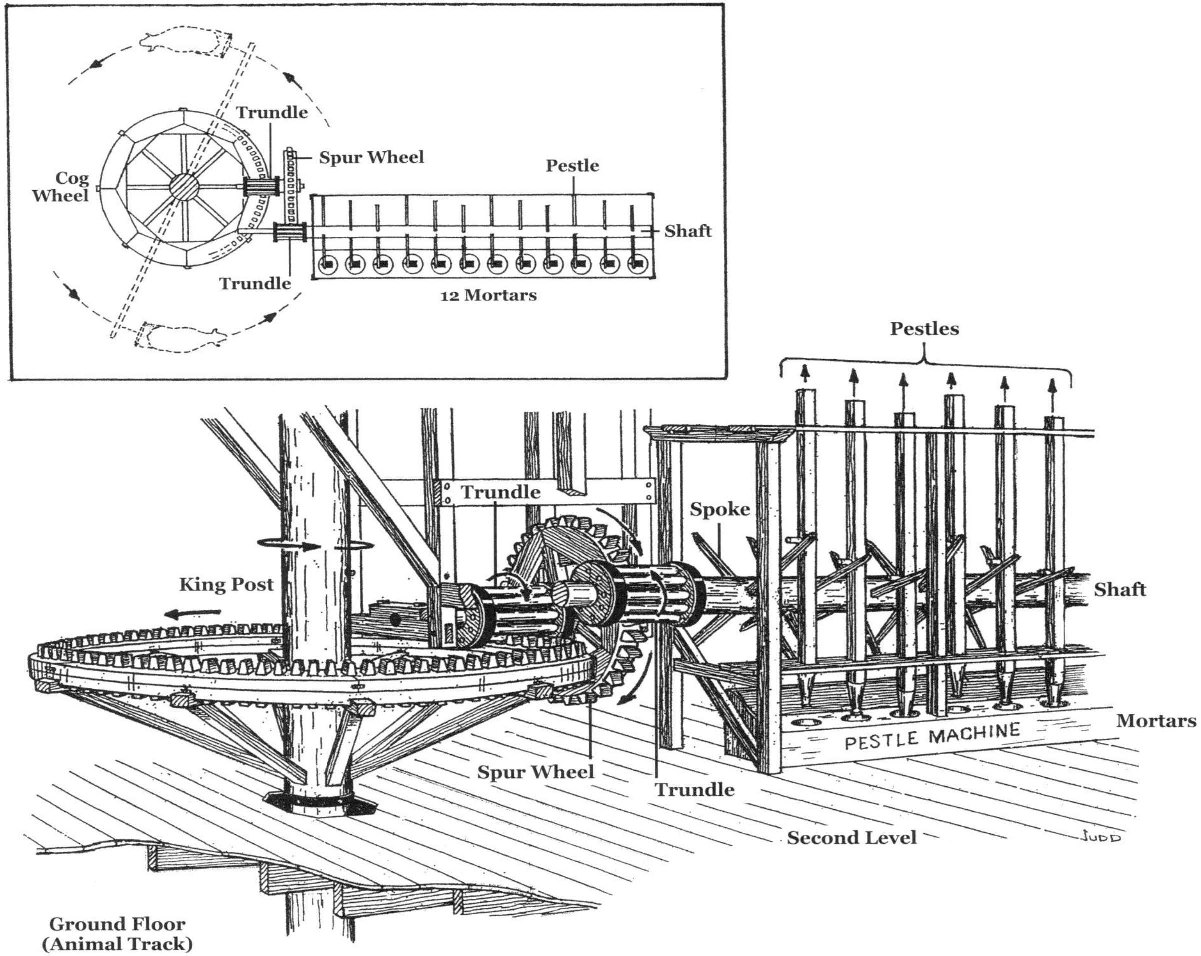

DIAGRAM 45. George Veitch's pestle mill. Illustration by William Robert Judd.

mortars. As the spokes continued to rotate, and as each spoke cleared a pestle's appendage, the pestles fell back into the mortars by gravity, only to be lifted again by the next rotating spoke.

Two methods of lifting pestles with spokes have been documented: an appendage attached to one side of the pestle and slots in the pestles. Which came first is unknown. By the time of the Lucas Water Rice Machine slots were the standard. Slots are used in the cog mill (diagram 42) and appendages in the Veitch mill (diagram 45) to demonstrate both these methods. Nothing in the archival descriptions of these two mills gives a clue to which was actually used.

Whether Veitch's pestle mill (1768) preceded or came after the cog mill described by Drayton is unknown because there is no recorded date for the cog mill. Since it is more complicated, the Veitch mill probably came after the cog mill. The only significant difference is Veitch's addition of the second trundle wheel. Veitch may have based his mill on the cog mill and added the second trundle wheel for more leverage to lift additional pestles.

Veitch's biggest improvement was substituting spokes for cogs. Spokes, being longer than cogs, allowed the pestles to be lifted higher, resulting in a greater force being applied to the grain when the pestles dropped into the mortars.

Although powered by animals, the Veitch mill and cog mill could have easily been adapted to water power. In fact a similar water-powered pestle machine was later incorporated into Jonathan Lucas's Water Rice Machine.

Water-Powered Mills

Water has been used as a source of power to run corn and wheat mills since the dawn of agriculture around ten thousand years ago. Water power in some form was used for centuries in Old World mills long before rice was introduced into Carolina. In 1086, in England alone, there were more than five thousand tidal-powered mills used to grind variety of grains.[54] Colonists who came to the New World brought the technology with them and soon built water-powered mills to process corn and wheat.

The waterwheel made its appearance in the first century B.C.E. in the eastern Mediterranean. Water power cost nothing to use and was not consumed. The waterwheel could develop up to fifty horse power at much less cost that any source available at the time. Water power incorporated three earlier technological innovations in its working mechanism: the vertical waterwheel, gearing, and a horizontal rotary mill. (All documented water-powered rice mills in the Rice Kingdom featured vertical waterwheels.) The earliest known description of a vertical water mill is found in the writings of Vitruvius, a Roman engineer from the first century B.C.E. (see diagram 46).

The waterwheel rotated in a vertical plane and drove a horizontal shaft. A vertical face wheel, its lower section operating within a pit, was attached at the opposite end of the horizontal shaft and engaged a lantern wheel, which changed the rotation from vertical

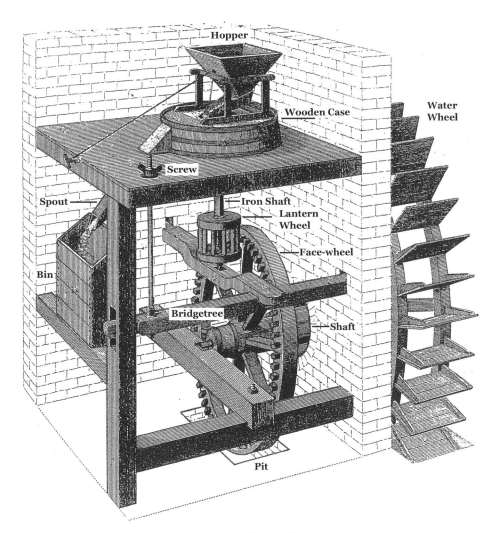

DIAGRAM 46. A Vitruvian water mill, circa first century B.C.E. Reproduced from Charles Howell and Allan Keller, *The Mill at Philipsburg Manor* (1977).

to horizontal. The lantern wheel was attached to the lower end of an iron shaft that rose up through a hole in the bed stone and was fastened to the runner stone. Both millstones were housed in a wooden case. The shaft rotated the runner stone. The lower end of the shaft sat on a bridge tree, which could be raised or lowered by a screw to adjust the distance between the runner and bed stone. The grain was processed between the two stones.

Above the case was a hopper that fed grain into the eye of the upper millstone. As the milled grain fell from the periphery of the millstones, it was swept around inside the wood case by the runner millstone to a spout that directed the grain to a holding bin.

Throughout history this first-century milling technology changed very little until the late nineteenth century.

When water power was first applied to a rice mill in the Rice Kingdom is uncertain. The earliest date may have been around 1701. Daniel Axtell and his partners employed water power to run a sawmill on Newington, a large plantation owned by a relative, Lady Rebecca Axtell. Daniel Axtell, in partnership with Lady Rebecca and Robert Fenwick, constructed a dam on the upper reaches of Dorchester Creek, which had its headwaters in a swamp north of Newington, near present-day Summerville in Dorchester County, South Carolina. Dorchester Creek flowed through Newington and into the Ashley River. The dam backed up water, creating a reservoir; water from the reservoir turned the waterwheel of his sawmill. No description of Axtell's mill has survived, and except for the dam, no traces of it exist today. Axtell was a New Englander, and his sawmill was imported from New England, where water-powered mills were common by the end of the seventeenth century.[55]

Rice was grown on Newington Plantation. Sales of upland rice at Newington were 6,000 pounds in 1702, 12,243 in 1703, 11,729 in 1704, 1,203 in 1705, and 8,791 in 1707.[56] These appear to be large crops for the time and may have been larger than a small slave force could process with mortars and pestles. It may be that Axtell used water power to run a mortar and pestle mill, reducing the number of slaves necessary to process Newington's crops. If so, it would be the first recorded water-powered rice mill. In any case, the documented existence of Axtell's mill shows that water power was being used run a mill at least by 1701.

The earliest extant record a water-powered rice mill in the Rice Kingdom appears in the journal of William Stephens of Chatham County, Georgia, on January 13, 1744, where he stated that "besides making a Mill of his own contriving, on a little Brook that runs thro his land, which turns a wheel, that with several coggs [cogs] upon it lifts up, and lets fall a number of heavy Pestles, where with he pounds Rice and fits it for Sale."[57]

This mill predates all animal-powered rice mills documented in the archival sources except the pecker mill. It may well be that the operation of the pestles of the animal-powered cog mill described by Drayton and in the Veitch mill were based on those in the mill Stephens described.

In 1745 the Salzburgers of Ebenezer, Georgia, constructed a water mill for pounding rice and barley (see figure 21). Johann Martin Boltzius, who reported on the daily activities of the Salzburgers, recorded: "The carpenter Kogler sent me word that the rice and barley stamp would function today for the first time. . . . In the afternoon I went out there, and steps were taken to begin stamping. In the long and thick beam there are seven holes in the form of an upright egg, as the holes usually are in stamps in Carolina. Every hole, in which a pestle falls contains more than a bushel of rice or barley, which can be stamped white in something more than an hour and prepared for use and sale. All seven pestles go as regularly one after the other as if seven men were threshing."[58]

FIGURE 21. The Salzburger water mills in Ebenezer, Georgia, 1747. Courtesy of Hargrett Rare Book and Manuscript Library/University of Georgia Libraries; hmap 1747s4.

Boltzius further stated that it was "an incomparably useful machine that is driven by a single waterwheel that stands next to the mill wheel." Boltzius did not explain how the pestles were operated in the Salzburger mill, but they may well have been operated by cogs, like those in the mill Stephens described. The Salzburger rice mill apparently shared its building and power source with a barley mill.

It is interesting to note that Boltzius said that "the holes usually are [like those] in the rice stamps in Carolina." This statement suggests the Salzburgers, and possibly Stephens, obtained the technology of pestles driven by a waterwheel from Carolina. Where in Carolina is unknown, but it suggests that somewhere in South Carolina prior to 1745, water power turning a waterwheel was used to drive a rice mill with pestles. Perhaps it was from Axtell's water-powered mill at Newington.

The possibility is also raised that in 1768 Veitch may have based the mortar-and-pestle design for his animal-powered mill on an existing Carolina water-powered pestle mill. This easily could have been accomplished by modifying the gearing to accept animal power.

When William Bartram visited Charleston in 1773, water-powered mills were already advanced. Bartram stated that "machines for cleaning the rice are worked by the force of water. They stand on the great reservoir which contains the waters that flood the rice-fields below."[59]

None of the water mills that graced rivers, streams, and reservoirs in the Rice Kingdom is left today; only partial ruins of a few millraces remain of the water mills that were an important step in the market preparation of Carolina rice.

Sources of Water Power

To drive rice mills planters took advantage of many sources of water in the Lowcountry. Unlike the New England area, where falling streams to power mills were abundant, Lowcountry planters had to rely more on trapping water in ponds or reservoirs to power mills, limiting the time the mills could run.

The simplest and possibly the earliest source of water to power mills was a *flowing stream or canal*. The mill building was situated on the bank of a stream or canal, and a waterwheel was located in a wooden or brick millrace. The waterwheel was connected to the mill's machinery by a shaft. Some method was employed to prevent the flow of water to the waterwheel (such as a bypass canal) to stop its rotation when the mill was not in operation.

An example of a water-powered mill on a stream was the one on Wantoot Plantation in St. John's Parish, Berkeley County. John Christian Senf's map of the Santee Canal shows a "Rice Machine" on a small stream (see plat 5). The map is dated 1787–1800, and the mill may have been in operation during that period, but it was constructed before 1787. Clearly there is no reservoir associated with the mill. Wantoot Plantation was inundated with the construction of Lake Moultrie in 1942, so there is no way to examine the mill site.

Rain-fed reservoirs were a more reliable source of water power for rice mills than a flowing stream. Reservoirs created by banking the upper area of an inland swamp to supply water for rice fields were common by the mid-1700s. It was a short jump to realize that these reservoirs could also drive a waterwheel, especially since the water was not needed to flood the fields after the crop was harvested. The 1797 map of Limerick Plantation in Berkeley County (see plat 6) shows the location of a water-powered mill (labeled "machine" on the map) below an earthen dam. Rainwater filled the reservoir and was held there by the dam through which a trunk box with a gate was installed. At milling time the gate was opened, and the water flowed through the trunk box into a millrace constructed of bricks; the water turned an undershot waterwheel, which drove the mill machinery. A gate in the bank acted as a bypass, allowing excess water in the reservoir to be removed and not pass through the millrace. Since there was no way to prevent a waterwheel from turning any time water was passing through the millrace, the bypass reduced excess strain on the wheel. As long as there was water in the reservoir, the mill could operate.

Tidal freshwater rice fields were a common source of water to drive mills. In 1791 Jonathan Lucas I built the first tidal rice mill in the Rice Kingdom. His mill used an adjacent rice field for water power. The field was flooded by the rising tide, flowing through the tide trunks used to irrigate and drain the fields. The field was never flooded through the millrace because this would have rotated the waterwheel backward, causing damage to the mill's machinery. When the tide ebbed, the gates automatically closed, trapping

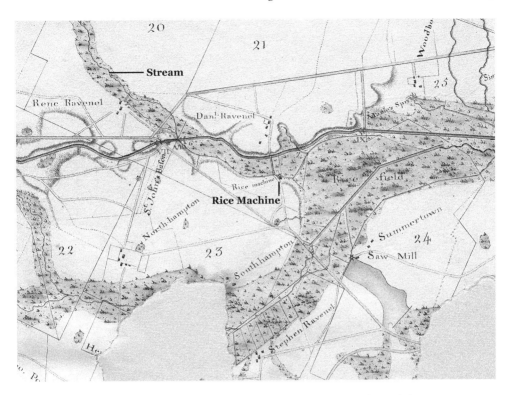

PLAT 5. Location of a water-powered rice mill in St. John's Parish, Berkeley County, South Carolina, 1800. From "General Plan of the Canal and its Environs between Santee and Cooper Rivers in the State of South Carolina. Commenced in the Year 1793 and finished in the Year 1800, by Christian Senf Colonel Engineer and Director in Chief of the Canal." Courtesy of the South Carolina Historical Society, Charleston.

---

the water in the field and creating a reservoir with about a four-to-five foot water head between the field and adjacent river at low tide. The gate at the head of the millrace was opened, and the flow of water from the field through the millrace turned an undershot waterwheel. The mill operated only when the tide was low and water could be flowed from the field. Because the mill's operation depended on the movement of the tides, operating hours differed every day with the daily variations in the times of high and low tides.

A description of this water source was given by Jacob Motte Alston, who planted Woodbourne on the Waccamaw River: "The mill was erected near the river, and the fields in the rear were flooded at high water. Then the floodgates were closed, and when the tide fell in the river, the water held back in the fields was some four or five feet higher than that in the river. This then was the motive power which set the machinery of the mill's waterwheel in motion; the huge stones to rotate and the heavy pestles to pound."[60]

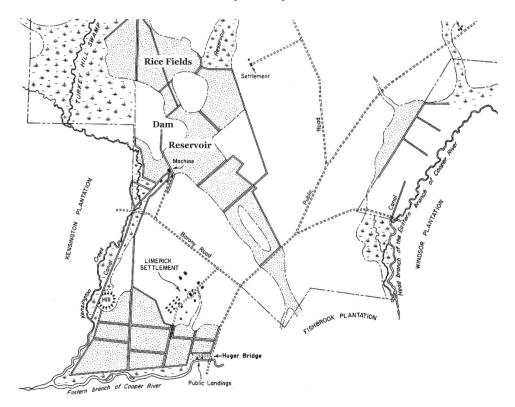

PLAT 6. "A Plan of Limerick, a Plantation Belonging to Elias Ball, Esq." Prepared by John Hardwick in 1797. Reproduced by permission of the South Carolina Department of Archives and History, Columbia.

Another source of power was a *saltwater tide*. Tibwin Plantation in Charleston County had a tide mill powered by a saltwater reservoir or millpond (see diagram 47). A tide trunk allowed salt water from Tibwin Creek to fill a pond behind the dam. The gate automatically closed as the tide ebbed, and the dam held the water back, creating a reservoir. At low tide a water head was created at the mill. The millrace gate was opened, and the flowing water turned an undershot waterwheel that drove the machinery in the mill. This system was identical to the those in freshwater rice fields except the saltwater millpond was never planted in rice.

Jonathan Lucas I built a mill on Shem Creek in present-day Mount Pleasant that used salt water from a millpond he constructed. His mill was a combination rice mill and sawmill, and its brick remains can be seen today at low tide.

Charles Drayton referred to a mill run by saltwater in a January 17, 1809, diary entry: "At Col. Cuthbert. his water-mills for preparing Rice & cotton are good—nothing extraordinary. they are impelled by Saltwater. He banked in a part of the salt water rush land. the 3d. Spring it was planted."[61]

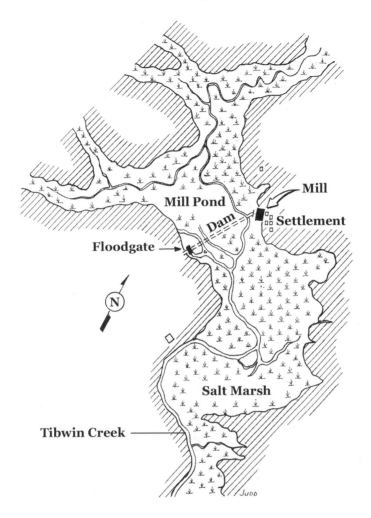

DIAGRAM 47. A rice mill powered by saltwater tides.
Illustration by William Robert Judd.

Another possibility was a *pump-fed reservoir*. In 1786 Henry Laurens of Mepkin Plantation contemplated an unusual method of providing water for a reservoir. While he was in Lymington, a port town in Hampshire, England, in 1774, he "saw several Pumps worked by the Wind_ the Wings fixed to the Pump were of Canvas about four feet long, Salt Water was by this means raised eight or ten feet high, & conducted through Pipes or Troughs to the salt works." Laurens added, "I have no thoughts of making Salt, my intention is to raise Water from an inexhaustible Source, over a rising ground of 16 or 18 feet high, to convey it into a Reservoir for turning a Pounding Machine for cleaning rice." Laurens believed it would relieve his servants from "that Branch of hard Labor." Laurens requested of certain gentlemen that, if they could assure him that a pump like the one he saw at the salt works would raise water as he envisioned, they send all the necessary articles to Mepkin, where he would construct the pump and windmill.[62] Laurens never recorded in his journal whether he ever constructed the pump and windmill; nonetheless it was an ingenious idea.

Construction and Operation of a Water-Powered Rice Mill

The first water-powered rice mills operated only mortars and pestles, which removed both hulls and bran. These mills are often referred to as "pestle rice mills." In 1787 Jonathan Lucas I added millstones to his Water Rice Machine to removal the hulls while retaining the mortars and pestles to remove the bran. The water-powered mechanism of both the pestle mill and the Lucas mill, however, operated alike and the description that follows applies to both.

All water-powered mills in the Rice Kingdom were set in motion by the undershot method because of the lay of the land. No waterfalls existed in the coastal area that could drive an overshot waterwheel, and reservoirs with a water-head high enough to drive an overshot waterwheel were impractical to build because of the generally flat land of the Lowcountry.

No rice mills with intact waterwheels survive, and only a few scattered remains of iron parts of waterwheels have been found. Several partial brick millraces survive, and although the lower sections are intact, the upper sections have deteriorated or were dismantled for their bricks, making adequate height measurements unreliable. The photograph of the mill at Middleburg Plantation (see figure 22) shows an intact millrace. The photograph was taken around 1911.

Housed inside the Middleburg building were the millstones, cleaning machines, and mortars. The entire building was made of wood except for the brick foundation. The abandoned millrace is located to the left of the building. The width of the millrace is 16½ feet. The inner millrace wall, which was the wall of the mill, is shown in its entirety. The gate that controlled the flow of water through the millrace is also visible. The outer wall is not visible; it was built approximately twelve feet high to accommodate a waterwheel twenty-two feet in diameter.

Only a partial diagram of a water-powered pestle mill has survived in archival sources (see diagram 48). Jonathan Lucas I diagramed a mill building for Elias Ball (1752–1810) of the Cooper River, depicting nine mortars in a line, offset slightly to the left of an opening in the wall labeled "Window for Mill Shaft."[63] The line of mortars represents a pestle machine placed directly on the ground within the building. The pestle machine was positioned to insure that the waterwheel and drive wheel rotated the spokes in an upward direction, lifting the pestles. The pestle machine worked the same for water power as it did for animal power.

Plantation journals and other archival sources record aspects of construction and dimensions of waterwheels, millraces, and rice mills in general. These sources and field surveys allowed the authors to diagram a reconstruction of a typical water-powered pestle

FIGURE 22. Two views of the rice mill at Middleburg Plantation on the Eastern Branch of the Cooper River in Berkeley County, South Carolina. From the Middleburg photograph album, courtesy of John Ernest Gibbs. Mr. Gibbs placed a date of around 1911 on the main photograph and a date of around 1900 on the insert.

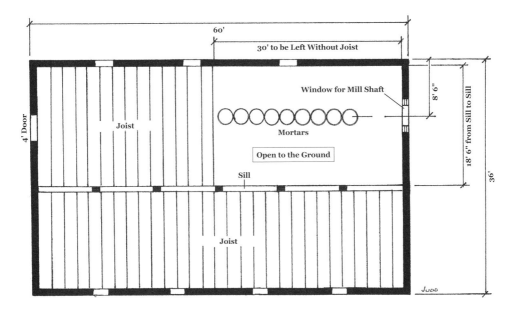

DIAGRAM 48. Floor plan of a water-powered mortar-and-pestle rice mill, 1793. Redrawn by William Robert Judd from "A Bill of Scantling For Elias Ball Esq.," Jonathan Lucas I to Elias Ball, Ball Family Papers, with permission of the South Caroliniana Library.

mill. The typical rice-mill building was a two- or three-story structure approximately thirty feet wide by forty feet long with a gable roof (see diagram 49). Its lower level was made of brick; the upper level(s) and loft were built of wood sided with clapboard or wood shingles. The building was situated on the bank of a canal or sometimes partly built over the canal. In many cases the bank was excavated to a level where the axle shaft of the waterwheel was approximately in line with the building's first floor. The level of the first floor, the millrace floor and the height of the outer millrace wall were determined by the radius of the waterwheel. The building's wall fronting the canal served as the inner millrace wall and sat directly upon the millrace flooring. The outer millrace wall was constructed of brick directly upon the millrace floor. Its outer location was determined by the width of the waterwheel. Early walls were approximately twelve feet apart.[64] Later walls were constructed wider apart to accept improved waterwheels that measured twenty-two feet in diameter by fourteen feet wide.[65]

The foundation and flooring of the millrace were built in the same manner as the floor of the floodgate, only longer. The floor of the millrace was constructed with great care. If not done properly, the entire millrace might have washed away. At Rose Hill Plantation on the Combahee River the remains of the millrace floor and its foundations are intact and can be seen and walked upon at low tide (see color plate 15).

Each end of the masonry millrace walls was extended with a horizontally planked retaining wall secured to vertical posts. These walls helped to prevent erosion of the embankments, which could cause silting within the millrace. The millrace walls served to direct the flow of water to the waterwheel. The outer brick wall supported and seated the outer end of the waterwheel's axle shaft, which rode in a greased stone bearing set into the wall's brickwork. The mill end of the axle shaft extended through an opening in the building's wall and rode in a similar bearing within the building. Fixed to the end of the wheel's shaft was a large driving wheel, which operated the mill's machinery. The driving wheel was sometimes referred to as a "pit wheel," so named because its lower half rotated within a brick-enclosed pit located just inside the building's wall. The remains of an inner wall and pit are seen in the ruins of the water mill at Rose Hill. The millrace gate, which held back a head of water, was located at the rear of the waterwheel. When time came to operate the mill, the gate was raised, and the water flowed through the millrace, striking the lower floats of the waterwheel and beginning its rotation.

The millrace gate underwent many changes in designs and methods of operation. The first gate design operated in a vertical position. Adjacent to the gate was a platform spanning the millrace, which allowed the operator to raise or lower the gate using a lever bar similar to the operation of the doors on a tide trunk. The gate was held at a desired height by dowels or pins. At some point builders found that "canting" the gate so that its bottom end was closer to the waterwheel increased the velocity of water striking the floats of the wheel, thereby increasing the wheel's revolutions per minute. The canted gate soon evolved into a curved gate, which "shot" the flow of water directly against the wheel's

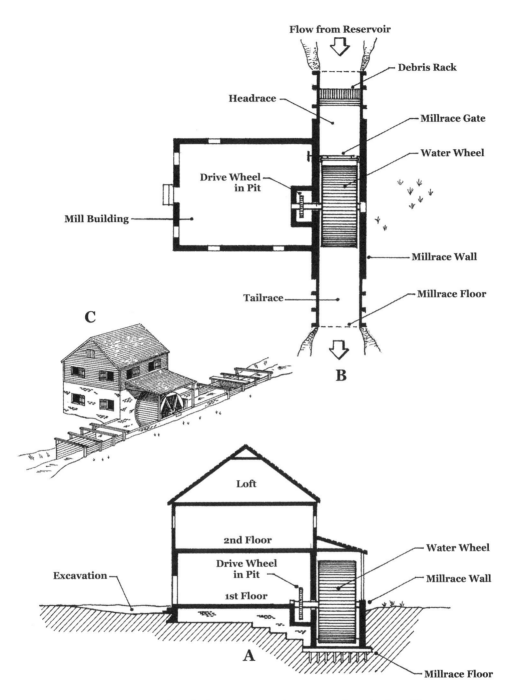

DIAGRAM 49. A typical water-powered mortar-and-pestle rice mill.
Illustration by William Robert Judd.

FIGURE 23. Stoke Rice Mill on Western Branch of the Cooper River, Berkeley County, South Carolina, early 1900s. Reproduced by permission of South Carolina Heritage Trust Program, Department of Natural Resources, Columbia. The photograph at upper left shows the mill as it appeared in the 1980s; the photograph at upper right, also from the 1980s, shows the brick pillars under the floor, where the mortars were located. Photographs by Richard Dwight Porcher, Jr.

floats. One such design was used at the Stoke Rice Mill on the Cooper River, as suggested by the curved slots in both walls of the millrace (see the insert in color plate 16).

Stoke Rice Mill (see figure 23) was probably typical of plantation water-powered mills of the late 1700s and early 1800s. The mill was situated next to the millrace. The lower walls were brick, and the upper floor and attic were enclosed by wood. The building measured forty feet by thirty feet with the short side next to the millrace. Nothing remains today except the brick shell that enclosed the lower floor (figure 23, insert left), the brick pillars that were placed below the floor under the mortars to support the floor from the pestles pounding into the mortars (insert right), and the millrace.

The lever method to raise and lower the gate was replaced by a windlass, which could be operated from within the mill building by rotating a spoked wheel clockwise or counterclockwise. As rice mills became more complex, cast-iron components were introduced into many phases of mill machinery. The old wooden windlass was replaced with cast-iron rack-and-pinion components, which operated more smoothly with less effort.

At the head of a millrace, spanning its width, was a wooden debris rack made of vertical boards spaced close together (see diagram 49B). Water flowed through, but debris such as floating limbs or branches that could damage the waterwheel, was trapped and then removed by workers.

Some planters protected their investment by building their mill over a canal with waterwheel in the center protected from the weather by the building's roof.[66] Other planters built a shed-type roof projecting from the main building. Unfortunately no diagrams or detailed descriptions of these mills exist.

A common size for a waterwheel was twenty-two feet in diameter and fourteen feet wide, with a massive framework of multiple sets of six to eight arms radiating outward from a thirty-two-inch-diameter axle shaft with outer ends secured to a wooden rim. Each rim was fastened with horizontal beams. Bucket boards, also called "floats," twelve to fifteen inches wide running the full length of the waterwheel, were installed around the circumference of the wheel. Water hitting the bucket boards rotated the waterwheel.

The waterwheel underwent a tremendous strain rotating the gearing within the building. This was especially true as milling evolved in complexity and additional gearing was installed to operate millstones and other machinery. To strengthen the wheel, cast-iron hubs were placed over the axle shaft at each end (see color plate 17). The flanges of each hub were bolted to the outer arms. Breakage still occurred because of the strain on the waterwheel. When a hub broke on Charles Manigault's waterwheel, James Haynes wrote to him:

> One of the flanges on the large waterwheel gave way. I sent immediately to Mr. Lacklison to have it repaired but from his men being sick and other casualties has not finished it. He proposed to cast a new as being the most expeditious way of replacing it as there would have been greater loss of time in taking down the whole structure saying nothing about expence both waterwheels would have to be taken

down as well as the large laboring wheel on the inside so as to uncouple the shafts in order to slip on the old flange but instead of doing all this he proposed to cast the new flange in two separate parts & put them together [with] lugs and bolts.[67]

Haynes's reference to "both waterwheels" is notable. With apparently few exceptions, water mills in the Rice Kingdom were powered by a single waterwheel. Manigault's mill, however, was driven by two side-by-side wheels, their axle shafts coupled together, turning the mill's driving wheel. Since Manigault's waterwheels drove both a thresher and mill, one wheel might not have had enough power to drive both systems.

Since the average waterwheel in the Rice Kingdom was fourteen feet wide, it would not have been feasible to construct a single waterwheel twenty-eight feet wide because its center weight might have caused it to sag and not run true or possibly break in the middle. But it was feasible to use two side-by-side waterwheels with their axle shafts coupled together. Near the center of the coupled shafts would have been a bearing mounted on a brick pier supporting the center of the combined axles' length. Two waterwheels would also require two gates, both being raised for maximum power.

Gowrie Mill on Argyle Island on the Savannah River was one of the more elaborate water-powered plantation mills. On March 22, 1867, Charles Manigault commented on the mill when describing the destruction of Gowrie during the Civil War: "The great loss in the settlement however was the huge Rice Pounding Mill burnt by the Yankees and from which I am told the flames extended to the other Buildings, but how true this we shall forever be at a loss to know. The Rice Mill was well known on Savannah River. It was built about the year 1809 by Mr. McLearn a Scotchman who also had constructed Mr. James Potter's Rice Mill, this latter however being not quite as large as the Gowrie Mill. Both of these large Buildings were upon solid brick foundations the rest of the Building being of wood. The motive power was water, and the peculiarity of these Mills was that the wheel was protected from the weather and in the center of the building. The frame work of the Gowrie Mill was cypress."[68]

Manigault also stated that the mill constructed by McLearn, who built mills in the Savannah area, for a Mr. Potter also contained the waterwheel inside. Evidently this was McLearn's innovation. Nothing remains of these two mills other than the brick foundations, and no photographs of them have survived.

A waterwheel acted on the same principle as a flywheel. As long as a force was applied to the wheel, setting it in motion, centrifugal force would contribute to its rotation. Power could not be increased by enlarging the diameter of an undershot waterwheel; more power could be obtained only by adding an additional waterwheel. Unlike the overshot wheels, which acted on gravity and the weight of the water collected in each bucket, the undershot wheel acted on percussion. Once a volume of water struck the waterwheel's floats, its power was spent. Power could be increased only by increasing the area (width of the waterwheel) struck by the water's momentum.

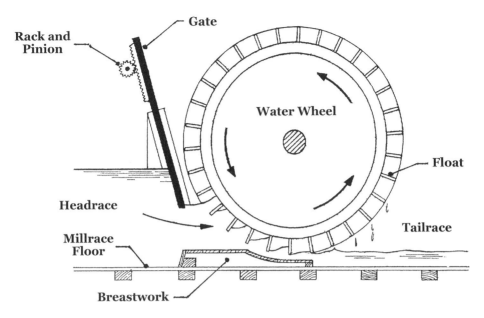

DIAGRAM 50. A waterwheel and breastwork. Illustration by William Robert Judd.

To reduce the drag created by the spent water surrounding the lower arc of the waterwheel, a low breastwork curving under the rear of the waterwheel was constructed across the millrace floor (see diagram 50). The lower framework of the millrace gate was attached to this breastwork. The height and curvature of the breastwork allowed the water, after striking the floats, to fall away quickly into the flowing tailrace, thereby reducing drag of the rotating waterwheel. Figure 24 shows the remains of a breastwork on the Combahee River.

FIGURE 24. Ruins of a breastwork at Nieuport Plantation. The breastwork has undergone significant deterioration since the photograph was taken in 1977. Photograph by William Robert Judd.

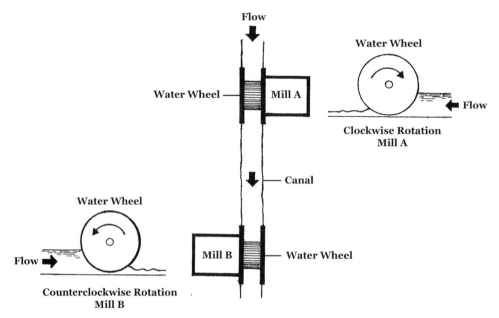

DIAGRAM 51. Positioning of machinery within a mill.
Illustration by William Robert Judd.

In a water-powered mill or an animal-powered mill correct gearing was essential to achieve the proper rotation of the mill's machinery. Almost all milling machinery was built to operate in one direction, and any reversed rotation could damage their components.

Positioning machinery within a mill building powered by water was dependent on the location of the building in relation to the flow of water through the millrace. In diagram 51, the rotation of the waterwheel is viewed from inside the mill. The flow of water through the millrace rotated the waterwheel of mill A in a clockwise direction. However, the waterwheel of mill B, on the opposite side of the bank, driven by the same flow of water, rotated in a counter-clockwise direction. Within each mill, fixed to the end of the waterwheel's shaft was a vertical driving spur wheel which set the mill's machinery in motion. The millwright had to position the machinery in the proper position to account for the different rotation of the driving wheel. In mill B this may have involved relocating a machine's drive wheel to its opposite end of its axle or adding an idle wheel between the mill's main drive wheel and the machine's drive wheel to maintain the machine's correct rotation.

At some point millwrights developed gearing to reverse the driving mechanism attached to the waterwheel so the mill machinery operated when the wheel turned in either direction. Regardless of the wheel's direction of rotation, the additional gearing allowed the mill machinery always to rotate in the same direction. This type of mill had a gate on

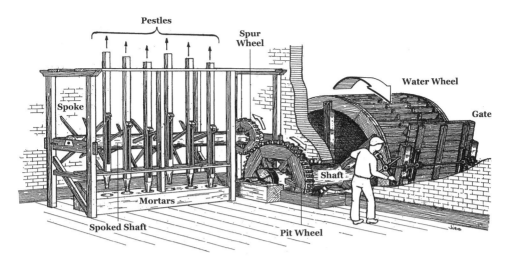

DIAGRAM 52. A water-powered mortar-and-pestle rice mill.
Illustration by William Robert Judd.

---

each side of the waterwheel. When one gate was in operation, the other gate was raised so as not to obstruct the flow of water. When the tidal fields emptied on the ebbing tide, the outer gate was closed. When the rising river water had almost reached its flood peak, the outer gate was opened, and water flowed through the millrace into the tidal fields, setting the wheel in motion. When the field was filled at flood tide, the inner gate was closed, trapping water in the field. When the tide reversed and a water head was created, the inner gate was opened, water flowed through the millrace and turned the waterwheel, but in a reverse direction. The mill could be operated for only a limited time—about five hours during each tide—but the ability for the wheel to rotate in either direction still doubled the time a mill could operate compared to those without the reverse mechanism in place.

Diagram 52 illustrates how a water-powered mill operated. Rice was placed into the mortars below each pestle. An attendant set the mill in motion by rotating a hand crank which raised the millrace gate, releasing the impounded water to begin the rotation of the waterwheel and its shaft. Fixed to the waterwheel's shaft within the building was a drive wheel or pit wheel. The pit wheel meshed with the spur wheel fixed to the end of the spoked shaft. As the spoked shaft revolved, each spoke rotated through slot in the pestles, lifting each pestle upward in sequence. As each rotating spoke slipped out of the slots in the pestles, the pestles fell by gravity into the mortars below. Each rotation of the spokes lifted and dropped the pestles in an alternating sequence into the mortars until the rice was cleaned. To stop the mill's operation, the attendant lowered the millrace gate, stopping the flow of water rotating the wheel.

In spite of the problems associated with maintaining the waterwheels and the mill's machinery, water proved an excellent and reliable source of power for rice mills for more than one hundred years. Water-powered mills became a mainstay in milling throughout the rice industry. Savannah and Charleston newspapers ran many sale advertisements for plantations with water-driven mills. In 1763 the *Georgia Gazette* carried an advertisement for a plantation on Hutchinson Island: "A tract, containing 1000 acres, known by the name of Vale Royal, of which 220 acre is exceeding good rice land; also 281 acres rice land opposite to it on Hutchinson's island, a large new barn, rice pounding machine by water, a good overseer's house, with several other improvements, a mile and a half from Savannah, as well situated for pleasure and profit as any tract in Georgia."[69]

In 1787 Childermas Croft advertised in the *State Gazette of South Carolina:* "70 acres of tide land are now in good order for planting, with a large dam just made, which will keep a sufficient head of back water to make a most excellent pounding machine or sawmill as any in these parts."[70] In 1799 Theodore Gaillard advertised for a "person to take charge of a Rice-Pounding Machine, which works by the tide; he must be sober, diligent and industrious; if he has some knowledge of the Carpenter's Trade, the better."[71] Advertisements such as these were common.

Jonathan Lucas I's Water Rice Machine

In 1787, millwright Jonathan Lucas I (1754–1821) (see diagram 53, insert) constructed a Water Rice Machine for a Mr. J. Bowman that revolutionized the market preparation of Carolina rice. Lucas constructed the mill on a reserve at Bowman's Peach Island Plantation on the South Santee River in Charleston County. Bowman's mill was the first of Lucas design incorporating millstones and was the basis for all mills to follow throughout the Rice Kingdom until after the Civil War.[72]

The son of a wealthy mill owner from near Whitehaven in England, Jonathan Lucas I was born in 1754 in Beckermet Parish, Cumberland County, England. Lucas came from a long line of millwrights and flour millers in England and married into another well-established family of millers. Lucas was knowledgeable about milling and building mills. He knew the correct stone to use for a particular grain and how to dress and maintain the stones, but he probably had no experience milling rice before coming to America in the 1780s, although he was familiar with the Peak stone used for milling barley and oats which he later used in his Water Rice Machine.

He settled in the Santee River area, where he realized that methods for milling were too antiquated to keep up with the steady rise in production of rice for export. Although plantation owners and local mechanics had been trying continually since the beginning of rice culture to improve the milling, only limited progress had been made by the time of the Revolution. Lucas involved himself in the problem. The Water Rice Machine was the result of his observations and knowledge of milling.[73]

Lucas was not the first to invent a water-powered rice mill. Water-powered mills were used in the Rice Kingdom before Lucas arrived. In fact, the complicated gearing of the

DIAGRAM 53. Jonathan Lucas I's Water Rice Machine, 1787. The original lettering has been replaced for clarity. Reproduced from John Drayton, *A View of South-Carolina* (1802). *Insert:* Jonathan Lucas I. Reproduced from William Dollard Lucas, *A Lucas Memorandum* (n.d.).

Lucas mill was documented in the Vitruvian water mill (diagram 46) long before rice cultivation began in Carolina. Even the mills designed by Oliver Evans were based in part on ancient mills such as the Vitruvian water mill.

The Vitruvian water mill operated with one-step gearing. The face wheel on the waterwheel shaft engaged a lantern wheel to rotate the grinding stone, which reduced the grain to meal or flour. Hulling and cleaning rice involved multiple processes. Lucas's water-powered rice mill required the face wheel on the waterwheel shaft to be replaced with a large spur wheel from which multiple gearing could receive power to operate machinery.

No diagrams of rice mills drawn by Lucas or his son Jonathan Lucas, Jr., have survived. Confederates burned all the Lucas buildings and Greenwich Mill on Shem Creek north of Charleston in 1865 to prevent supplies held in them from falling into the hands of the advancing Union army. Most of the Lucas library and milling records were probably destroyed in the fire. John Drayton, however, included an engraving of a Lucas mill in *A View of South-Carolina* (1802). Drayton sketched the Lucas rice mill located on the plantation of Philip Tidyman at North Santee. In a notation written on his sketch, Drayton stated: "This was drawn by John Drayton from the Rice Mill of Doctor Philip Tidyman at North Santee—The proportions however were not sufficiently correct according to the scale of drawing, and it was accordingly altered by Mr. Latrobe, at Philadelphia."

The engraving in Drayton's book (diagram 53) was redrawn from his sketch by the British-born architect Benjamin Henry Boneval Latrobe (1764–1820), called the "Father of American Architecture,"[74] who came to the United States in 1796.

Lucas's contributions to rice milling were monumental and ingenious. He incorporated previous mechanical advancements into his mill, in some cases modifying them, and added some of his own inventions. His mills set a course of technical innovation that led to extensive mechanization of milling in the 1800s. Lucas was the first millwright to use millstones to remove rice hulls. He preferred millstones from his English homeland. Lucas's millstones had the same purpose as the wooden rotary quern: removal of the hulls. Milling methods prior to 1787 did not clean the rice very well, and resulted in considerable breakage. Millstones cleaned rice better, and breakage was reduced. After the millstones stripped the hulls from the rice, the bran and germ were removed by mortar and pestle; then brushes removed the polish.

Lucas later automated his mills. He added steam power in 1817, and his mills became the standard of the industry. During 1791–93 Lucas built mills of his design on fifteen rice plantations in the Carolina rice-growing area. Undoubtedly, his milling experience in England contributed to the success of his first rice mill and subsequent mills.

Once Lucas developed his Water Rice Mill, other millwrights built mills based on his design. Lucas's mills outperformed these mills, and Lucas was often called on to repair and modify mills built by others. Plowden Weston wrote to Lucas on February 20, 1793: "I am really uneasy for fear I may have a bad mill for want of your presence occasionally to give instructions for I am persuaded Mr. Marshall is not equal to the task. I have lost a deal of time already by Mr. Marshall's mistakes."[75]

Diagram 53 illustrates how the Lucas mill operated. (The lettering and the legend at the bottom of the engraving have been enhanced.) The mill was situated on a millrace through which water flowed from a nearby millpond. To operate the mill a windlass (A) was turned to open the gate holding the water in the millpond. As the water flowed through the millrace, it turned a waterwheel (hidden behind the far side of the mill wall). The main driving cog wheel (C) was fixed on the waterwheel's shaft.

The main driving cog wheel was recessed in a pit and turned a large wheel (D) by engaging a small wheel (Y) on the same axle as the large wheel. The large wheel (D) engaged a small lantern wheel (E). Connected to the lantern wheel was an iron shaft that passed through a hole in the bed stone above (hidden behind the horizontal beam), which was fixed onto the underside of the runner stone (F). The lantern wheel turned the runner stone.

Rice passed through a funnel (H) from the loft of the mill into a hopper (G). The bed stone and runner stone lay below the hopper. The grain passed from the hopper through a hole in the center of the runner stone, down and around the iron shaft and drive plate (which set the stone in motion) into four channels carved on the underside of the runner stone (see diagram 54). These channels directed grain between the stones. Centrifugal force from the turning of the runner stone forced the rice toward the stones' edge. The distance between the stones was set a little less that the length of the unhulled rice. As the rice turned end over end in a vertical position, the hulls were stripped by the pressure on the ends of the rice. Lucas used Peak millstones from England because they were pock-marked and somewhat abrasive (color plate 18-A). The rough surface grasped the sharp ends of the rice, causing them to assume a vertical position. Hulled rice and the hulls exited the millstones all around the outer edge, falling down a chute attached to one side under the millstone cover. The chute sent the rice and hulls to the wind fan.

The distance between the stones was set by a screw (K) that played into the end of the moveable beam (U) upon which rested another moveable beam (V). As the screw turned, it raised (or lowered) the moveable beam (U), which in turn raised movable beam V. The small lantern wheel (E) was anchored into the moveable beam. Since the trundles of the lantern wheel were vertical, it could move up or down and still be engaged by the large wheel D. The lantern shaft was anchored into the upper millstone, and as it was raised, it lifted the stone (or lowered the stone as the screw was turned the other direction).

Being able to set the distance between the two stones was important for two reasons. First, the stones wore thinner with use. Consequently, the distance between the stones could be adjusted to compensate for the wearing. Second, all rice grains were not the same length, and the distance between the two stones could be adjusted depending on their length.

Lucas installed a wind fan (based on Kogar's wind fan?) to blow away the hulls. The fan was turned by a band (W) from the main wheel (D). A strap (L) worked in an eccentric motion vibrated a riddle (not seen) inside the wooden frame that housed the wind fan. The rice and hulls were discharged from the stones into the wind fan hopper (I) which fed the rice and hulls onto the riddle. The rice and hulls passed through the riddle into the blast of the fan. The lighter hulls were blown through a door (M) to the outside of the mill barn, while hulled rice was discharged into a holding bin (O).

Lucas retained the mortars and pestles to remove the bran. The hulled rice was taken by hand from the bins and placed in the mortars. The main cog wheel (C) turned by the waterwheel engaged the spur wheel (P) which rotated the axle (S). Attached to the axle was a series of spokes, one set of four spokes for each pestle (Q). As the axle rotated in a counterclockwise direction, the spokes fit into a slot in the pestles, lifting the pestles. As the axle continued to rotate, the spokes ultimately disengaged from the slot in the pestles, and the pestles fell by gravity into the mortars. The action of the pestles is similar to the machines already described: friction between the rice rubbed off the bran. Each revolution of the spur wheel (P) raised and dropped the pestle four times. The sets of spokes were staggered so that all the pestles were not raised and dropped at the same time, thereby distributing the force applied to the mill floor.

A more detailed diagram and explanation of the operation of millstones is shown in diagram 54. (The drawing is not to scale and represents only pertinent components.) The underside of the runner millstone had four slots placed around the eye to receive a stiff, four-clawed rynd (an iron bar supporting the upper or runner millstone) fixed to a vertical shaft that through a wooden neck in the eye of the stationary bed stone. The lower end of the shaft pivoted in a socket plate fastened to a bridgetree. To achieve the correct spacing between the stones, an adjusting nut on a rod was turned clockwise or counterclockwise to raise or lower the outer end of a pivoting beam which supported the outer end of the bridgetree. This, in turn, raised or lowered the drive flange shaft and the runner stone, thereby increasing or decreasing the space between the millstones.

A large face wheel meshed with the lantern wheel attached to the drive flange shaft which gave the runner stone its rotary motion. (The face wheel is not shown; see Water Rice Machine, diagram 53.) The ratio of the gears varied according to individual millwrights but was often around five revolutions of the millstone to one revolution of the waterwheel.

Not only did the miller have to keep the correct distance between the stones but also he had to keep the runner stone level, or balanced. Periodically, a millwright was summoned to a plantation to "true" the balance of the stones. Charles Manigault's overseer, K. Washington Skinner, wrote to Manigault on October 31st, 1852: "Last week I found upon examination that your Pounding Mill Stone (top one) is not true, & don't run level, one side when at the right place, or height to grind, leaves the other side twice as high. So (in a few words) they must be put in first-rate order by a first-rate mill-wright, before I can have rice ground. Stephen Tommy & Primus have been endeavoring to fix them

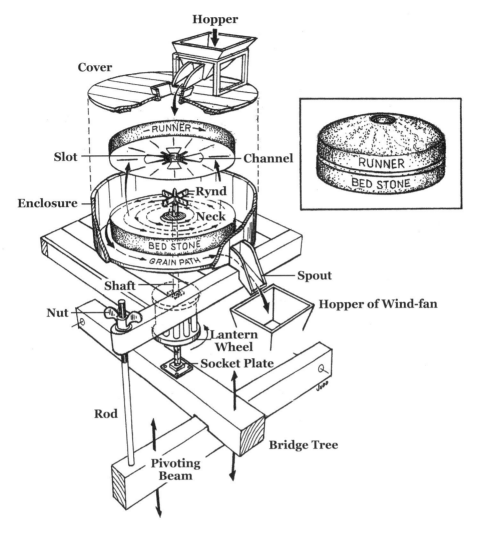

DIAGRAM 54. Operation of millstones for hulling rice.
Illustration by William Robert Judd.

and cannot do. Stephen says that it will take a mill-wright to fix them properly."[76] Small plantation mills probably used lead weights in the area of the eye to balance the runner stone to keep it level.

The milling process began when rice was poured into the hopper (fixed to the cover of the millstone's top enclosure), which fed the rice into the eye of the runner stone. Rice passed into the space between the stones through four channels cut into the underside of the runner stone (see diagram 54). A combination of small depressions (pocks) and/or furrows caused the rice to flip end over end. Since the stones were set at a distance slightly less than the length of the rice, the hulls were stripped off. Centrifugal force moved the rice and hulls in a spiral path from the eye to the periphery of the stones, where they fell into the area between the stones and the side of the tub enclosure. As the quantity of rice and hulls exiting the stones accumulated, the rotating runner stone swept them around, and they exited through a spout into the hopper of a wind fan. (The wind fan is not shown; see diagram 53.) The wind fan blew away the lighter hulls while the rice fell into a holding bin to be placed in the mortars later.

Some runner stones were fitted with a tag known as a "sweeper," which projected from the edge of the runner stone. As the runner stone turned, the tag swept the space between the edge of the stone and casing and carried the rice to the spout opening. Color plate 18B shows a piece of metal embedded in the edge of a runner stone that functioned as the sweeper.

Whether the Water Rice Machine had screens and sieves is unknown. They do not show up in the diagram and are not mentioned in any description. Sieves and screens would have been known to Lucas because they were commonly used in earlier water-powered pestle mills. Sieves and rolling screens are included in later Lucas mills.

Mortar and Pestle Evolution

As pounding pressure from the pestles became more forceful, mortars and pestles underwent an evolution in construction. The early mortars used by slaves pounding rice by hand were made of a single piece of native cypress or heart pine hollowed out at the top to produce a concave depression where the rice was poured. The hand-operated pestle was a wooden implement either of cypress or heart pine. As water- and steam-driven mills with increased pounding force evolved, changes were inevitable.

There are no surviving documents of any major changes to the mortar and pestle, however, until the diagram of the Lucas's 1787 Water Rice Machine. Lucas apparently used a solid beam and simply hollowed out eight concave depressions for the mortars (see diagram 53). The pestles were made of wood. No reference was found as to what kind of wood Lucas used, but pine was probably the logical choice because it was available and had proved reliable in pestle mills. A close examination of the tip of the pestles reveals that the working end had some kind of cap, probably cast iron. Some type of pestle cap was standard on pestles in later mills.

In 1846 Robert Allston reported on a newer design for a mortar and pestle: "These mortars, improved by Mr. Kidd's design, are constructed beautifully of four pieces of the heart of pine, seasoned. They are in figure a little more than a semi-ellipsoid, and are made to contain four and a half bushels of ground rice each."[77] Mortars constructed of multiple pieces of wood produced a stronger unit because of a better selection and arrangement of wood grain as opposed to the grain of a single block of wood. The pestles were also constructed of heart pine and were "sheathed at foot with sheet-iron, partially perforated in many places from within by some blunt instrument, so as to resemble, on a very coarse scale, the rough surface of a grater."[78]

In a footnote to his 1846 article on "Rice," Allston described a mill on the Savannah River "carrying eleven pestles," which "are shod with cast-iron about one foot in length, and secured to the wood by a long bolt driven from the bottom, and fastened by a screw and nut. The mortars are of cast-iron, weighing 600 lbs."[79] This is the first surviving primary reference documenting mortars made entirely of cast iron. Judging by archival sources and field studies, prior to this time, wooden mortars with cast-iron bottoms were standard. Doar reported that "the great stones used for grinders lie scattered around and the concave iron bottoms of the mortars can be seen in front yards of the negro houses in service as wash pots."[80]

Field surveys located three cast-iron mortar bottoms. One was in the yard Middleburg Plantation (see color plate 19).[81] Another was at a Lucas mill site on Wambaw Creek. This Lucas mortar bottom is presently housed at the Village Museum in McClellanville. A third cast-iron bottom, also from a Lucas mill, serves as a flower pot in a yard in McClellanville. Since all three mortars came from mills Lucas built, he may have introduced cast-iron mortar bottoms.

Millstones

Millstones have been around for centuries and were first used to grind grains such as corn into meal and wheat into flour. Millstones were among the items on ships that brought English settlers to Carolina. During the early years of the colony they were used on corn and wheat, and it seems unusual that no one tried to use them to remove rice hulls prior to Jonathan Lucas I's Water Rice Mill, even if only on a small scale. Yet no documentary evidence of any earlier use was found.

Archival sources relating to the Rice Kingdom contain sparse material on the operation of millstones. We do know, however, that, when Louisiana began growing rice in 1886, it borrowed millstone technology from the Rice Kingdom. Consequently, the operation of millstones in Louisiana was similar to that in the Rice Kingdom, and we can assume that a description of Louisiana technology will demonstrate how millstones operated in the Rice Kingdom. General accounts of milling in Louisiana stated that the millstones were flat on each working surface, and set at a distance apart a little shorter than the length of the rice. As the rice turned end over end, the hulls were stripped off.

Writing in 1893, Amory Austin gave a detailed description of how millstones operated. Austin, however, misspoke when he said the rice revolved on its *shortest* axis and should have said its *longest* axis:

> This is effected by passing the grain between burr-stones, by which the husk is literally ground off. These stones are generally about 5 feet in diameter and make 200 revolutions per minute. They are not grooved like ordinary millstones, since the object is not to crush the grain, but simply to crack and rub off the husk; therefore the faces of a pair of stones are made smooth and level and are nicely adjusted at a distance apart equal to the length of a grain of rice in its husk. A concavity in the

center of the upper stone admits the grain, which, impelled by centrifugal force, revolves upon its shorter [longer] axis and passes between the stones, the husk being thus stripped off, while the kernel is left unbroken. Shorter grains escape unhusked.[82]

A 1905 in-house document of the Rice Millers Association stated that the "rice grains fall down into these stones in such a way that they pass from center to circumference end over end."[83]

Seaman Asahel Knapp reported on rice milling in the Southwest in 1910. His lengthy article on rice culture covered every aspect of getting the crop to market, even commenting on modern millstones: "The hulls, or chaff, are removed by rapidly revolving milling stones set about two-thirds of the length of a rice grain apart." Afterward the rice was cleaned of the bran in mortars and polished between wire gauze and sheepskin.[84]

Beth McLean described the operation of millstones in her 1934 article on rice culture in the Southwest:

> The cleaned rough rice is passed between the face of large revolving "shelling" stones. The centrifugal motion causes the grain to stand in a perpendicular position; the shelling stones are placed just close enough together to crack the ends of the hulls free from the grain without crushing the grain. Since each variety has a characteristic length and breadth of the kernel, besides the individual variations which naturally occur within each variety, some of the grains are unhulled on the first shelling; these grains must be returned to a second, and sometimes a third pair of shelling stones set more closely together that the first stones thru which they passed in order to completely shell all of the grains.[85]

We examined many discarded millstones from former Lowcountry rice plantations. The majority of the working surfaces were smooth, which supports the assumption that Lowcountry rice-milling practices were like later ones in the Southwest. Several millstones have a few grooves cut into the working surfaces (see color plate 18C). Presumably these grooves helped flip the rice end over end the same way the rough surface of the stone did.

Millstones used in the plantation mills in the Rice Kingdom were generally a standard size. According to Robert Allston, "The stones which are used for grinding rice should be 5–6 feet two inches diameter, and 18 inches thick at the center. There is said to be a quarry in Northumberland, affording stones of such excellent substance, that they will grind rough rice enough for packing 1,000 barrels without being taken up."[86] The millstones measured during fieldwork were generally the size Allston described.

The bed stone was eight inches thick with a flat surface, top and bottom. The runner stone's milling surface was flat. Its top surface was convex or dome-shaped measuring eleven to thirteen inches thick at the center of the eye and decreasing to eight inches thick along its outer edge. The purpose of the dome was twofold. First, the added dome replaced the stone material cut away on its milling surface to create the grain channels and the deep slots to receive the flanges of the drive plate, thereby maintaining the millstone's strength. Second, the dome added balance to the revolving stone. The runner stone was supported in the center on a rotating drive shaft that spun the stone at approximately one hundred rotations per minute. The dome created a greater centered mass to the revolving millstone, reducing centrifugal weight at its outer edge. A greater balanced weight at the center allowed the stone to run truer, reducing the wobble effect.

The average millstone weighed approximately two thousand pounds. Installing or replacing a millstone was a monumental task. Charles Drayton recorded in his diary on April 4, 1811, that millstones for his animal-powered mill were "raised into the barn loft, on an inclined plane, by my schooner's tackle. It required the whole day to effect, which, thanks be to God was accomplished safely and without accident. It required 16 fellows and Mr. Thomson's management."[87]

Basing his conclusions on observations at an abandoned quarry in Northumberland, England, George Jobey described the dressing of millstones: "Central holes in the eyes of the stones were cut on site, though they may well have required further attention when the final feathering of the grinding surfaces was carried out, presumably at the mills since no completely dressed millstone has been found at a local quarry. The practice of cutting a cross on the rough faces of a millstone, as an indication of the level to be achieved in subsequent dressing, has never been encountered locally, but this is not to deny a practice which seems to have been followed in both Derbyshire and the Anglesey millstone quarries."[88] Many Lowcountry planters secured their millstones from Derbyshire. More than likely, however, planters had their millstones shipped undressed to avoid damage to a dressed face. Upon their arrival at the mill, the millstones were dressed by local millwrights who traveled from plantation to plantation.

A millstone at Rochelle Plantation on the North Santee River is an example of a millstone shipped undressed. The millstone is an unfinished runner stone, five feet in diameter and approximately eighteen inches thick. Cut onto the rough, working surface are several linear channels a few centimeters deep in the shape of a cross (see color plate 18D). The surfaces of the channels were cut to the same level. The level of the channels represented the surface to which the rest of the millstone was to have been cut. It appears the cross was cut by a local millwright. After the millstone shattered at its edge during cutting of one channel (bottom right), the dressing was never completed, and the millstone was discarded.

The use of millstones to remove rice hulls created a problem: All rice was not the same size; consequently, when the stones were set at a distance apart to mill the majority of rice, slightly smaller rice was not milled. The unhulled rice could be sent back to the stones, and the stones set a shorter distance apart, which milled the unhulled rice. First, however, a means had to be developed to separate the unhulled rice from the hulled rice. On April 1, 1807, Lewis Dupre of South Carolina received a patent for his "Winnowing Screen Pendulum." The patent was lost in the 1836 fire at the Patent Office, and no description or diagram of his invention exists. Charles E. Rowland, however, wrote about Dupre's machine in 1828, explaining how "after the rough rice has been ground in the usual way [millstones], it is rayed in one of Mr. Duprix's [Dupre's] pendulum screens, which takes out the dust and mill cut,—separating, at the same time, the rough from the ground rice. The rough rice is carried back to the mill to be ground over. The rice goes much purer to the mortar. After coming from the mortar, it is run twice through the rolling screen and brush."[89] This is the first mention in any archival source of a means to separate the unhulled rice from the hulled rice.[90] Dupre's screen must have had openings of several sizes to allow the dust, mill cut, and hulled rice to pass through, while the larger unhulled rice exited through the bigger openings. Since it was called a pendulum screen, a method must have been devised that caused the screen to rock back and forth, striking something to jar the contents and facilitate passage of the materials through the screen. The unhulled grains were sent back to the stones and hulled after the distance between the two stones was changed to accommodate the shorter length of the unhulled rice.

Diagram 55 is a conjectural view that incorporates the concept of pendulum motion and a rice screen. Milled rice was placed into the sieve, which fit into a wooden frame

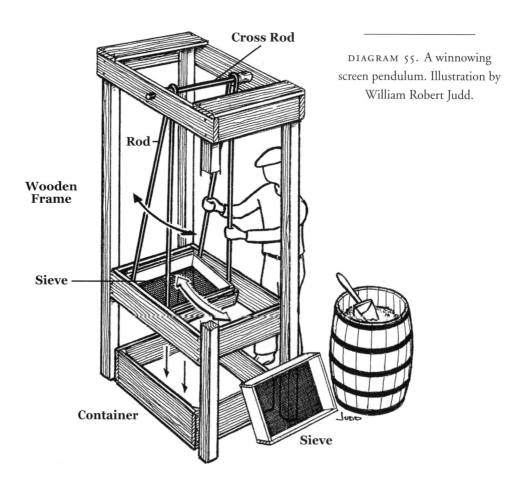

DIAGRAM 55. A winnowing screen pendulum. Illustration by William Robert Judd.

suspended by rods from a cross rod so that the frame could be manually rocked back and forth. A wooden frame housed the machine. Below the sieve was a container that received the contents of the sieve. The unhulled rice was retained in the sieve and sent back to the millstones. A sieve of one size could be changed out for another.

Peak millstones used by Lucas and other millwrights came from England.[91] Excellent rock for processing millstones was found in southwest Yorkshire and the northeastern perimeter of Derbyshire. This rock was appropriately known as Millstone Grit, also called Peak or gray stones, used by British millers for hulling barley or oats. Peak stone is a monolithic stone made from a conglomerate. It is a type of sandstone containing a high density of small rounded quartz pebbles, usually under 1 centimeter in maximum dimension.[92] The Peak District was the most famous of the millstone-making areas in Britain, and its output exceeded that of all other areas combined. Peak stone was used widely in the grain-milling industry until about sixty years ago. Advertisements appeared regularly in local newspapers offering millstones from England for sale. For example, the *City Gazette* in 1804 carried this notice: "Mill Stones. Ten pair large MILL STONES, of a very superior sort to any hitherto imported, of a fine soft grit, clear of Pebbles and according to the directions of Mr. Lucas, imported in the Rolla, from Liverpool. For sale by Lockey, Murley & Naylor."[93]

The company of W.J. & T. Child of Whitkirk, Leeds, claimed the largest stock of Derbyshire Peak millstones. It advertised Peak millstones for shelling oats and rice in the March 2, 1903, issue of the *Miller*. Unfortunately, the advertisement does not show a photograph of the millstone used for rice, but it documents that in the early 1900s there must have been a market somewhere for Peak millstones to shell rice.[94]

Millstones made from French buhr stone were primarily used for milling wheat and corn. Buhr stone was almost invariably mined from the quarries at La Ferté-sous-Jouarre, about 50 kilometers east of Paris. Buhr stone was found only in small pieces ranging from about 12 to 18 inches long, 6 to 10 inches wide, by 5 to 10 inches thick, so a millstone made from buhr stone was a composite.

Buhr stone was a hard stone suitable for milling, superior to native British stone. Buhrstones (as millstones made from buhr were called) were made by cementing together from ten to as many as thirty individual pieces of French buhr stone. Iron bands were shrunk around the periphery to give strength and prevent shattering under centrifugal force. A plaster back was provided to give a smooth finish to the nongrinding surface, or bed stone, so that the stone would lie flat on its base. The centerpiece was made of a local grit stone, sandstone, or other suitable rock because little grinding took place near the center. Buhr stone could then be reserved for the outer parts where grinding occurred. In England there were many manufacturers of buhrstones, London being the center. In Scotland they were manufactured only in Glasgow and Edinburgh.[95]

Buhrstones were common in America. The first pair of buhrstones was brought to Virginia in 1620 for a wind-powered mill. The United States became the largest importer of buhrstones in the world. B. F. Star of Baltimore, an importer of buhrstones, claimed in their advertisements that they had two thousand men working in French quarries making blocks to be exported to America. George Washington's Grist Mill at Mount Vernon used buhrstones. After about 1750, buhrstones were used throughout the United States for grinding wheat into flour and corn into meal.

The United States had a buhrstone crisis during the War of 1812, when trade was cut off with France. A substitute for French buhrstone was needed, and Christopher Fitzsimons, an Irish-born resident of Charleston, South Carolina, found buhrstone beneath the soil at his Old Town Plantation, near Louisville, Georgia, which he had acquired in 1809. He sent this Georgia buhrstone to millwright Oliver Evans in Philadelphia, who was impressed with it, and in 1811 his sent his workers to Georgia to acquire a quantity of the stones. More than one thousand buhrstones were taken from Georgia quarries by 1815.

With trade reopened with France in 1815, the Georgia buhr-stone trade ceased. It resumed in 1849, with buhr stone from a plantation in Burke County. The Lafayette Burr Millstone Company of Jefferson and Burke Counties produced more than one thousand sets of Georgia buhrstone for buyers throughout the South. But Georgia could not compete with the lead that French buhrstone had established. France had name recognition and ships that carried cotton and other products from America to France had abundant room to carry buhrstones from France back to the United States. The Georgia buhrstone industry died around 1854.[96]

Archival records do not indicate whether buhr stone, either French or Georgia, was used to make millstones in the Rice Kingdom. Some type of buhr stone, however, was used in the rice industry in the Midwest. In his 1893 article Amory Austin stated that the husk was removed "by passing the grain between burr-stones [that is, buhrstones]." By that time buhrstones had become common in the Southwest.

Indirect evidence that buhrstones were used in the New World for rice milling appeared in advertisements by John Reid and Peter Reid in Scotland. Their firm advertised in 1877–88 "French burrstones for shelling rice for Home or Abroad."[97] How long before 1877 they offered buhrstones for use with rice could not be determined. It is certainly possible that someone tried French buhrstones in the Rice Kingdom.

Belts and Drills

Elevator belts with attached buckets and drills were used to move the different milling products from one floor to the next. These two mechanical devices were first developed and employed by Oliver Evans in 1785 for a flour mill he and his two brothers erected on Red Clay Creek in Delaware.[98] They were later adapted by the rice industry. Evans also introduced other labor-saving devices that greatly facilitated getting grain to market and were incorporated into the market preparation of Carolina rice.

The first mention of elevator belts in the Rice Kingdom was in 1794, when Jonathan Lucas I built his Improved Water Rice Machine for Henry Laurens. The system (see diagram 56) consisted of a continuous canvas belt approximately nine inches wide. Small

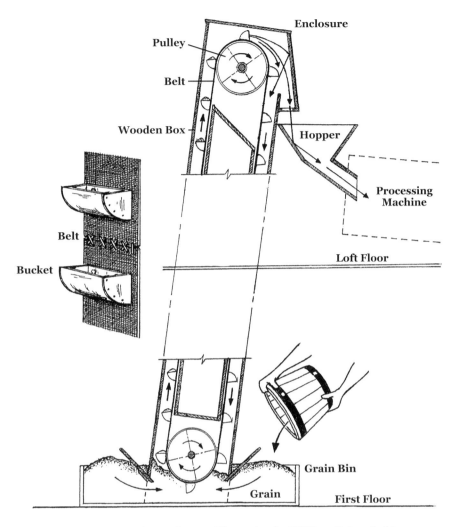

DIAGRAM 56. A rice elevator. Illustration by William Robert Judd.

cups or buckets fashioned from wood, cast iron, or formed tin were riveted to the belt every eighteen inches. The buckets came in two sizes: one pint and one quart.[99] The length of the belt was determined by the distance between the holding bin and the machinery to which the rice would be sent for processing.

The belt was enclosed within a long, vertical wooden box having a pulley mounted at the top and bottom. The box was slightly inclined, allowing the buckets to discharge the rice better. The box insured that any rice spilled from the buckets would fall back into the bin and be scooped again. The top pulley set the belt in motion. This arrangement allowed the belt to hang tighter because of the weight of the buckets, thereby preventing the belt from slipping. The bottom end of the box was placed in a bin holding the rice.

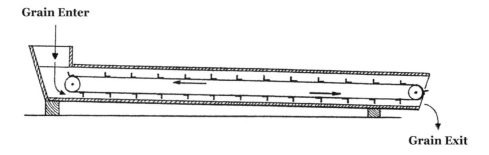

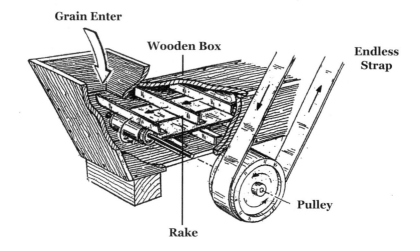

DIAGRAM 57. A drill conveyor. Illustration by William Robert Judd.

Each bucket scooped rice from the bin and elevated it. As the bucket passed over the top pulley, rice fell into a hopper that passed it to a processing machine.

The drill (see diagram 57) was an endless strap revolving over two pulleys enclosed within a wooden box. It was like the elevator belt but moved grain horizontally. Instead of buckets, small wooden or metal rakes fixed to the strap drew the grain along the bottom of the box. If the enclosure was set a little descending, it would move the rice with ease. In some drills, instead of an endless strap, rakes were fixed to metal links, spaced apart and resembling a wide bicycle chain.

Jonathan Lucas I's Improved Water Rice Machine

Jonathan Lucas I and his son Jonathan Lucas, Jr. (1775–1832),[100] made many improvements to the original Water Rice Machine. No diagrams or drawings have survived of the improved mills Lucas and his son built after the first mill, built for Mr. J. Bowman, and the mill built for Dr. Philip Tidyman on the North Santee. Several descriptions of the workings of the improved mills, however, were recorded. Whether these accounts are from the same original source is not certain. Lucas built his first improved water mill around 1794, when he modified a water mill he had built for Henry Laurens at Mepkin Plantation on the Cooper River.[101]

The Lucas improved mill was fully automated and could be operated by three workers. Rough rice from the threshing barn was deposited into a bin, then elevated to a rolling screen that initially cleaned the rice. From there it was carried throughout the milling system automatically and was never handled by workers. The cleaned market-ready rice was discharged into a barrel ready for shipment. Workers only monitored the working of the milling system, making repairs and keeping the system moving.

Historians question whether Lucas based part of his improved mill on the works of Oliver Evans, who by 1785 had built a flour mill in which he created the first continuous production line in any industry.[102] All movement was automatic. Power was supplied by waterwheels, and grain was passed by conveyors and chutes through the stages of milling and refining to emerge as finished flour. This is not to detract from Lucas's accomplishment. Lucas still had to modify the flour mill's machinery to a rice mill. But it is certainly possible that Lucas knew of Evans and incorporated Evans's work into his improved rice mill just as he incorporated and modified the works of others in his initial Water Rice Machine.[103]

In *A View of South-Carolina* (1802), John Drayton gave the earliest account of Lucas's improved water mill, comparing it to the initial Lucas Water Rice Machine diagramed by Latrobe: "Of late years, some have been erected with complicated mechanism; whole movements proceed with perfect harmony; carrying the grain through a variety of changes, until it be finally delivered into the barrel, and is there packed for market." Drayton went on to describe the operation of the new mill:

> The rough rice, is carried by a set of elevating buckets, from the lower, into the upper story of the machine house, from whence it falls into a rolling screen; which separates the sand and gravel from it; and pours it clean into the hopper. From the hopper it passes to the mill stones, where the chaff [hulls] is [are] separated from the grain, and is afterwards blown away by a wind fan. The milled rice is then discharged into a bin, placed above the mortars; having funnels communicating there from to the mortars. The rice is then introduced into the mortars by the funnels, and is there beaten by pestles weighing about 230 lbs. weight; which strikes the rice from 34 to 44 times a minute. When the rice is sufficiently beaten, it is taken out and thrown into an hopper; from whence, by a set of elevating buckets, it is carried up to another rolling screen, where the small rice and flour [bran] are separated from it. The whole rice then passes through a funnel, under the friction of a brush, which takes off any flour [rice polish] which may still adhere to the grain; it thence falls into a wind fan, which winnows it clean, and discharges it into a bin. From whence, by funnels, it is received into barrels; and in some mills, it is even packed in them by mechanical operation.[104]

FIGURE 25. A rice-polishing brush from the threshing barn at Cockfield Plantation. When the barn was converted to a residence, the threshing machines and brush were moved. When Richard Dwight Porcher, Jr., took this photograph in the 1980s, the sheepskin was intact. The brush could not be relocated in 2010.

In 1803 Lucas introduced the brush to polish rice (see figure 25). The brushes removed the very inner layers of the bran and the aleurone layer (the polish), which were not completely removed by mortar and pestle. Brushing gave rice its pearly luster, which greatly improved in its appearance—and consequently its market value.

Polishing was effected by friction between the rice and sheepskin, which was tanned and worked to a high degree of softness. The sheepskin was loosely attached to a large drum, which worked inside another drum made of wire gauze on a wooden frame. The brushing cylinder was worked in an upright position. Rice was put into the brushing cylinder at the top, and the polish from the surface of the grains was forced out through the wire gauze.

Robert Allston's 1846 description of the milling process undoubtedly refers to the Lucas mills. His description is similar to Drayton's forty-four years earlier; there are certain refinements in the process that evolved after Lucas made improvements over his initial Water Rice Machine in 1787 and the improvements cited by Drayton in 1802. According to Allston,

> The rough rice is taken by means of elevators . . . up to the highest apartment in the building, to be passed through a sand-screen revolving nearly horizontal, which, in

sifting out the grit and small grain rice, separates also all foreign bodies, and such heads of rice as were not duly thrashed.

From the sand-screen the sifted rough rice is conveyed directly to the stones on the same floor, where the husk [hull] is broken and ground off, thence to a wind-fan below, where the chaff [hull] is separated and blown off. The grain is now deposited in a long bin, placed over the pestle-shaft, and corresponding in length with it, whence the ground rice is delivered by wooden conductors into the mortars on the ground floor,—ten, twelve, fourteen, or twenty-four in number, as the power applied may justify.

The pestles . . . are sheathed at foot with sheet-iron, partially perforated in many places from within by some blunt instrument, so as to resemble . . . the rough surface of a grater. A mortar of rice is disposed of, or sufficiently pounded in one hour and forty minutes to two hours. The grain thus pounded is again elevated to the upper floor, to be passed through a long horizontal rolling-screen, slightly depressed at one end, where by means of a system of wire-sieves grading coarser and coarser towards the lower end, are separated, first the flour, second the "small rice" (the eyes and smaller particles of the broken grains), third the "middling rice," or the smaller and the half broken grains, fourth and last, the "prime rice," and fifth, the larger and chiefly unbroken grains, which fall through the largest wire, and forthwith descend to the "polishing screen," whence it descends through a fan into the barrel on the first floor, where it is packed, and the preparation is completed. The head rice, or largest grains of all, together with the rough unbroken by the stones, passes off at the lower end of the screen, to be pounded over.[105]

One feature worth noting in Allston's description of the improved Lucas mill is the method of separating the unhulled rice from the hulled rice so the unhulled rice could be sent back to the stones. The rolling screen had five different sizes of perforations to separate the flour, small broken rice, middling rice, prime rice, and at the very end of the screen, the rough, unhulled rice. Whether Lucas added the separation of the unhulled rice based on Dupre's pendulum screen is unknown; however, it served the same purpose as Dupre's screen.

The Lucas mill was quite simply an astonishing and ingenious piece of work. One mill could turn out approximately one hundred barrels of rice each day, with a work force of only three. Each barrel weighted six hundred pounds. No one better summed up the Lucas Water Rice Mill than the editor of *Southern Cabinet* in 1840:

With respect to our other great staple, rice, we were most fortunate, many years ago, in having directed to its preparation, the great mechanical skill of the elder Mr. Lucas, who not only invented, but perfected the mill, which is now and has been for so many years used almost exclusively in preparing the paddy rice for market. Many have been the alterations, and new inventions brought forward, to supersede either in whole or in part, this beautiful combination of machinery, which as a whole, is perhaps the most perfect which has ever been introduced for similar purposes. Nearly every alteration, or proposed improvement, has, after comparatively short trials, been abandoned, and the rice-mill as invented and erected by Mr. Lucas, is now in almost the exact state in which he left it.[106]

After the death of Jonathan Lucas I in 1821, Jonathan Lucas, Jr., carried on his father's mill-construction business. During this period significant losses were incurred from breakage or spoilage in the shipment of milled rice overseas. Rough rice kept much better during shipping, and was more esteemed when it was milled at its destination overseas. To force its former colonies to send rough rice to England after the War of 1812, the British imposed a four-dollar-a-barrel tariff for rice milled in the former British colonies (sixty-six dollars in 2009 money) but not on rough rice. In 1823 Jonathan Lucas, Jr., left South Carolina for England at the request of the British government to build mills to process the increased volume of rough rice being shipped to England because of the tariff. Another reason for his leaving was that he felt his improvements in rice milling were accepted better in England and that mill owners at home, in as much as their mills were doing well, were reluctant to change. Jonathan Lucas, Jr., built two mills in Liverpool and two in London, as well as one each in Copenhagen, Denmark; Bremen and Flensburgh, Germany; Amsterdam, Holland; Lisbon, Portugal; and Bordeaux, France.[107]

The circle was complete: millwright Jonathan Lucas I, a native of England, created mechanized rice mills in the American Rice Kingdom, starting in 1787, which could mill the increased crops resulting from tidal culture. In 1823 his son Jonathan Lucas, Jr., carried his father's milling technology back to England, where rough rice from the states was milled. The son planned to return to America, but he died in England in 1832.

Jonathan Lucas, Jr.'s Rice Cleaner

Like his father, Jonathan Lucas, Jr., was an inventive genius. Not only did he build rice mills in the states and abroad, he made improvements to mills, including his father's. In July 1808 he was granted a patent, signed by President Thomas Jefferson, for a machine to clean rice. (His patent was destroyed in the 1836 fire at the U.S. Patent Office and not restored.) Although some references state his patent was for "hulling and cleaning rice," this invention was not a huller. Lucas's invention replaced only the mortar and pestle in the milling process. His cleaning machine removed the bran by friction between two revolving cylinders. According to Lucas, his rice cleaner

> consists of two cylinders, one to run within the other, the inner cylinder to have a motion of much greater velocity that the outer one, & both of them to be faced with sheet iron punched like a grater, or the inner cylinder may be covered with cork or any other soft wood fluted, or sand or any other sharp scouring substance cemented on the surface might answer.

The rice is intended to pass between the two cylinders which will run at or within the distance of one half inch of each other as experience may advise. These cylinders are conical or tapering, so that the inner one may be moved upward or downward at pleasure to widen or diminish, as may be necessary, the vacancy thro' which the rice passes to take off effectually the inside skin [bran] of the rice & scour it clean ready for the rolling screen or sifter & thence to the brushing machine. The two cylinders are to turn in contrary or opposite directions and to be placed perpendicularly or otherwise & their length sufficient to prepare the rice for the sifter in its passage thro' to be about twenty five feet more or less.[108]

Lucas's theory is easily understood. As the rice passed between the two cylinders, the bran was rubbed off as the rice grains were forced into contact with the abrasive surfaces of the two cylinders.

There is no record of how Lucas's cleaning machine was inserted into the milling sequence. Since it replaced the mortar and pestle, there must have been some type of funnel that fed the hulled rice from the second story of the mill barn into the cleaner after it left the stones and was winnowed. There also must have been a method to collect the rice from the cylinders and deliver the rice upstairs to the rolling screen and then to the brushing machine. How power was applied to the cylinders is also unknown.

Nor is there any record of whether or not Lucas's invention was successful and used by planters in their mills. No references were found in any documents to a planter having ordered or built a mill using Lucas's rice cleaner. One might speculate that Lucas installed his cleaning machine in the mills he built overseas.

The Deforest Rice Mill

In the early 1800s, J. Deforest of Georgia invented a mill that employed four sets of three millstones each. It is documented only by two crude diagrams and notes that James Hamilton Couper (1794–1866) of Hopeton Plantation on the Altamaha River made in 1835, so it is difficult to understand its construction and operation fully (see diagram 58).

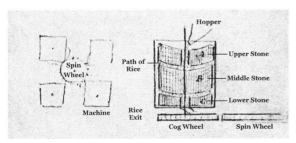

DIAGRAM 58. The Deforest rice mill. Reproduced with permission of the Southern Historical Collection, Wilson Library, University of North Carolina at Chapel Hill, J. Hamilton Couper Plantation Records, "Agricultural Notes," vol. 4, p. 55, #185-Z. Letters added.

Deforest's mill consisted of four individual machines run by a single spin wheel located at the base and middle of the four machines. Each machine housed three millstones, stood five feet high, and was three feet square. Based on the diagram by Couper, each machine consisted of a wooden frame that enclosed the three millstones. The upper and lower millstones were stationary; the middle stone was turned by a rod that was connected at the base to a cog wheel, which was engaged by the spin wheel. The spin wheel engaged all four machines simultaneously.[109]

Rough rice from the hopper passed to each of the four mills, then through a hole in the upper stone of each mill to the space between the upper and middle stones. By centrifugal force, rice was passed to the edges of the stones, where it fell down the sides and passed between the middle stone and lower stone. The clean rice, hulls, broken rice, and flour exited at the base, where they presumably were sent to a sieve and rolling screen for separation. Since Couper used the term flour [bran] in his account of the products of the mill, the passage between the stones must have removed hulls and bran.

Many questions remain about the operation and construction of Deforest's mill: How was distance set and maintained between the stones? What was the source of the stones? Why were two milling surfaces superior to one milling surface? Did the passage between the upper stone and stationary stone remove only the hulls, and the passage between the lower stone and stationary stone remove the bran? Or did both passages contribute to the removal of both hulls and bran? How did the rice that spilled from the passage between the upper and middle stones pass between the middle stone and lower stone?

James Hamilton Couper installed Deforest's mill at Hopeton and compared its production to his pestle mill. In his "Agricultural Notes" he recorded that one set of four machines milled about twenty-five bushels of rough rice per hour. This output was superior to his pestle mill in terms of speed but was inferior in that it produced a much higher percentage of flour and small rice. This obviously reduced the net value of his yield of whole, market rice. Whether Deforest's mill was used on other rice plantations in Georgia or elsewhere in the Rice Kingdom is unrecorded.

The most intriguing aspect of this mill is the power source. Couper did not mention a power source. The location of the spin wheel suggests that the mill was powered from underneath. If run by animal power, the mill would have been placed on the second floor of a mill building. Animal power at ground level turning a king post attached to the spin wheel could have turned the spin wheel. This technology had already been in use as evidenced by the cog mill described in diagram 42.

The possibility that water powered Deforest's mill is raised by the presence of a brick structure in the canal at Butler's Island, adjacent to the site of the abandoned steam mill.[110] Butler's Island is on the Altamaha River, just downriver from Hopeton Plantation. Someone placed the label "Tide Mill" on a cylindrical brick structure in the canal. Yet no archival source identifies this brick structure as a tide mill. If it were a tide mill, it would have had to have had a horizontal waterwheel contained within the base and turned by water flowing in the canal. No physical structures at the site indicate that a vertical waterwheel was used. Mills turned by a water-powered horizontal wheel date to the Roman empire and predate vertical water mills. Deforest's mill would have been placed in a barn projecting outward from the bank over the canal. A king post attached to the horizontal waterwheel could have turned the spin wheel that turned the four individual mills.

Were water-driven mills with a horizontal waterwheel known in the vicinity of rice plantations on the Altamaha River in the early 1800s? One source indicates they were. In 1832 John D. Legaré, editor of the *Southern Agriculturist,* toured the south of Georgia and visited Sapelo Island, home of Thomas Spalding (1774–1851). Legaré reported that Spalding "found from long experience that a vertical cattle-mill would not express the juices from the blue cane sufficiently, and not finding it convenient to procure a steam-engine for the purpose, it was determined to erect a horizontal water-mill, excavating for the purpose about five acres of marsh, so as to be able to work neap-tides, a few hours. This mill is connected to the water-wheel simply by a coupling box."[111]

Spalding's Sapelo Island plantation was ten miles east of Butler's Island and Hopeton Plantation. Planters socialized together and discussed aspects of agriculture practices, either informally or as part of agricultural societies, and overseers or managers of plantations were in frequent contact. Roswell King (1765–1844), who managed Butler's Island, and his son Roswell King, Jr. (1796–1854), who succeeded him, would have known about the Deforest mill on Hopeton and Spalding's horizontal water mill on Sapelo Island. They could have used this knowledge to construct a horizontal water mill on Butler's Island.

The description of Deforest's mill is vague and incomplete; the description or diagram of the Deforest mill is in Couper's "Agricultural Notes." The possibility that it was driven by a horizontal waterwheel brings a new dimension into rice-milling technology.

The Ravenel Rice-Hulling Machine

On May 7, 1828, John Ravenel of Charleston was granted a patent for a rice hulling machine.[112] Ravenel's patent was one of the many destroyed by the 1836 fire in the U.S. Patent Office and not restored, so no diagram exists of his invention. A gentleman identified as J. G., however, gave a written description of Ravenel's invention, claiming: "The introduction of Mr. Ravenel's principle in preparing Rice, will be, we hope, the commencement of a new era in the annals of this old staple of Carolina." Ravenel's improvement in rice mills, according to J. G., consisted "in getting rid of the enormous pestle-shaft and lifters, and the heavy pestles weighing 360 lbs.; with the expensive foundation of solid brick work required for the mortars to lie upon; a part of the machinery requiring very considerable experience to manage; always attended with danger; propelled (even by water) with difficulty, and hence inapplicable to mills moved by animal power, without sacrifices almost ruinous."[113]

Ravenel substituted pestles moved by cranks, which rose and fell only a few inches, each pestle "3 × four inches, and about four feet in length." According to J. G., the pestle did not strike the rice in the mortar with a force or weight equal to some thousand pounds and did not pound or beat the rice, instead it hulled the rice with "rapidity of friction, what is done on the old system by weight and hard blows." Ravenel also used wire mortars, but nothing in J. G.'s article explained how the wire mortars cleaned the rice better. The wire mortars acted as a grate to strip the hulls from the rice.

J. G. claimed that Ravenel's mill would "in a great measure get rid of one of the heaviest items of loss to the Rice-planter; we mean small Rice," for the "old mode of pounding, will perceive that each stroke of the pestle must have broken the rice more or less." With Ravenel's mill the planter no longer had to send his rice to a toll mill. He could mill rice at his plantation because animal power could be used on a small mill, and planters with extensive operations could drive more pestles with the same water power as on previous mills. How successful Ravenel's mill became is not known. His mill was probably run by water power since steam power was in its infancy in 1828. Steam power, however, could have run the mill later.

Steam-Powered Mills, 1830s

Major changes in rice milling occurred in the third decade of the 1800s. Steam engines and more efficient milling machines were coming into common use, giving the planters more and better options in marketing their rice. The primary goal of rice planters, however, was still the production of clean rice highly polished and packaged in wooden casks, tierces, or barrels. With the advent of the newer machines for milling, it was possible to produce a better quality, clean rice. A planter who did not produce clean rice was beaten out of the market. Rice was purchased by merchants and agents of northern and European importers and exporters and shipped out of the ports of Charleston and Savannah. Some areas of the world owned their own milling facilities, such as Great Britain (which is why Great Britain sought Jonathan Lucas, Jr., to come overseas), and wanted only rough rice.

By the third decade of the 1800s, four different methods had evolved to prepare rice for market. First, rice was milled on the plantation and shipped to the factors in the cities. The factor arranged for shipment from the plantation to the city and to overseas markets.

Charles Manigault gave a description of a typical work force for a large plantation mill: "Mr. Taylor's Steam Mill opposite my Plantation occupies 7 Prime men, 10 Boys & girls, a white Miller & 5 men who attend constantly rolling Rice & other wharf work & to attend to his flats & flat his rice to Savannah, 23 in all. His Mill has a 22 horse power engine, & 20 Pestles."[114] This method could be used only on large plantations that could afford complex and costly milling machines.

Second, smaller plantations sent their rice to neighboring plantations where they paid a toll to have it milled. The cost of complex mills driven by steam was generally too expensive for small plantations. The larger plantations served as toll mills, and the planter assumed the cost of transportation and the toll for milling his rice. Middleburg Plantation served as a toll mill for plantations along the Eastern Branch of the Cooper River.

The Committee on Foreign Mills, established by the Agricultural Society of South Carolina, reported, however, that the "loss to many planters by sending to Toll Mills, has been so great; the disappointments and delays so vexatious, and the mortification so deep that in many cases those who have frequented these mills, and who would still have resorted to them from being unable or unwilling to undertake beating [milling] at home, have preferred selling in the rough, for exportation, as the least of the many evils they are liable to."[115] The committee realized that not using toll mills represented a great loss of money to the area and suggested that "it might be worthy of the attention, and the calculation of every individual proprietor of a Toll Mill whether he will not lose his whole trade or employment, unless he bends all his attention to reducing, in every way, the losses and charges, which those who send their rice from a distance, are subject to at present."[116] No further mention of toll mills appears in the Agricultural Society's minutes.

Third, rough rice was sent to a city factor, who handled all milling, freight, and marketing and charged the cost back to the planter when the rice was sold to a merchant. One advantage of this method was that the newly milled grain looked and kept better. This method was used by most small plantations.

Fourth, planters shipped their rough rice to a factor, who purchased it outright. The planter had no more involvement with the transaction. It was sold as rough rice by the factor in the world market. The planter paid freight costs from the plantation to the city and was charged the usual factor's commission. In the long run the planter did not receive as much for his rice as he would have in the first three cases; however, for a planter who did not want to bother with all the problems of cleaning rice or for plantation managers who were serving an absentee owner, it was an acceptable trade off.

This fourth method garnered the attention of the Agricultural Society of South Carolina, the leading agricultural society for rice planters. A committee was formed in 1827 with the object "to call attention of foreign capitalists to the new branch of commerce, the exportation of rough rice from Carolina, with the view of being prepared in other countries; and to offer all the information necessary to enable them to form a correct judgement on this interesting subject."[117] The committee's investigation documented that a planter would sacrifice 27¾ percent of his total sale of rough rice. A planter who raised 400 barrels of rough rice would give 111 barrels of his crop yearly to "strangers he never saw." But the committee concluded that for the planters who could not mill at home, "this may have hitherto appeared to be attended with the smallest sacrifice of feeling, or interest."

The Early History of Steam Power

The steam engine was the first practical engine that converted heat energy into motion to drive machines. Heat produces steam pressure in a boiler, increasing the steam's bulk some 1,600-fold at atmospheric pressure over the volume of water. Steam engines harness this pressure by moving a piston back and forth within a cylinder to run a machine. The steam engines powered the machines of the Industrial Revolution and the many factories and small mills that sprang up in America. The development of steam power in the Rice Kingdom may be divided into three overlapping stages, or periods:

1. Steam engines in forms developed by Thomas Newcomen, Matthew Boulton, and James Watt came from England to America during 1750–1815. These early condensing, low-pressure engines were used for pumping water from mines and for urban water supplies. Later they were used to power steamboats and city mills.
2. Between 1800 and 1825 Oliver Evans (1755–1819) created and developed new stationary, high-pressure, noncondensing steam engines with a vertical cylinder, which quickly found a niche in the country's industrial expansion. His engines generally replaced the low-pressure, condensing engines except in steamboats, where Evans's engines played a secondary and diminishing role. Operationally, the Evans high-pressure engine had minimal water requirements, and steam pressure could be altered at the boiler. The high-pressure engine required only feed water, the amount required to replace the water lost to make steam. On the other hand, the low-pressure engine required up to thirty-five times as much feed water to supply its condenser. In fact it was the advent of successful steamboats that may have influenced rice planters who were looking for a better source of power for their mills. Not only did steamboats provide a means of transporting rice to market, but they also demonstrated the capability of a new, more reliable source of power as an alternative to animal and water power. Experience gained in the building and repair of steamboat engines was generally applicable to stationary engines, and both types were often built in the same shops. For example the Phoenix Iron Works on Pritchard Street in Charleston advertised "Marine, Stationary and Portable Steam Engines and Boilers." Although differing in size, capacity, and sturdiness, the basic problems of design, construction, and operation were essentially the same for steamboat and stationary engines.
3. Between 1825 and 1850 many improvements were made to the high-pressure noncondensing engine. In the eventual useful form that became the norm for mills, the cylinder was placed horizontally on a flat bed.

The engines James Watt and Matthew Boulton developed in England from the Newcomen engine, which were later adapted by Robert Fulton with certain modifications for his steamboats, were condensing steam engines. A condensing engine worked on atmospheric pressure (see diagram 59). With the engine at rest, the piston was at the top of the steam cylinder since the combined weight of the end of the working beam where the piston and piston rod were attached was less than that of the end of the beam where the pump (not seen in diagram) and mine-pump rod were attached. The engine cycle started when steam was admitted into the cylinder through a valve. The pressure of the steam within the cylinder drove out the air formerly there, and the valve then closed. A spray of cold water from the reservoir was injected into the cylinder, which condensed

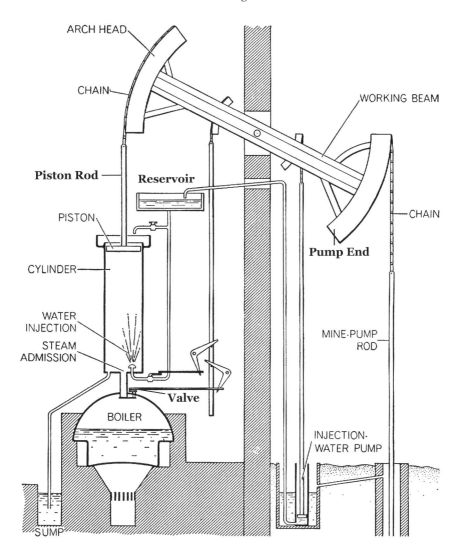

DIAGRAM 59. Operation of the condensing steam engine. Reproduced from Eugene S. Ferguson, "The Origins of the Steam Engine" (1964). Reproduced with permission. © 1964, Scientific American, Inc. All rights reserved.

the steam and produced a partial vacuum. The pressure of the atmosphere above the cylinder (about 15 pounds per square inch) added to the weight of the arch head of the working beam, piston, and piston rod made this end of the beam heavier and drove it down, raising the pump end and with it the injection water pump.

The cycle was completed when the valve into the cylinder was opened, relieving the vacuum in the cylinder beneath the piston. As the pressure approached that of the atmosphere, the pump end of the working beam (again relatively heavier) descended, raising the piston to its original position next to the top of the cylinder.[118] Because water was injected into the cylinder after each stroke, a large nearby water supply was necessary. It was often harder to obtain an adequate supply of water than it was to acquire coal to heat the boilers.

The next step in steam-engine evolution was taken by Scottish inventor James Watt. When he was asked to repair a Newcomen engine in 1763, he realized how he could improve its wasteful design. Watt made three major changes. First, he added a separate condenser that eliminated the injection of cold water into the steam cylinder so the cylinder was not cooled at each condensation, saving fuel that had previously been used in reheating the cylinder after each condensation. The piston, however, was still "pulled" by the vacuum in the cylinder. Second, Watt changed the reciprocating motion of the engine into a rotary motion. And third, he enclosed the top of the cylinder and applied steam to each side of the piston alternately, which increased engine speed. Watt completed his design in 1775 and produced an engine that more efficient and better suited to driving machinery.

The low steam pressure employed in the Newcomen and Boulton & Watt condensing engines required a large, vertical cylinder and piston to eliminate the heavy friction and wear of the piston on the cylinder wall. The condensing engines were too large for plantation mills and required a large supply of water; however, the mills in Charleston used one of these "walking beam" engines. Clearly visible on the roof of the West Point Rice Mill (see figure 26) are eight large cylinders that contained freshwater to fill the reservoir as it became depleted.

FIGURE 26. Rear view of West Point Rice Mill in Charleston, South Carolina, in the early 1900s. The cylinders on the roof are believed to have held freshwater for the condensing steam engine. Courtesy of the Charleston Museum, Charleston, South Carolina.

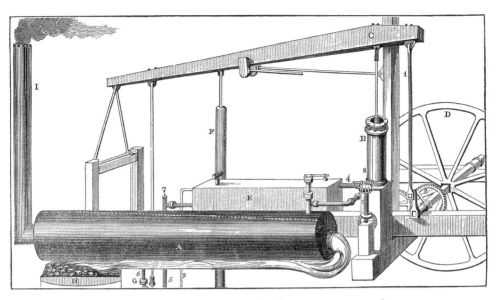

DIAGRAM 60. Oliver Evans's Columbian steam engine of 1811.
Reproduced from P. R. Hodge, *The Steam Engine* (1840).

In 1811 Oliver Evans produced independently in America the noncondensing, high pressure Columbian engine (see diagram 60). The transformation from a condensing steam engine to a noncondensing engine was a major step in steam-engine evolution. The Columbian engine first made use of the expansive nature of steam, the major nineteenth-century advance in steam engines. Although Watt had already anticipated the development of using expansive steam to drive an engine, it is not clear whether the first engines made in America used his theory. Evans, however, championed using expansive steam in his high-pressure engines. His Columbian engine featured a vertical steam cylinder driven by expansive steam, a boiler with a single return flue through the boiler, a furnace below the boiler to produce gases of combustion, a smoke stack to carry away the products of combustion, a large flywheel on the crank shaft, and an overhead beam.

Evans's engine was built with a range of capacities from six to twenty-four horsepower. Evans constructed his boiler of riveted wrought iron, which allowed high-pressure steam for the first time (from 100 to 125 pounds per square inch and even higher). During the period 1814–20, Evans abandoned the flue through the boiler and drew the combustion gases along the underside of the boiler by the draft created by a chimney. This system was adapted by the manufacturers of mill engines.

> Louis C. Hunter described the expansive nature of steam in engines:
> In conventional full-stroke practice, steam was admitted through almost the entire length of the piston stroke . . . the pressure upon the piston was constant

throughout the length of the stroke. On passing into the condenser, on engines so equipped, or into the open air, the exhaust steam contained much of the original energy.... In expansive working, the admission of steam to the cylinder was stopped, cutoff, at some immediate point, ranging from one-half or more of the stroke in early years, to one-fourth, one-eighth, or even less in the more advanced practice and with the higher steam pressures employed during the later years. Following cutoff, the steam within the cylinder expanded as the piston advanced, extending a gradually diminishing pressure upon the piston and performing work in proportionate ratio but bringing a net gain in terms of steam expended.[119]

Evans employed this cylinder cutoff in his high-pressure Columbian engine (see diagram 61). High-pressure steam from the boiler was routed by a rotary valve through a pipe into the cylinder above the piston and drove the piston down with great force. The rotary valve shut off the entrance of steam into the cylinder when the piston had moved one eighth the distance of a full stroke. Under atmospheric power, the steam expanded and decreased in power but pushed the piston to the end of the cylinder with steadily diminishing pressure through the remainder of the stroke. At the end of the stroke,

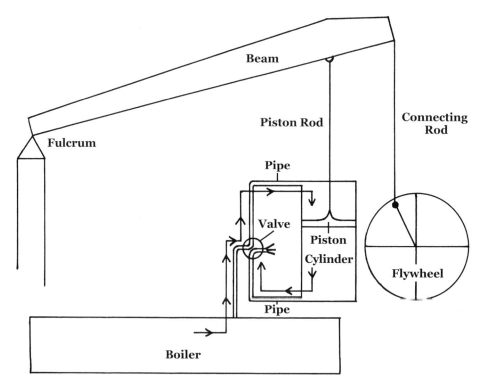

DIAGRAM 61. Operation of the Columbian noncondensing steam engine. Reproduced from Carroll W. Pursell, Jr., *Early Stationary Steam Engines in America* (1969), with permission of Carroll W. Pursell, Jr.

atmospheric pressure inside and outside the cylinder was equalized. The spent steam below the piston was forced through a pipe and the rotary valve and escaped into the air.

When the piston reached the bottom of the cylinder, the rotary valve was turned clockwise ninety degrees, allowing steam from the boiler to enter the bottom of the cylinder through the lower pipe and push the piston upward. Again the rotary valve shut off the entrance of steam into the cylinder when the piston had moved one eighth the distance of a full stroke. The steam expanded and decreased in power, but pushed the piston to the end of the cylinder with steadily diminishing pressure through the remainder of the stroke. The spent steam above the piston was exhausted through the pipe and the rotary valve into the air.[120] The main advantage of the cutoff was the limited use of steam, which resulted in a fuel saving.

Rotary motion was converted into motive power by the beam. As the piston was raised and lowered, the beam, which rested on a fulcrum, was raised and lowered by the piston rod. The beam turned the flywheel by means of a crank and connecting rod. Motive power was then applied to machines from the shaft of the flywheel.

The Columbian engine did not have a condenser; therefore its water requirements were limited to feedwater, the amount of water lost as steam from the boiler, which was far less than that used in a condensing engine. Consequently the Columbian engine was not limited to a location where there was a large water supply. Increase in power to the engine was effected by increasing the pressure of the steam from the boilers, which, because of the Evans design, could take the additional pressure without exploding. The Evans noncondensing steam engine was the forerunner of the mill engine that powered the threshing barns and mills that prepared rice for market.

The Mill Engine

The last major change in what came to be the common mill engine in the United States was the placement of the cylinder in a horizontal position. John Fitch was the first American inventor to attempt to build an engine with the steam cylinder placed horizontally instead of upright, but he ultimately abandoned the use of the horizontal cylinder. Why is not known for certain, but he may have listened to those who said a horizontal piston within the cylinder would place too much pressure on the lower part of the cylinder and cause uneven wear and leakage of pressure. This proved unfounded.

Pursell explained the main advantage of the horizontal cylinder: "The horizontal placement of a steam cylinder permitted a direct linkage between the piston rod and an eccentric arm connected to the flywheel."[121] The direct link eliminated need for the beam and its many bearings and friction. Eliminating the beam also allowed for a more sturdy engine mount and eliminated a costly and cumbersome structure to secure the beam engine in a room. Within the limited space of a small plantation mill, the smaller and more compact horizontal cylinder engine was clearly preferred.

By midcentury, the horizontal placement of the cylinder was the standard construction for steam engines and for the small rice-mill engines. No one inventor or manufacturer was responsible for acceptance of the horizontal cylinder; rather, many companies experimented and tampered with improved piston packing and better construction material, and the horizontal cylinder was established.

The engines of the Industrial Revolution, which were generally thought to have bypassed the agricultural Old South, were actually much in evidence there. Manufacture of steam engines and boilers (plus many other implements and machines for other agricultural industries of South Carolina) required skills of foundries, machinists, and machine shops. Ernest M. Lander, Jr., documented that in the decade before the Civil War, Charleston, although small by northern standards, had a varied and complex manufacturing base linked to cotton, rice, and lumber.[122] Foundries such as Eason and Dotterer, Eagle Foundry, Smith and Porter, Vulcan Iron Works, and Phoenix Iron Works (renamed from Cameron and McDermid, which was destroyed by fire in 1850), manufactured steam engines and boilers, rice and grist mills, threshers, railway locomotives, railway cars, and lumber mills. Charleston's industrial production reached its peak in 1856, possibly topping $3 million ($78.3 million in 2009 money) for the entire district. However, between 1856 and 1861 a series of fires destroyed the three largest rice mills, three iron foundries, a railway-car factory, the largest lumber mill, and two flour mills. Although some of this industrial base recovered, manufacturing in Charleston never reached its previous production level. The Civil War destroyed much that remained.

Savannah developed an industrial base similar to Charleston's, and the Georgia city manufactured much of the equipment for rice and other industries. Like Charleston's, Savannah's industrial development was linked to cotton, rice, and lumber. Shipyards built vessels; rice was milled for market; shops constructed locomotives and railcars; pine casks for rice were produced; foundries built steam engines and boilers; a saw and planing mill produced a variety of finished lumber products.

The steam engine that emerged around 1825 in the Rice Kingdom was technically called the "American, stationary, horizontal, noncondensing, high-pressure steam engine," or mill engine, and employed steam expansion with cylinder cutoff. The mill engine was developed in America and became the workhorse of the rice plantations, gradually replacing water-powered threshing barns and mills and making threshing and milling independent of the tides.

The mill engine did not create industries in the South. Industries such as rice mills, sugar-cane refineries, cotton gins, and sawmills were already established and run by animal or water power. In the main the mill engine was applied to already existing industries and greatly expanded their operations.

When the first mill engine was used on a rice plantation is uncertain. Millwright Jonathan Lucas I is generally given credit for introducing steam to a rice mill in 1817; however, his first engine was probably a beam engine used for a mill in Charleston and not the plantation mill engine. Beam engines were used in the larger city mills. Historians base Lucas's introduction of a steam engine on a letter Thomas Naylor wrote to William Lucas on September 11, 1820, concerning the cost of copper boilers: "They are very expensive but in the end will be a saving to him. Mr. Bowman says he advised them to be sent out when the engine was made."[123] No further record is found on this early Lucas steam mill or the copper boiler. However, horizontal steam engines quickly became standard motive power in the Rice Kingdom, but they were powered by boilers made of rolled iron, not copper.

Carroll Pursell reported that in 1838 there were forty stationary steam engines in operation in South Carolina and twenty-three in Georgia.[124] Robert Allston reported that by 1846 "almost every planter of four hundred acres and upwards, is provided with a Tidewater or a Steam-pounding mill for preparing his own crops for market."[125] Through archival and field research, we located upward of fifty abandoned rice-threshing barns and mills that were powered by steam engines.

Planters converted their water-powered mills and threshing barns to steam. They did not necessarily build new mills or barns. Instead they installed the steam system in the same mills or threshing barns and abandoned their waterwheels. On the other hand, if a planter had no water-powered mill or threshing barn, and he was just getting into rice cultivation, he probably installed a steam system.

Two abandoned mills demonstrate the conversion of a water mill to a steam mill. At Stoke Rice Mill on the Cooper River, the engine was placed inside the mill next to the wall on the side of the millrace. The boiler was mounted outside the building and a chimney constructed in the mill yard. The engine, boiler and chimney have disappeared; only the engine mount remains.

At Middleburg Plantation on the Eastern Branch of the Cooper River, the engine and boiler were situated on the opposite side of the mill from the millrace (see figure 22, insert) and a room was installed to house the engine and boiler. The photograph, taken around 1900, shows a two-story building with a large attic or loft. The structure attached to the front of the main building may have been used for storage of rice before it was shipped to Charleston for export.

Only four mill engines survive in the Rice Kingdom. The loss of engines (and other machinery) began as the Civil War drew to a close and Sherman's army entered South Carolina. Williams Middleton wrote to his sister Eliza M. Fisher on September 25, 1865: "I have just heard that the Hobonny steam engine has been carried off, and also, that a man went to Old Combahee to take down the engine there & remove it."[126] And John H. Screven at Pocotaligo wrote to Williams Middleton on February 14, 1866: "Will you please in form me whether you will Sell the boilers & pump at 'Old Cumbahee' Plantation, and for what sum cash? I have not Seen them—but as I expect to want them—write in time. I learn through Mr. Henry Hucks (who informs me that he is not in the market for them) that all the rest of the machinery, Engine, etc—except the driving wheel, have been carried away—I will take every thing in the shape of your machinery left there—if we can agree on a price—so as to make up that which the tender mercies of our Enemies left <u>me</u>."[127]

Many engines were dismantled and melted down for the World War II effort. In 1978 the Henry Ford Museum in Dearborn, Michigan, purchased from Elise M. Pinckney

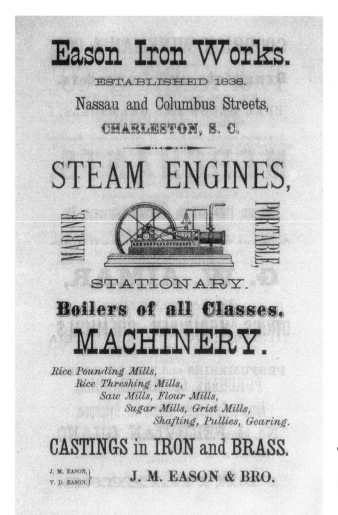

FIGURE 27. A typical mill engine in a flyer for the Eason Iron Works in Charleston, South Carolina. Collection of Richard Dwight Porcher, Jr.

a Cameron McDermid Mustard engine that was located at Fairfield Plantation on the South Santee River;[128] an engine built by Thomas D. Dotterer when he was a partner in the Charleston firm of Eason and Dotterer, Machinists and Engineers, was purchased from Fairfield Plantation on the Waccamaw River.

The four engines still remaining in South Carolina are at Mansfield Plantation in Georgetown County (color plate 20), Cedar Hill Plantation (color plate 21), and Middleburg Plantation on the Eastern Branch of the Cooper River in Berkeley County (color plate 22), and a Cameron McDermid Mustard engine from Fairlawn Plantation in Charleston County, which was moved to the Charleston Museum.

Mill engines were constructed on the same basic plan. In reality, however, manufacturers in Charleston (and manufacturers in other cities of the Rice Kingdom) designed and created their own engines (see figure 27). The photographs of the three extant engines show the diversity of engine design. The engines differed in types of governors, horsepower, steam-chest design, feedwater pumps, and flywheel size and design. Often parts such as governors were obtained from northern manufacturers and installed on local engines. Some engines were bought from northern manufacturers, so there was a mix of local engines and those from the North.

The mill engine was mounted on a brick base and secured by six iron rods. The base was essentially the same design for all steam engines, but varied in size based on the engine's horsepower. Larger horsepower engines required a larger base. The brick base provided a place where the other engine components were secured to hold them in a fixed position. A detached brick base supported the shaft and shaft bearing of the flywheel.

Diagram 62 depicts the basic horizontal, noncondensing, high-pressure mill engine. To begin operation, the throttle valve was opened and steam entered the steam chest. A governor regulated the amount of steam through the throttle valve into the steam chest. The most common type of governor that emerged as horizontal engines evolved in complexity and sophistication was the fly-ball governor. The governor maintained a uniform velocity of the engine, which in turn maintained a uniform velocity of the milling and threshing machines powered by the engine. A belt attached to the engine's shaft operated the governor. Any acceleration of speed of the engine above a given setting caused the

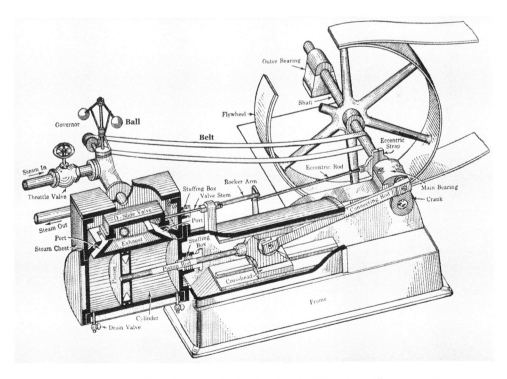

DIAGRAM 62. A basic horizontal-cylinder, simple slide-valve mill steam engine. Reproduced from C. M. Leonard and V. L. Maleev, *Heat Power Fundamentals* (1949).

governor balls to swing outward and partially close a valve in the throttle, reducing the amount of steam passage. Any decrease in the speed of the engine below what was desired to run the milling and threshing machines adequately caused the governor balls to fall inward, increasing the passage of steam into the steam chest.

The slide valve was the key to the operation of the high-pressure engine that worked on steam expansion as championed by Oliver Evans. The slide valve took its motion from the valve stem (see diagram 63) connected to a rocker arm operated off an eccentric collar assembly keyed to the axle of the flywheel. The collar assembly consisted of a wheel with an offset center hole. Around the rim of the wheel was a machined convex bead that acted as a track, which fit into a mating groove cut into a two-piece collar that clamped around the diameter of the wheel. The eccentric wheel rotated within the collar which was held in a somewhat stationary position by the rocker arm operating the valve stem of the slide valve. Diagram 64 illustrates the reciprocating action of the slide valve as the eccentric collar assembly was rotated through one revolution of the flywheel. (The steam chest in which the slide valve operated has been omitted.)

In step 1 the eccentric wheel's offset (A) is shown to the left of the flywheel axle. The stationary collar has pushed the slide valve all the way to the left—the end of its stroke. To begin the cycle of the slide valve, the flywheel axle rotated the eccentric wheel's offset toward a straight down position as shown in step 2, causing the collar to pull the slide valve to the right. As the offset began to rotate (step 3) toward a straight-up position in step 4, the collar began to push the slide valve back to the left. As the wheel continued to rotate, the offset moved toward the left of the flywheel axle, causing the collar to push the slide valve further left, back to its original position (step 1), completing its cycle. Steam entering the engine's cylinder pushed the piston back, which rotated the flywheel axle to begin another cycle of the slide valve.

The slide valve in the steam chest operated as follows (see diagram 64): Steam entered the engine through the steam chest which contained the slide valve. To begin the engine's operation, the slide valve (V) in step 1 was moved to the right by the action of the valve stem uncovering the head-end cylinder port (H). High pressure steam (S) quickly flowed through the port into the head end of the cylinder pushing the piston (P) to the right. After the piston had moved one eighth of the full stroke (the cutoff point), the slide valve (step 2) moved to the left, blocking any more steam from entering the head end of the cylinder, while at the same time opening the cylinder port (C) at the crank end. The piston continued to move to the right because of the pressure of the expanding steam (ES), which pushed the air (A) in the crank end of the cylinder (first start up) out through the cylinder port (C) to the exhaust port (E). As the piston neared the end of its stroke to the right in step 3, the slide valve moved further to the left, closing the cylinder port (C) to the exhaust. The remaining air (A) served as a cushion for the piston as it approached the end of its stroke.

In step 4 the slide valve moved still further to the left, opening cylinder ports H and C at each end of the cylinder. Steam quickly flowed through the open port (C) at the

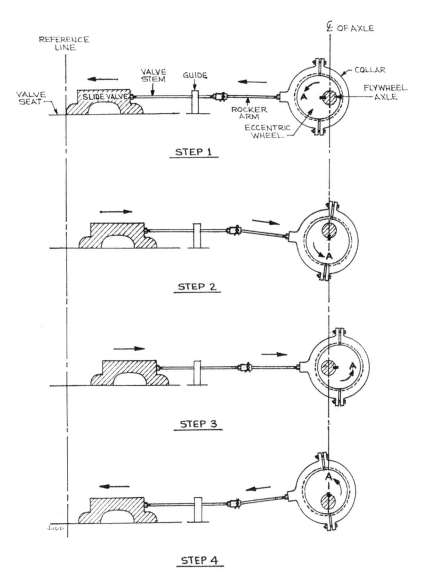

DIAGRAM 63. Operation of the valve stem.
Illustration by William Robert Judd.

crank end, pushing the piston to the left in the reverse direction. When the piston had moved one eighth of s full stroke (again, the cutoff point), the slide valve (step 5) was moved to the right, blocking the steam from entering the crank end of the cylinder and at the same time opening the head-end cylinder port (H) to the exhaust port (E). As the expanding steam (ES) pushed the piston to the left, the spent steam (SS) in the head end of the cylinder was expelled through the cylinder port (H) into exhaust port E. The slide valve (step 6) was moved further to the right closing the head-end cylinder port (H)

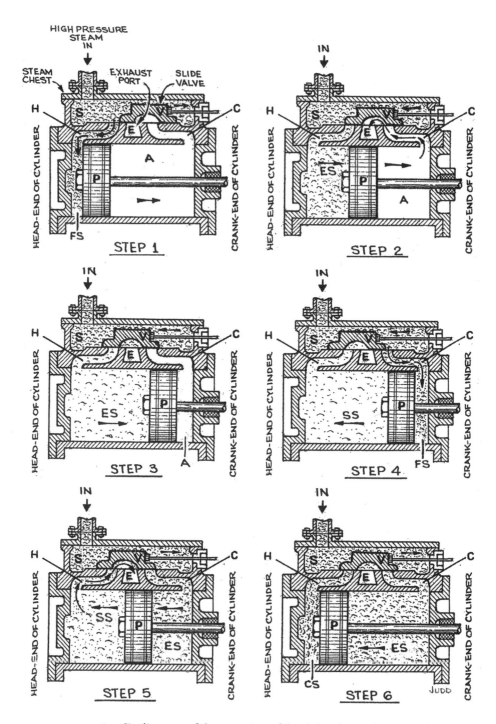

DIAGRAM 64. Six diagrams of the operation of the slide valve in the steam chest. Illustration adapted by William Robert Judd from four diagrams in Terrell Croft and E. J. Tangerman, *Steam Engine Principles and Practice* (2nd edition, 1939).

to the exhaust. The remaining spent steam (SS) in the head end of the cylinder became compressed steam (CS) acting as a cushion for the piston at the end of its stroke to the left, completing one full cycle of the engine.

To begin another cycle the slide valve was moved still further to the right, opening cylinder port H at the head end of the cylinder as in step 1 and allowing fresh steam (FS) to send the piston back in the opposite direction toward the crank end of the cylinder. With the slide valve taking its motion through the valve stem from the rotating flywheel, the reciprocating motion of the piston was automatic and continued as long as the throttle valve was open.

The motion of the piston was transmitted through the piston rod to the crosshead (see diagram 62). The crosshead was attached by a connecting rod to a crank on the flywheel shaft, and the reciprocating motion of the piston was converted into rotary motion of the flywheel. A belt was attached either around the flywheel or to a pulley mounted on the shaft on the outside of the flywheel. The belt ran to an overhead pulley on the outside of the barn whose shaft operated a pulley assembly that ran all the machinery inside (see diagram 65). The machinery shown in diagram 65 is conjectural and is shown only to present the capabilities of the axle/pulley assembly to transfer power.

Attached at one end to a shaft that was supported by an outer bearing on a brick foundation, the heavy flywheel was attached at its other end to the engine and equalized its movement. The flywheel's inertia opposed any sudden acceleration of speed and prevented any sudden decrease in speed. At the end of each stroke of the piston, before the piston reversed direction, there was no motion for a brief period. The flywheel carried the piston past this dead period.

Unfortunately no mill engines are operational anywhere in the former Rice Kingdom. The mill engine was the dominant energy power that ran the two great agricultural industries of the South for one hundred years: Carolina rice and cotton (both short staple and Sea Island). The Henry Ford Museum, in Dearborn, Michigan, has a Cameron McDermid Mustard engine on display run by electrical power but no working steam engine.

Portable Engines

Just before the Civil War, the Charleston manufacturers Smith & Porter (successors to William Lebby) and Taylor Iron Works Manufacturing Company built portable steam engines. Smith & Porter built the most successful portable engine. A restored Smith & Porter portable engine built around 1858–60 is located in the Roundhouse Museum in Savannah, Georgia (see color plate 23).

Portable engines gained most use in the lumbering trade because, as timber was used up at a site, the engine could be hauled to the next site to run a sawmill. They probably had limited use in the rice industry because a rice plantation grew rice in the same fields year after year. It is possible that someone who was starting a rice mill or threshing barn just after the Civil War and wanted steam power might have used a portable engine. A

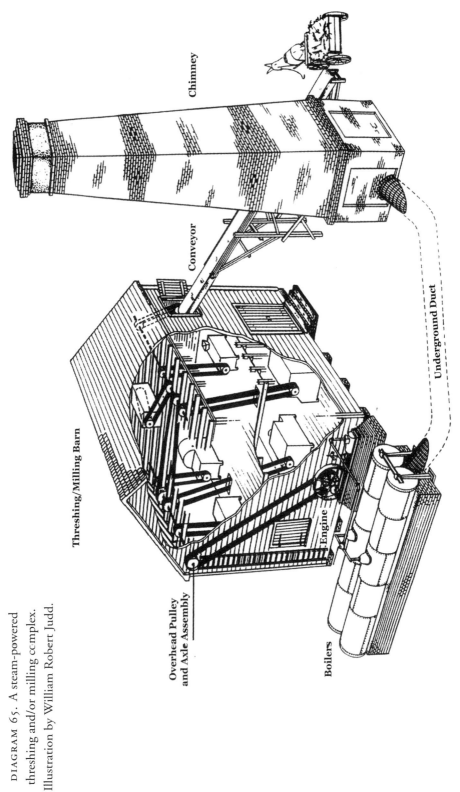

DIAGRAM 65. A steam-powered threshing and/or milling complex. Illustration by William Robert Judd.

FIGURE 28. A portable steam engine on the Eastern Branch of the Cooper River in Berkeley County, South Carolina, circa 1900. From the Middleburg photograph album, courtesy of John Ernest Gibbs.

portable engine was easy to install and cheaper than bringing in boilers, mounting a stationary steam engine, and constructing a brick chimney. Figure 28 is an unknown rice mill. Plainly visible is a tall metal smoke stack that was part of a portable steam engine.

The Plain Cylindrical Boiler

Because of its simplicity and low initial cost, the plain cylindrical boiler was favored where fuel was cheap and efficiency secondary. Consequently it was the boiler of choice to power the mill engine. Fuel from nearby forests was plentiful, and straw from threshing could also serve as fuel. In addition the energy requirement of the steam engines used in the threshing barns and mills was not great. Plain cylindrical boilers were easy to make and simple to maintain, clean, and repair. They had a large water capacity in relation to their steam-making capacity, which enabled maintenance of a steady steam pressure with less frequent attendance. Because of their sturdy construction, they could produce the high pressure needed to run the mill engines. [129]

The cylindrical boiler had its origin in England. Boilers for low-pressure Newcomen, Boulton, and Watts steam engines were first constructed of copper. Although expensive, copper sheets were smooth, making lap joints tight when riveted, and heat transferred quickly. With the advent of steam engines to power boats, copper was used in the boilers because it was not subject to corrosion in the salt atmosphere.

At least one cylindrical boiler was made of copper. Thomas Naylor wrote to William Lucas on September 11, 1820: "I was anxious to have sent your brother's account long ago but would not until I had shipped the copper boilers and paid for them which I have now done, with all the expenses on the same. They are very expensive but in the end will be a saving to him. Mr. Bowman says he advised them to be sent out when the engine was made but your brother John opposed it, when they are worn out I am told they will be worth nearly half their cost as old copper."[130]

With the increase in popularity of the stationary steam engine and later the mill engine, cast iron became the choice over copper for boiler construction because of its low cost and ease of fabrication. Cast-iron boilers, however, were short lived because fire playing directly on cast iron causes the material to become brittle and crack, and punching and reaming rivet holes caused small cracks that deteriorate under tremendous heat. Black iron sheets were next used for boiler construction. When the sheets were hammered flat and riveted to make the boiler, however, they were not smooth enough to make a tight seal at the joints.

In the early 1800s, wrought-iron sheets became available and many of the fabrication problems disappeared. Wrought iron could expand and contract and was malleable, allowing the edges to be hammered smooth for a tight seal; rivet holes could be drilled instead of punched and reamed. Wrought iron was rolled in sheets and riveted together in alternating, overlapping sections (see diagram 66, color plates 24 and 25, and figure 29). Sections of the boiler could be fabricated and shipped in pieces to the site or completely assembled in the shop and shipped to the plantation. One thing is certain: A single boiler weighed around 7000 pounds, and it must have been a difficult task to transport an assembled boiler from the city to the plantation and install it.

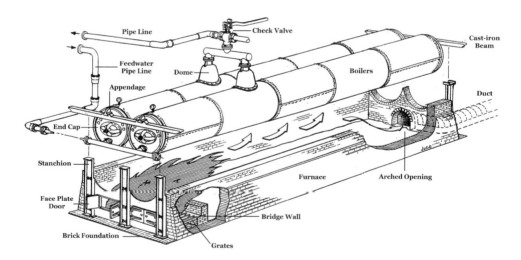

DIAGRAM 66. Boilers and a furnace. The diagram of the boilers is based on the remains of those at Middleburg Plantation. Boiler systems were essentially the same in all mills, but there were minor modifications based on the preference of manufacturers and whether there were one or two boilers. Illustration by William Robert Judd.

FIGURE 29. A boiler at Social Hall Plantation on the Ashepoo River in Colleton County, South Carolina. Photograph by Richard Dwight Porcher, Jr.

To assemble the boiler, red hot rivets were inserted through the aligned holes of the plates from inside the boiler. A large hammer or block of iron was held against the head of the rivet while a worker on the outside peened the molten rivet to create a head on the outside. The lapped edges of the plates were riveted approximately every two inches.

Only four plain cylindrical boiler units survive in the former Rice Kingdom (all in South Carolina). They are at Middleburg Plantation on the East Branch of the Cooper River (color plate 24), Nemours Plantation of the Combahee River, Social Hall on the Ashepoo River (figure 29), and Atkinson Creek on the North Santee River in the Santee Delta (color plate 25). The mills at Middleburg and Nemours had double boilers; Social Hall and Atkinson Creek had single boilers. The brick setting of Middleburg is partially intact; the boiler settings at Nemours, Social Hall and Atkinson Creek are gone. The description and diagram of a boiler unit is based on the remains of these four boiler units and descriptions in archival sources.

Boilers supplied steam to the engines, which induced motive power to operate machinery. Smaller horsepower engines (10 to 20 hp) required only one boiler; larger engines (25 to 30 hp) required two boilers. The boilers in the Rice Kingdom were circular in design (the strongest type possible) and ranged from twenty-four inches in diameter to a diameter not exceeding thirty-six inches.[131]

Boilers ranged in length from twenty to thirty feet.[132] The end of each boiler was fitted with a formed cast-iron cap riveted in place (see diagram 67). The front cap had an elliptical opening, called a manhole, which allowed access into the boiler for cleaning. The opening was sealed from the inside with a cast-iron assembly made up of three parts: a ribbed convex cover plate with an inserted cast shaft threaded on its outer end, a formed compression bar, and a large nut. The cover plate, with a half-inch-thick rubber gasket, was inserted through the opening and pulled against the inside surface surrounding the opening. The formed compression bar, two arms cast to a hub, was slipped on the threaded shaft from the outside and pushed against the outer embossed lip of the

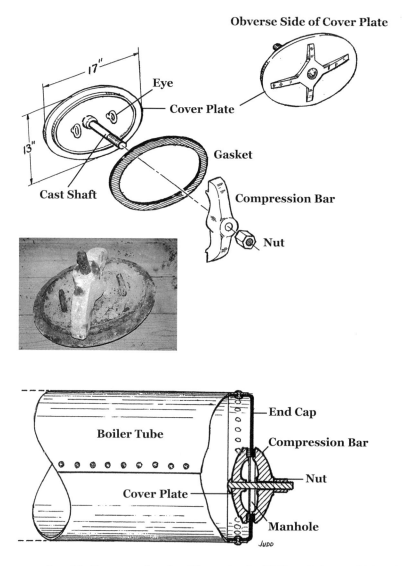

DIAGRAM 67. A boiler cap and gasket. Illustration by William Robert Judd based on two cast-iron cover assemblies found in the threshing barn at Chicora Wood Plantation.

manhole opening. The nut was then threaded on the cover's shaft. Projecting outward from the concave side of the cover plate are two elliptical eyes, one to the left and one to the right of the threaded shaft. These eyes possibly received a rope loop or tool to assist the installer in holding the cover plate in place against the surface of the opening while snugging the assembly together. The nut was then tightened, sandwiching the boiler's end cap between, thus sealing the opening.[133]

Boilers were installed in place independent of support of the brick setting (see diagram 66). The boilers were first positioned to the proper height and slightly inclined for drainage. Cast-iron stanchions of a predetermined height (three stanchions at the firebox end and two at the duct end) were secured to a brick foundation and topped with a cast-iron horizontal beam that supported the boiler's weight. Cast seats on each horizontal beam accepted an appendage cast on each end cap of the boiler tube. These appendages carried the weight of the entire boiler.

The furnace, built of common brick, was constructed to provide a sealed area below the boilers from which hot gases from the burning fuel could act on the bottom of the boilers and provide the heat to create steam. Grates were installed beneath the front end of the boilers and a cast-iron faceplate with doors was secured to the front stanchions to allow fuel to be placed in the firebox. A bridge wall separated the firebox from the furnace area, confining the contents of the firebox. The draft from the chimney drew oxygen into the firebox and lifted the hot gases[134] from the burning fuel up over the bridge wall. The height and position of the bridge wall forced the gases and particles from the burning fuel along the underside of the boilers (see arrows), increasing the heating efficiency. At the opposite end of the furnace, an arched opening led to an underground duct through which particles and gases from the burning fuel were drawn to the chimney and expelled.

Heat from the furnace converted water within the boilers into steam. Expanding pressure forced the steam through the boiler's dome into a pipeline controlled by a check valve. This pipeline ran to the engine's governor, which controlled the pressure to the engine. A feedwater pipeline from the engine's feedwater pump replaced the water lost as steam from the boilers. The boilers (and steam engines) were protected from the elements by a roof or shed extending off the side of the barn that housed the threshing or milling machinery.

It was important to keep the proper water level in a boiler. There had to be enough room for steam to form—the upper one-third to one-fourth of the boiler diameter—and the water level had to be maintained high enough in the boiler shell to prevent heat from weakening the iron plates. Two types of water-level gauges were used during the 1800s before the Civil War: try cocks and a glass-tube gauge (see diagram 68). Of the four boilers left in the Rice Kingdom, only the Middleburg boiler has a remnant of the water-level gauge, a glass-tube gauge.

The glass-tube gauge was a small-diameter glass tube joining upper and lower gauge cocks so that the water level in the boiler was extended to the glass tube outside. The tube was positioned on the faceplate so that the proper water level filled the tube halfway. If

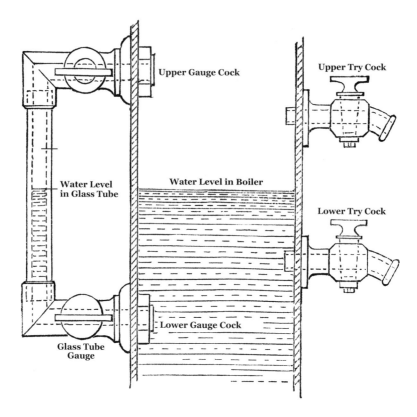

DIAGRAM 68. Water-level gauges for a steam boiler.
Reproduced from P. R. Hodge, *The Steam Engine (1840)*.

---

the tube was empty, the water level was too low, and if the glass tube was completely filled with water, the level was too high.

Try cocks were fastened just above and just below the desired water level in the boiler. If steam issued when both cocks were opened, the water level was too low; if water issued from both, the water level was too high. If steam issued from the upper cock and water from the lower cock, the water level was adequate. The try cocks and glass-tube water gauge were not completely adequate to measure low-water in a boiler. Since low-water was a major cause of boiler explosions, other low-water detectors were devised. These relied on elaborate arrangements of floats, levers, rods, gears, and wheels—all designed to give a warning signal through a visible indicator, or a whistle or bell when the water level fell below a designated danger point. In some cases a safety device activated the water pump to raise the water level. If any of these warning devices were used in Rice Kingdom boilers and mill engines, it is not recorded. With the advanced level of technology prior to the Civil War, however, at least some Charleston manufactures must have experimented with these warning devices.

Boiler explosions did occur, with tragic results. William Daniell, a Savannah doctor and prominent Savannah River rice planter, installed a modern steam mill on his plantation, and as Savannah rice factor Robert Habersham reported in January 1852, "Dr. Daniell's mill boiler exploded yesterday, killing one negro & numbing several others, destroying some of the Mill, in a slight measure, and burning 13000 bushels of rice in the Ricks. We sent a [fire] engine & the Dr. thinks the [illegible] may be saved perhaps a good deal of the rice. All hope so. The explosion was seen & heard for miles and from the boiler, parts of it, being hurled 80 yards all the ricks 20 or more were instantly fired, so that they could not stop the fire before it got well ahead.[135]

Since spent steam was lost from the engine with each stroke of the piston, water had to be put back into the boiler to replace spent steam. Initially a feedwater force pump was used to force feedwater into the boiler against the pressure of the boiler. The force pump was powered from some moving part of the engine—beam, crosshead, or main shaft—and forced a small amount of water through a one-way "check valve" against boiler pressure into the boiler with each stroke of the engine. Color plate 21 of the Cedar Hill engine shows a device on the side of the engine that was probably the force pump.

Feedwater drawn from brick-lined wells or cisterns was probably the standard. Brick wells were located at two mill sites with a lead pipe projecting through the brick wall into the well. The pipe probably extended to the force pump of the engine, and the force pump drew its water supply from the well and then forced the water into the boiler. How the force pump regulated the amount of water injected into the boiler is unknown. Some method must have insured that the boiler was not filled to the top or the water fallen too low. Both events would exact serious consequences.

Like all mechanical parts, the force pumps gave problems. K. Washington Skinner wrote to Charles Manigault from East Hermitage on October 29, 1852: "Everything about engine being in readiness, John made steam on Wednesday, but could not get the water up by pumping,—had to fill the Boilers with pails, buckets &cs., then got up the steam easily, & John put the Engine in motion with the greatest facility, and she moved finely. The pumps were soon put on, but they failed to keep up the water. So we had to stop and fix the pumps,—got up steam yesterday morning when the pumps failed again."[136] How John filled the boiler without using the pump is unknown. None of the boilers documented in the field had an access or removable cover to allow manually filling the boiler with buckets of water. The manhole could not have been used because the bottom of the opening was so low that the boiler could not have been filled but a few inches. One possibility might have been to remove the blow-off valve assembly mounted on top of the dome. This allowed a four-inch opening down through the dome into the boiler. Once the water reached the correct level in the boiler, the blow-off valve would have been reinstalled, and the boiler could be fired.

Evidently there was no standard type of force pump in the engines made locally or imported from the North. This was especially true early on when construction of the mill engines was in its infancy. Engine makers experimented with their own force pumps.

Unfortunately the four engines mentioned above have either suffered damage to the force pumps or the pumps are gone, making it difficult to determine how these force pumps operated. During the 1860s, two new methods to add feedwater to the boiler emerged: the steam injector and the direct-acting steam pump.[137] No evidence was found that either of these two pumps was employed locally.

From the beginning of steam power, fuel for the boiler furnace was wood from the pine trees (lightwood) and hardwood of the nearby forests. The wood was either purchased off the plantation if there was not an adequate supply or cut on the plantation by slaves. The planters also made use of the straw from the threshing machines. Straw, mixed with pine, or burned alone, gave an intense heat. This once-discarded refuse provided an endless supply of fuel at no cost to the plantation. As Charles Manigault told Louis Manigault in 1856, "Venters [the overseer] writes that he will start the Thresher the middle of this month, I told him to go over with 'Jimmy' & 'Isaac' to see how they feed & carry on entirely with straw. D. Heyward tells me they use not a stick of Lightwood, & that straw is better in every respect. So I got no lightwood this season."[138]

The construction of the boiler shell (riveted together in alternating, overlapping sections) allowed a boiler to be lengthened in place. If a planter found that his original boiler did not supply sufficient steam for his engine, additional sections could be added to the boiler, and it was not necessary to buy a new boiler or have the boiler sent to city shops to be retrofitted. In the same 1856 letter Charles Manigault wrote to Louis Manigault: "I have had much talk with Eason & Dotterer about our steam thresher, Engine, &c., & with others, & believe this is our best plan. Add 5 feet to the length of our two Boilers, by disengaging the Heads of the Boilers & advancing them out 5 feet in front, which will take 2 sections more of Boiler Iron, the front wall, fire place, & grating ash hole & all to be also brot. forward."[139] The cylindrical boiler provided a reliable source of steam to drive the mill engine, and remained basically unchanged until portable steam engines replaced the mill engine.

Rice Chimneys

The rice chimneys of the abandoned steam-powered threshing barns and mills stand in silent testimony to the ingenuity and professionalism of the designer-builders and black artisans, and as solitary witnesses to the bygone activity of once-thriving rice plantations. Of all the artifacts from rice culture, the chimneys stand preeminent as the reminder of this extensive agricultural enterprise. Field surveys covered the entire coast of South Carolina and excursions were made into North Carolina and Georgia, documenting fifteen extant chimneys in South Carolina and three in Georgia. An extensive photographic collection of the chimneys, many of which are included in this book, resulted from the field surveys.

Little material was found in archival sources on the construction and origin of the rice chimneys. Most of the descriptions of rice chimneys come from examinations of extant chimneys and archival sources dated well after their construction and use. The basic concept of steam-engine chimneys, however, is similar, allowing a reasonably accurate history of the construction and function of the rice chimneys.

The rice chimney owed its origin to Great Britain and the Industrial Revolution. In Great Britain a chimney was first employed for the steam engine invented by Thomas Savery in 1689. John Farey, in his 1827 book *A Treatise on the Steam Engine,* included an engraving of Savery's steam engine for raising water.[140] The engraving includes a drawing of a chimney (see diagram 69) that in all outward appearances is similar to the rice chimneys. Whether it had all the individual structural features of the rice chimneys is unknown. Both had the same function.

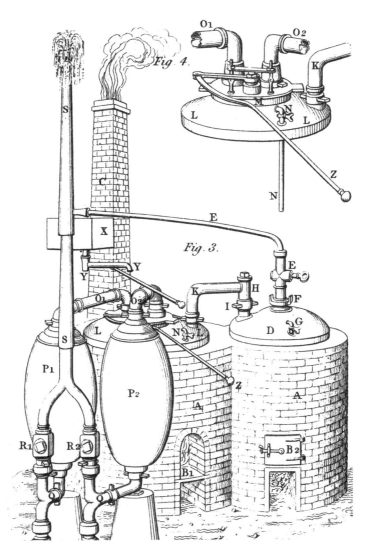

DIAGRAM 69. Thomas Savery's brick chimney, Great Britain, 1689. Reproduced from John Farey, *A Treatise on the Steam Engine* (1827), which republished the engraving from Thomas Savery, *A Description of the Draft of the Engine for Raising Water by Fire* (1702).

By the time horizontal steam engines were first used to power threshing and milling machines, the technology of chimney construction was already advanced in Great Britain. Who first brought the technology of chimney construction to the Rice Kingdom in the early 1800s is lost to history. Millwright Jonathan Lucas I, a native of England who built the first steam-powered rice mill in Charleston on Mill Street in 1817, may have introduced the rice chimney, but nothing in the Lucas papers supports this possibility.[141]

An 1831 article in the *Southern Agriculturist* contains the only description of chimney construction dating to the beginning of the appearance of chimneys on rice plantations. The article, written by James Hamilton Couper of Hopeton Plantation, on the Altamaha River in Georgia, describes a chimney used for a steam engine that drove the sugar-mill machinery on West Indies and Louisiana plantations. The engine pulled the cars containing the sugarcanes onto an elevated plane where the rollers were located and drove the rollers that pressed out the sugar. Couper stated the engine was "intended to drive rice-pounding and threshing-mills when not used for grinding canes; the power is greater that is required for the sugar mill alone."[142]

Couper's statement suggests that a steam engine could have been used to power a rice mill. The engine in question was "on the low pressure principle and of the portable form, from the manufactory of Boulton & Watt [of England], and is of a fourteen horse power."[143] The engine was enclosed in a separate room set off from the cooling and curing room of the sugar mill, and the boiler was mounted in a small wing on the outside of the main building. The chimney was fifty feet tall and placed adjacent to the boiler. The spatial arrangement of the engine, boiler, and chimney was similar to the engine-boiler-chimney complex that developed on rice plantations, with the exception that most rice chimneys were placed away from the barns and mills. Whether the rice mills and threshing barns were copied from the sugar mills or established independently is unknown. Since Lucas built his first steam-powered rice mill in 1817, an independent origin seems probable.

Because steam engines were invented in Great Britain, it was there that a search for literature on the chimneys (and steam engines) began. In May 2006 a trip to Great Britain resulted in a visit with H. Alan McEwen of Keighley, an expert on steam-engine restoration who is also knowledgeable about the history of steam engineering and chimneys.[144] The following notes were recorded from our discussions on steam mills:

1. The function of the chimney was to pull gases of combustion under the bottom of the boiler to increase surface area of the boiler exposed to heat, and to induce oxygen for combustion into the firebox.
2. The crown on the chimney top was to lift and diffuse the smoke and not allow the smoke to fall back down the chimney or on the surrounding area. The crown also added strength to the top of the chimney.
3. The inner chimney was to protect the outer chimney and to increase the strength of the draft in the lower section of the chimney. (A narrow diameter creates a stronger draft.)

Milling 251

The brick chimneys of the rice plantations, like the factory chimneys, designed to create a draft to draw oxygen into the furnace, to pull the gases of combustion along the underside of the boiler, and to disperse the gases, smoke, and cinders from fuel combustion. A well-designed chimney consisted of a base, stack, flue, and crown (see diagram 70 and color plate 11). The base was the foundation on which the stack rested. The lower portion of the base and flue extended four to five feet below grade to ensure stability in

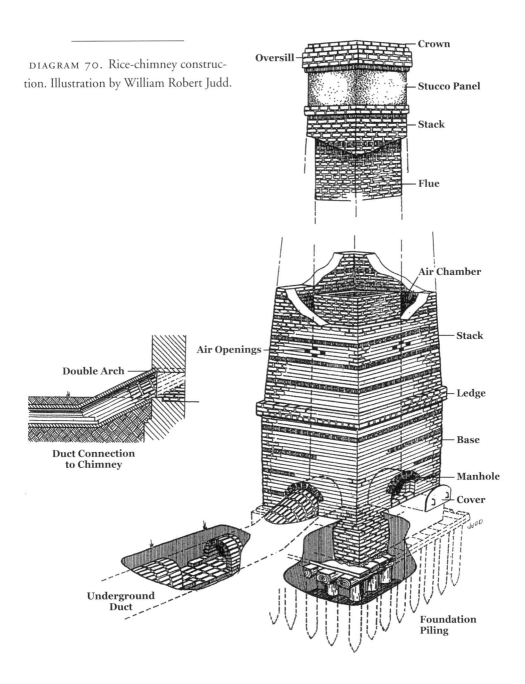

DIAGRAM 70. Rice-chimney construction. Illustration by William Robert Judd.

the soft soil. In marsh areas, where chimneys were occasionally built, a foundation of underground pilings was constructed to add support to the chimney. Two chimneys, at Fairfield and Black Out Plantations,[145] have no defined bases and may have been of earlier construction (see color plates 26 and 27). One destroyed chimney in Georgia has no defined base. With one exception the bases were square, and the shape of the base had no bearing on the shape of the stack. Both round stacks and square stacks on square bases occur locally, as well as in the factories of Great Britain.

All the brick chimneys we surveyed were made of common brick measuring 2½ × 4 × 9 inches and known as "Carolina grey brick" along the South Carolina coast and "gray brick" in Savannah, Georgia. The walls of the stack and base were two bricks thick tied together every sixth course with a horizontal header course. This pattern is called a common bond. In some chimneys this pattern is every fourth course while others follow no uniform sequence to the common bond pattern. All mortar joints are plain and struck flush with the face of the bricks.

All the stacks had batter, a tapering inward of the stack toward the top. Batter is generally found in all chimney stacks of the industrial world, both past and present. Batter was a structural feature that counteracted horizontal force (wind) from an opposing side.

Although the factory chimneys in Great Britain were made on a much larger scale and were of more complex construction, they shared similarities with the rice chimneys. In Great Britain there were three basic stack shapes: round, octagonal, and square. These three shapes occur on rice plantations; one chimney, however, found at Laurel Hill Plantation, was built in a star pattern (see figure 30).

Flues of the chimneys copied the shape of the stack. (The one exception is the star-shaped chimney stack at Laurel Hill which has a round flue.) The round flue was more effective because it is the natural form of an ascending column of gases from a fire; however, the cost of construction of a round stack with a round flue was more expensive. The chimney at Fairfield (see color plate 26), with an octagonal stack and flue, is next most effective, and its flue and stack the next most expensive to construct. The Fairfield chimney is the only extant octagonal chimney, although the bases of two others have been found. The square stack and flue were the least effective functionally, but this type of flue and stack was the least expensive to construct because of the rectangular shape of common bricks.

Most rice chimneys had square stacks and square flues (see diagram 70).[146] Functional efficiency was sacrificed for cost; however, since the chimneys were not in everyday use (as were factory chimneys in Great Britain) and were used with smaller steam engines, this was an acceptable sacrifice. Square chimneys were also less wind resistant; high winds, however, were not common, except during severe storms, so wind damage was minimal.

The only extant chimney with a round stack and flue is at Willtown Plantation (see color plate 28). Even though its construction is round, rectangular bricks—not specially shaped bricks—were used. Constructing a round chimney and flue using rectangular-

FIGURE 30. The rice chimney at Laurel Hill Plantation on the Waccamaw River in Georgetown County, South Carolina. Photograph by Richard Dwight Porcher, Jr.

shaped bricks was more complex than the octagonal or square design. Each graduating course was laid in a circle radiating outward from an imaginary center point. Unlike the vertical brickwork of the flue, special attention had to be given to the stack as the diameter of each course was smaller than the previous course, thereby creating batter to the top of the chimney. This was a costly design and more difficult to construct.

The junction between the base and stack could be made abruptly, without any attention to an agreeable line. In rice chimneys, however, a brick ledge separated most stacks from the base, adding to the overall aesthetic appearance (see diagram 70). There was considerable variation in the number of brick courses of the ledges, how far the ledges extended from the base, and how each course related to the others.

A rice chimney was topped with a masonry crown, which was both ornamental and functional (see color plate 27). A chimney is essentially a cylinder, and a cylinder erected in the air creates air currents that caused smoke and gases to eddy and trail back down the exterior of the chimney instead of being dispersed into the atmosphere. Since the chimneys were near the barns, this could create a serious problem for the workers. When correctly built, chimney crowns were supported with one or more projections called oversillers, which changed the direction of the air currents, directing them upward where winds carried the smoke and gases away from the barns. With one exception, crowns or remnants of crowns with oversillers are found on all the rice chimneys.

With the exception of the chimney at Black Out, all the chimneys surveyed had an inner stack called the "flue" (see diagram 70). Some chimney flues reached to the top of the outer stack; others went from halfway up to three fourths. Apparently there was no standard height for the flue, and decisions about height rested with the designers. Although flues of factory chimneys were made of fire brick, no flues made of fire brick were found in the Rice Kingdom (although some of the many destroyed chimneys may have been).

The flue was independent of the outside stack and separated from it by an air space that tapered upward along its entire length. The flue had two main functions: to create a draft and to protect the main stack from radiated heat from the gases being discharged. The narrower diameter of the flue created a stronger draft. Without the flue alternating cooling and heating of the outside stack would cause expansion strain and could cause damage to the mortar, causing the stack to crack or collapse.

Over time both flue and stack suffered from the elements and began to deteriorate. Charles Manigault suggested a way to stabilize a weakened stack, writing to Louis Manigault: "With regard to the Chimney, it stands up well in line, & is solid, except the cracks on each side. There is a way of fixing this by putting 4 square pieces of flat iron all round it. We will see about it."[147] The chimney on Hope Plantation (see color plate 29) on the South Edisto River was braced with flat iron.

The heights of the chimneys vary from thirty to sixty feet. The chimney at Laurel Hill (see figure 30) is the tallest, and the one at Black Out is the shortest (see color plate 27).[148] The height of the chimney served an important purpose: to ensure that the chimney did not emit any sparks which could have started a fire. Both threshing barns and mills held flammable material that could be ignited by sparks. To prevent fires, planters used slate, tin, or wood shingles painted with fire-proof paint for the roofs. Particles of fuel from combustion in the boiler furnace tended to burn up as they progressed through the underground duct and up the chimney. As long as the chimney was tall enough, all the fuel particles burned out before they could be emitted from the stack. Touring Jehossee Island in 1850, Solon Robinson noted: "The flue [duct] is carried off fifty or sixty feet along the ground and there rises in a tall stack that never emits any sparks."[149] Whether there was a mathematical calculation to determine the correct chimney height for a particular boiler, or whether chimney height was determined by trial and error, is not recorded. But it is certain that all builders and planters knew that the chimney had to be high enough for sparks to be extinguished.

The chimney was connected to the boiler by an underground duct (see diagrams 65 and 70). Products of combustion were drawn from underneath the boiler through the duct by the chimney draft and were emitted from the top of the chimney. Most of the ducts collapsed or were dug up for the bricks, but enough remnants remain to document their construction. The underground duct was made of brick with an arched roof to give added strength since it was probably not possible to keep workers and carts off the duct. More important, being placed underground insulated the duct, preserving the heat of the gases from the boiler, which contributed to the chimney's draft.

The distance from the chimney to the boiler's furnace and the path of the duct varied for each chimney. Some ducts ran in a straight line, some angled off left or right, and others had curving paths. The ducts ran twenty to fifty feet before each end exited the ground, at approximately thirty degrees. One end entered the arched opening at the base of the chimney; the other end entered the boiler's furnace.

Most chimneys have one or two arched openings in the base (see figure 30) on the side or sides not taken up by the duct. Solon Robinson detailed the function of the opening in the Jehossee Island chimney: "Governor Aikin [Aiken], however, has one improvement that I recollect mentioning to Mr. B., that he would require; that is, a 'man hole' into this flue, to enable him to clean out the great accumulation of cinders at the bottom of the stack. In Gov. A.'s, there are two which are closed by iron covers."[150]

No iron covers were found at any of the fifteen chimneys; they may have been lost or sold as salvage. Or perhaps the use of iron covers at Jehossee may have been a concept introduced by the builder or the overseer and not used elsewhere. Manholes may have been closed with loose brick, which could be easily removed for access to the flue. Additionally being able to change the number of bricks meant one could change the size of the manhole opening, which could alter the strength of the flue's draft.

The manholes entered into the flue, and if necessary, an individual could crawl through into the chimney to clean out the ashes. The ashes, however, could have easily been raked from the interior through the manhole. Cleaning was probably an annual event done in the off-season. Chimneys that had no manholes may never have been cleaned out, or perhaps the connection of the brick duct was disassembled to give access for cleaning. The bricks could be replaced after the cleaning was completed, restoring the seal between the duct and flue.

Two aspects of smoke-duct construction were important.[151] First, the duct leading to the chimney could not be so narrow that it constricted the passage of gases from the boiler. Second, it could not be so large that the passage of gases was so rapid that their heat was not communicated to the bottom of the boiler. Nothing in archival sources on construction of chimneys documented that these facts were known to the planters or builders. The chimneys worked, however, so it must be assumed that this knowledge was available and applied.

For many years some presumed that a fire was started in the flue of the chimney to begin the draft. Access to the interior of the flue to start the fire would have been through the manholes. No evidence was found in archival records or during field studies to substantiate this presumption. Arguing *against* this concept is that some chimneys have no manhole openings. A revised theory is that sufficient draft was created by air passing over the opening at the top of the chimney to create a natural draft back to the boiler's furnace. The draft then was increased by the flow of hot gases from the furnace.

Several chimneys have a series of air openings of various patterns in the stack (see color plate 11). The openings lead into the air chamber between the stack and flue and were created by the omission of half bricks. The openings are placed from one to three intervals up the exterior height of the stack. On some chimneys these openings appear on all four sides. On other stacks they appear on the front and rear face, while some stacks have no openings.

The function of the openings is not recorded. Some suggest that they were "put-log holes" to hold horizontal timbers that supported the builder's scaffolding. Normally put-log holes would have been filled after the chimney was built. Two arguments against this idea are that not all chimneys have the openings and that the placement of the openings forms decorative patterns that do not correspond to how a scaffolding would have been supported.

These openings may have been air vents leading into the air chamber between the stack and the flue. While decorative in appearance, they contributed to the chimney draft and helped cool the brickwork of the flue through convection. As the heated air in the chamber began to rise, cooler air was drawn in through the openings to take its place. The cooler air was then heated, and the cycle repeated again and again. The rising air current accelerated up the tapered air chamber, creating a suction effect as it exited the chimney, increasing the chimney's draft all the way back to the furnace.[152]

No reference to side openings was found in the descriptions of factory chimneys of Great Britain, and no openings are apparent in the many photographs in Walter Pickles's *Our Grimy Heritage*.[153] The absence of openings in factory chimneys may be attributed to the height of the factory chimneys. Taller chimneys created a greater draft and did not need extra draft created by the side openings. Rice chimneys were much shorter and had less draft; consequently the extra draft created by side openings compensated for the reduced draft of a short chimney. Why some chimneys do not have openings remains a mystery.

There are many permutations of chimney designs. In fact no two chimneys in the Rice Kingdom are identical. Since there are probably fifty or so that have been destroyed (an estimate based on our field surveys), the number of different chimney designs might have been considerable. Why were there so many different designs? Surely, for ease of construction, a chimney builder would have offered a stock design that he was familiar with. One possible reason is that each planter wanted a distinct chimney for easy recognition of his plantation. Most plantation threshing barns and mills were located on waterways to provide access for boats on which to ship rice to city markets. Each plantation owner may have required the builder to create a different pattern so his plantation could be easily identified from the waterway.

Individuality and personal pride were a hallmark of the planter's psyche. They were not unlike their English forebears. Alan McEwen related that some chimneys in urban areas of Keighley where he lived were signatures of the owners, and Pickles stated, "The height of the chimney was another source of pride amongst the mill-owners [in Great Britain] who even went so far as deliberately to increase this height so that their chimneys would tower over their neighbors."[154] As planters became wealthier from their rice crops, a tall and highly ornate chimney became a way of saying that they had "arrived" and were "of the manor born."

The chimney at Laurel Hill must be an extreme example of a planter's pride. This chimney towers above all other chimneys documented in the Rice Kingdom, and its unusual star-shaped stack readily identifies the plantation to travelers on the Waccamaw River. Some planters went to elaborate efforts to create painted panels on the sides of their chimney. The chimney at Middleburg Plantation supports four panels with remnants of stucco and were probably painted.

Perhaps the most ostentatious display was the crown on top of the chimney. Although many crowns have deteriorated, enough remain to establish the extent to which the planters went to decorate their chimneys and make them distinct. Although the oversiller had a vital function, it is obvious that most of the ornamentation of the oversillers and crowns were a form of ostentation. They were embellished with decorative brick patterns and some finished with a band of whitewashed stucco. When a chimney lost its top, it appeared impoverished when compared with the ornamented and crowned chimneys of neighboring plantations.

Carpenters, Blacksmiths, and Mechanics

The successful operation of a rice plantation depended not only on the manual labor of the slaves but also on their ability to master the skills of the carpenter, blacksmith, and mechanic. Slaves were trunk minders, hydraulic engineers, canal and floodgate builders, and barrel makers. Slave watermen traveled the rivers and sounds between the plantations and cities, carrying staples to market and returning with plantation supplies. The early mills run by animal and water power were built and repaired by plantation carpenters.[155]

As mill machinery became more complex before the Civil War and steam-driven mechanical threshers and rice mills became common, manufacturers in the cities began to produce mill machines and sell them to plantations. Many foundries and machine-manufacturing companies were established to supply plantations with the new machinery. These machines were either assembled in the city factories and shipped to the plantations or were shipped dismantled and assembled in place.

The plantation carpenter, blacksmith, and mechanic remained important. Many plantations were remote from the city repair shops and machine shops. Plantations had to be self-sufficient. Only as a last resort did planters request outside help or take a machine to the city for repair. Often manufacturers used plantation workers to assemble a machine in place, and it was often a plantation carpenter or blacksmith who had to repair broken or worn-out machines or to replace parts sent from the cities. Overseer James Haynes wrote to Charles Manigault from Gowrie in 1847 concerning running a driving band of the beater directly to the pulley of the beater and said: "This work can be done by the carpenters at home rakes and all."[156]

Traveling through South Carolina in February 1853, Frederick Law Olmsted wrote:

> From the settlement, we drove to the "mill"—not a flouring mill, though I believe there is a run of stones in it—but a monster barn, with more extensive and better machinery for threshing and storing rice, driven by a large-steam engine, than I have ever seen used for grain before. Adjoining the mill-house were shops and sheds, in which blacksmiths, carpenters, and other mechanics—all slaves, belonging to Mr. X.—were at work. He called my attention to the excellence of their workmanship, and said that they exercised as much ingenuity and skill as the ordinary mechanics that he was used to employ in New England. He pointed out to me some carpenter's work, a part of which had been executed by a New England mechanic, and a part by one of his own hands, which indicated that the latter was much the better workman.[157]

Charles Heyward of Rose Hill on the Combahee River commented on January 12, 1860, that he "only had a 'White Engineer' to work the thrashing mill the first winter it was put up. My young Black-smith Albert was placed in charge afterwards and performed the duties very satisfactorily."[158]

South Carolina planter David Doar also noted the significant role of slaves in maintaining the rice machinery and the plantation in general: "Those men, many of them, were no mean artisans. Generally, they were some intelligent boys or sons of some favorite slave who were then sent to the city and put to trade with such firms as Cameron and Mustard, where they served 6 to 7 years; on completing the term, they could do almost any kind of work in their line. They could make repairs and run the engine of the rice mill, and I have seen them patch a boiler, replace a cog broken out of a spur wheel and various other jobs."[159]

Anne Simons Deas wrote about the workers at the Stoke Rice Mill on the Cooper River:

> There were always skilled mechanics of various kinds on the plantation. Besides the barrels for the pounding-mill, the cooper's shop furnished cypress tubs, buckets, pails, and piggins, of every size and sort; to that a "bought basket" was rare. The carpenters could not only construct the flood-gates and rice-field trunks, and build the negro houses, but make the plantation wagons and carts, and do work, requiring great neatness of finish. The blacksmiths made and mended what ever of wrought iron was used on the place. . . . Among these mechanics was always to be found one, capable of being miller to the pounding-mill, or engineer to the thresher; and simple repairs to the machinery were done at home.[160]

These testimonies, written by people who actually observed slaves' work, are a permanent and vivid record of the skills of slaves who kept the plantation.

The Napier Rice Cleaner

In 1846 Robert Allston reported that a Mr. Napier on the Cooper River substituted "wire cards" for the mortar and pestle. Unfortunately no diagram or description exists to explains how the cards worked. Allston said that the wire card "imparts a slightly bluish tinge to the grain, though it is supposed to keep longer than rice prepared in the ordinary way. Rice thus prepared will not command as high a price per cwt. as that from the pestle or similar quality, but it is said to be the interest of the planter to patronize the 'Cards,' inasmuch as the yield in whole rice from a given quantity of Rough is invariably greater, the offal being less. In the year 1842–3, this mill prepared seven thousand barrels, and seems to have given satisfaction to patrons."[161]

Nothing in any archival source indicates that wire cards were used elsewhere in milling Carolina rice, and no source explains where how the cards fit into the milling sequence. Napier may have been the only planter to use them.

The Williams Cam Pestle Mill

At a regular meeting of the Agricultural Society of South Carolina on July 18, 1843, Mr. S. K. Williams showed the society his model of an improved rice-pounding mill, which was thoroughly examined by the members. A report on his machine by a committee appointed at a previous meeting was then read, adopted, and ordered published.[162]

The report suggested that Williams's invention "may be important to the rice planting interest." Williams substituted a cam (see diagram 43) in place of spokes, which resulted in "an increased speed of the pestles than in the common mill; and, also, the quality of the rice would be improved by better brushing: an increased speed to the Pestles would necessarily impart additional warmth to the pounded rice, and all experienced and intelligent millers must know, that rice cannot be effectually brushed, if not conveyed warm from the mortars to the brushing screen; and planters as well as factors are well satisfied as the very decided advantage of well brushed rice in the market."[163]

In the common mills of the day, which used spokes to operate the pestles, the speed of the pestles could not be increased beyond forty-three to forty-eight strokes per minute without slamming the pestles against the spokes and breaking the lifters. The Williams cam was intended to increase the speed of the pestles beyond that range.

It is debatable how unique or new Williams's cam was. The ancient Chinese used the cam with waterwheels to operate trip hammers for hulling rice and crushing stone, converting rotary motion into linear motion, and William Stephens, of Chatham County, Georgia, used cogs that functioned basically like the Williams cam.

In spite of initial praise by the Agricultural Society, no information was found that the Williams cam gained commercial use, and all photographs of the inner workings of mills show spokes, not cams. With so many new milling machines being invented and advertised, planters or city mill operators may have been reluctant to spend the cost of converting their lifters to cams.

McKinlay's Improvement in Cleaning Rice

On April 1, 1851, Peter McKinlay of Charleston, South Carolina, was granted patent no. 8,010 for an "Improvement in Mills for Cleaning Rice." McKinlay stated that the object of his invention was "to separate the thin skin or film [bran] which remains on the grain after the hull or rough outside shell has been taken off and blown away."

His invention modified the motivation of the pestle. In all previous cleaning machines (and the ones to follow), the pestle was motivated from above, lifted, and then dropped by gravity into the mortar. McKinlay attached the pestle to a rod passing through the bottom of the mortar, and the pestle received motion through a crank or its equivalent placed below. He stated that because of the more rapid motion of the pestles, the rice was cleaned more quickly and cleanly, that the rice could be drawn off from the bottom of the mortar, which saved time, and that the pestle could be raised to any height above the mortar, which gave greater or less friction to the rice as required.

In McKinlay's mill (see diagram 71), hulled rice was placed in the mortar. The power source imparted rotary motion to a shaft, which gave a reciprocating rectilinear motion to the pestle by an eccentric connecting rod. The pestle moved at a rate of 120 to 150

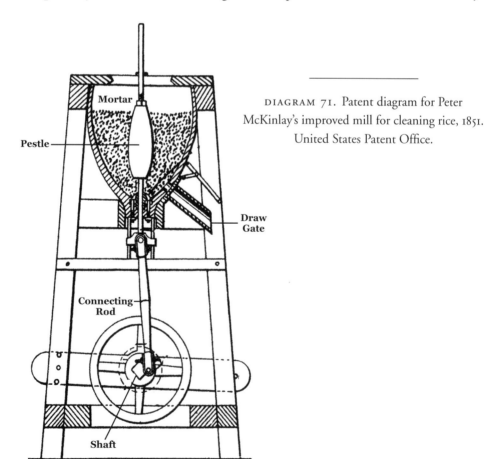

DIAGRAM 71. Patent diagram for Peter McKinlay's improved mill for cleaning rice, 1851. United States Patent Office.

FIGURE 31. "Rice Mills on Savannah River."
Reproduced from *Harper's Weekly*, July 5, 1884.

strokes per minute, which forced the rice, under considerable pressure, from beneath the pestle to the sides of the mortar, creating friction between grains of rice, and the bran was rubbed off. When the rice was cleaned, a draw gate at the bottom of the mortar was opened, and the rice poured out.

J. M. Eason & Brothers advertised McKinlay's "Crank Pestle Mortar" for cleaning rice in 1859 and again 1875 and stated it was "the best Machine now in use."[164] There is no record of how successful McKinlay's cleaner was or how often it was used. But the fact that it was advertised over sixteen years indicates that some planters must have used it. At least one planter ordered McKinlay's machine. In a bill of agreement dated May 1857, P. McKinlay granted to the Estate of H. Middleton "the Right to use five pair of my patent rice pounders @ $200 per pair."[165]

Either McKinlay's machine or a similar one was apparently used in the rice mill in Savannah, Georgia. A series of mortars appear in an 1884 depiction of a Savannah mill (see figure 31).[166] There are no spokes or pestles showing above the mortars, but there are power belts running to the bottom of each mortar indicating that the pestles were moved from below. Certainly the Savannah mill's mortars and pestles operated similarly to McKinlay's improvement and may have been influenced by his invention.

McKinlay's Rotary Rice Cleaner

Peter McKinlay was awarded patent no. 8,841, dated March 30, 1852, for a "Rotary Rice Cleaner." His machine (see diagram 72) served the same function as his improved mill for cleaning rice. It removed the bran after the hulls were removed by millstones. He stated in his patent: "The nature of my invention consists in having a circular conical chamber into which the rice is put after having the outside shell [hulls] taken off by a pair of stones operated on in the usual way. The rice after having the outer shell taken off has still a thin film or skin [bran] on it of a dark color. The object of my invention is to take off this skin or film by rotary friction without breaking the grain."

In McKinlay's cleaner an opening with a movable cover led through a passage into a circular conical chamber formed by two side plates screwed together with a stone or metal plate between. The surface of the side plates toward the chamber was rough. Flutes radiated from the center of the side plates to the inner diameter of the chamber. Rotary arms were secured to a shaft, which passed through the chamber and rested on bearings. Pulleys on the shaft imparted motion to the rotary arms from the power source (probably steam by 1852). The back of the rotary arms was flat. The working front of each rotary arm was made fast to a sharp edge from the back to the front to within five or six inches of the end. The end of the rotary arms came within three-fourths of an inch to the metal plate.

Hulled rice was poured into the chamber through the passage. As the arms rotated, the rice was forced against the sharp edges of the flutes and between the ends of the rotary arms and the rough surface of the metal plate. By this means a large amount of friction was brought on the rice, and the bran was rubbed off. Since no percussive force (as was the case with mortar and pestle) was brought against the rice, less breakage occurred.

24. Boiler at Middleburg Plantation. Photograph by Richard Dwight Porcher, Jr., 2010.

25. Boiler at Atkinson Creek in the Santee Delta, Georgetown County, South Carolina. Photograph by Richard Dwight Porcher, Jr.

26. Rice chimney at Fairfield Plantation on the Waccamaw River in Georgetown County, South Carolina. Photograph by Richard Dwight Porcher, Jr., 2010. *Insert:* the original crown as shown on a postcard from the early 1900s. When the chimney was restored, the crown was not replaced.

27. Rice chimney at Black Out Plantation on the North Santee River in Georgetown County, South Carolina. Photograph by Richard Dwight Porcher, Jr., 2010.

28. Rice chimney at Willtown Plantation on the South Edisto River in Charleston County, South Carolina. Photograph by Richard Dwight Porcher, Jr.

29. Chimney stabilization at Hope Plantation on the South Edisto River in Colleton County, South Carolina. Photograph by Richard Dwight Porcher, Jr., 2010.

30. Daniel Hadley's model of an 1850 Carolina rice mill, located at the Rice Museum in Georgetown, South Carolina. Photograph by Richard Dwight Porcher, Jr., with permission of the Rice Museum.

31. Abandoned rice fields on Quenby Creek in Berkeley County, South Carolina. Photograph courtesy of Bernard Joseph Kelley.

32. Abandoned slave village on Crow Island in the Santee Delta, Georgetown County, South Carolina. Photograph by Richard Dwight Porcher, Jr., 2010. The village was inside the earthen bank. *Insert:* remains of a slave cabin chimney in the 1980s, now gone. Photograph by Richard Dwight Porcher, Jr.

33. A storm tower in the Santee Delta, Georgetown County, South Carolina. This storm tower was converted to a hunting lodge; the roof, door, and steps are not original. Photograph by Richard Dwight Porcher, Jr., 2010.

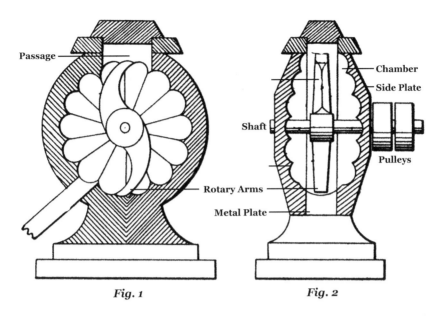

DIAGRAM 72. Patent diagram for Peter McKinlay's rotary rice cleaner, 1852. United States Patent Office.

McKinlay claimed that his machine was simple to operate, would save power, and would not get out of order or become damaged. All the same, no record has been found that it was commercially used.

The Taylor Machine for Cleaning Rice

In 1857 three men from Charleston and one from Georgetown invented three machines, two for cleaning rice and one for brushing rice. All three machines were intended for plantation mills. The decade before the Civil War was the zenith of rice culture. Planters endeavored to present the cleanest rice possible to the overseas market, and they were eager to produce new machines that might gain widespread use and return a handsome sum for their efforts.

On June 23, 1857, John F. Taylor, of Charleston, South Carolina, employed by the Phoenix Iron Works, was granted patent no. 17,646 for a "Machine for Cleaning Rice." In his patent he stated: "The object of this invention is to rub or take off in an expeditious and perfect manner the flour [bran] which encompasses the kernels or grains of rice. . . . Rice has hitherto been cleaned; that is, deprived of this flour by means of pestles or mortars. . . . This cleaning of the rice by means of pestles or mortars, and known as rice pounding, is a slow operation, and the kernels or grains are liable to be broken and bruised."

Taylor's machine (diagram 73) consisted of a large, spherical vessel with a bar on top. A vertical shaft fit through the bar with a crank on the end. The shaft extended to the

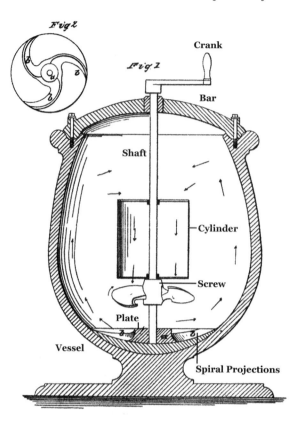

DIAGRAM 73. Patent diagram for John F. Taylor's machine for cleaning rice, 1857. United States Patent Office.

bottom of the vessel where it was anchored. At the bottom of the vessel was a dish-shaped plate fitted with stationary spiral projections (labeled "b"). A screw with two flanges was attached to the shaft a short distance above the plate. A cylinder placed just above the screw was permanently attached to the shaft.

Taylor's machine did not remove hulls, only the bran and germ. Hulled rice from the millstones was placed in the vessel, and a crank rotated the shaft. The rice below the screw was forced downward. The spiral projections forced the rice outward, and the continuous action of the screw forced the rice upward on the sides of the vessel and into the cylinder, where it was forced downward and again underneath the screw. The arrows in fig. 1 show the direction of the rice's movement. The movement of the rice through the vessel caused sufficient attrition between the grains to remove the bran and germ.

How the inventor fit the machine into the automated mills of the time is unknown, and how the power source was applied is unknown. The machine could have been used singly on a plantation for domestic use and operated by hand with the crank. Whether mechanical power was applied to the machine as a labor-saving device is unrecorded. Presumably the rice was winnowed and sifted later. Although Taylor's machine demonstrated technological ingenuity, no references were found to indicate if it was a commercial success or was employed on plantations for domestic use.

The Lachicotte and Bowman Improvement in Cleaning Rice

On July 28, 1857, Philip Rossignol Lachicotte and T. B. Bowman, of Charleston, South Carolina, were granted patent no. 17,832 for an "Improvement in Machines for Cleaning Rice." Lachicotte and Bowman's invention was similar to Taylor's, but it had certain enhancements. It was identical in purpose. Their machine (see diagram 74) replaced the mortar and pestle to remove the bran. The inventors claimed in their patent that it would clean rice in "one-eighth the time required by the usual mortar and pestle."

Their invention consisted of a hollow, inner cylinder with screw flanges attached to its outer side. This assemblage was placed in a cylindrical case (figs. 1 and 2). A shaft passed down through the cover of the outer cylinder and was anchored at the center of a bar that formed the base of the machine. The inner cylinder was secured to the shaft by a yoke.

Rice was placed in the upper part of the chamber of the outer cylinder case, and the inner cylinder was rotated manually. The rice was pressed downward (arrows, fig. 1) by the flanges of the inner cylinder as it rotated, pushed underneath the lower edge of the

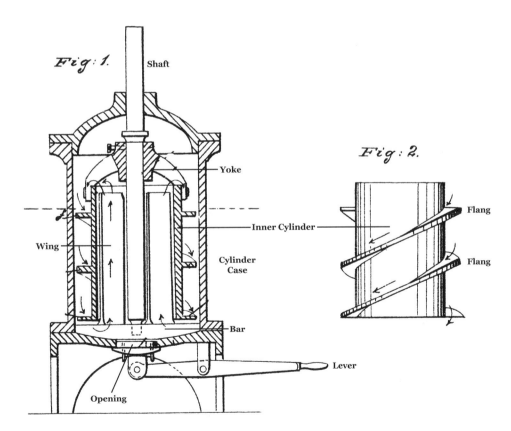

DIAGRAM 74. Patent diagram for Philip Rossignol Lachicotte and T. B. Bowman's improved rice-cleaning machine, 1857. United States Patent Office.

inner cylinder, and then forced upward through the inner cylinder and over its upper edge. Then the rice was forced downward again by the flanges. The forced movement caused friction between the grains which rubbed off the bran. The inventors also claimed that two wings in the inner cylinder prevented "the mass of rice as it passes upward within the cylinder from turning with the cylinder, thereby increasing the friction of the grains or kernels one against the other." The cleaned rice exited through an opening at the bottom of the machine.

The inventors credited John F. Taylor with the concept of the screw within a cylinder to move the rice; however, the vertical movement of rice in their machine caused by the two wings created more friction between the grains, resulting in better cleaning. No record was found that this machine gained any commercial use.

Oliver J. Butts's Machine for Brushing Rice

Oliver J. Butts was a prominent member of the Georgetown mercantile community and also the son of Jehiel Butts of Charleston who invented a thresher described above. Oliver Butts served as a private in Company A of the Georgetown Rifle Guards, but because of his mechanical knowledge, he was discharged and transferred to the Government Machine Shop. After the Civil War he was a charter member of the Georgetown Rice Milling Company in 1879 and served as miller and superintendent in 1880. His background in the rice industry and his mechanical ability must have inspired him to work on rice machines.

On October 27, 1857, Butts was granted patent no. 18,496 for a "Machine for Brushing Rice." He said his machine would take "less power than the old cylinder brush and gives the rice a much higher polish with no breakage whatever." He claimed as his invention "the application of a flat brush for brushing rice consisting of a flat runner dressed with sheepskin and basils in connection with a wire bed."

The Butts machine (see diagram 75) featured a brush made of a cast-iron plate (fig. 4) faced on the bottom with wood. Attached to the wood facing were strips of wool and basils (that is, "basil leather," which is sheepskin tanned with agents of plant origin). The wool was cut into segments, and each segment was tacked on one edge to the wood facing (figs. 2 and 4) and around the eye. The basils were also cut into segments and tacked in a like manner onto the wood facing over the wool. The plate acted as the runner, which rotated over a stationary wire bed (fig. 3 and 4).

To operate Butts's machine, rice from the mortars and pestles was placed in the hopper (fig. 1), and by gravity the rice passed through the eye or center of the brush, where it was thrown by centrifugal force onto the wire bed by the rotation of the plate. The plate was rotated by a spindle turned by the driving wheel, which was powered by either water or steam.

The friction of the revolving wool and basils against the rice rubbed off the polish and imparted a high gloss. The flat surface of the wire bed caused the rice to pass upon its side through the brush, instead of end over end, and reduced less breakage than in

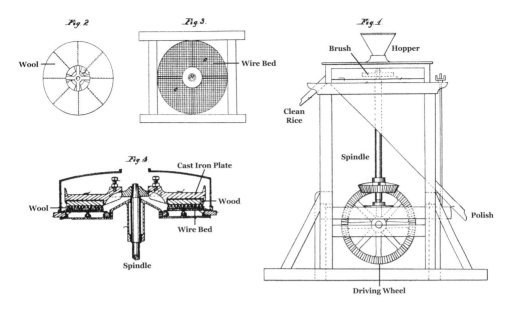

DIAGRAM 75. Patent diagram for Oliver J. Butts's machine for brushing rice, 1857. United States Patent Office.

other cylinder brushes where rice tumbled end over end. The polish passed through the wire bed into a spout, and the polished rice was discharged from the machine through the clean rice spout.

As Butts claimed in his patent, "The advantage of the basils is that they prevent the wool from becoming clogged with flour [polish], and in presenting a more even and firmer surface gives the rice a higher and more uniform polish—at the same time being more durable that the wool by itself." Furthermore "attaching the wool and basils to the runner only around the eye and one edge of each segment they are made to drag loosely upon the wire bed, and the rice is thereby kept in constant contact with them."

Since Butts was a charter member of the Georgetown Rice Mill, he probably installed his brushing screen in this mill; however, no records of the mill survive, so it is impossible to determine for certain if it was used. Butts's machine was nonetheless an ingenious invention and might have proved useful if rice cultivation had not diminished after the Civil War.

Many inventions for rice milling and preparing rice for market became available after the Civil War. Most were invented for the burgeoning rice culture in the Southwest. With the exception of the Engelberg Huller, most were not commercially used after the war because rice culture waned, and most planters could not afford new machines. By this time most rice was sent to the city as rough rice for milling or shipped as rough rice overseas. Milling on the plantations was rare except for local consumption. Steam was the likely power source for these machines.

McKinlay's Machine for Cleaning Rice

On November 6, 1866, Peter McKinlay of Charleston was granted patent no. 59,432 for a "Machine for Cleaning Rice." It was touted in an 1873 issue of the *Rural Carolinian* as "McKinlay's Portable Rice Mill." The magazine pointed out that, since the close of the war, the culture of rice in the interior has become more general than previously and that one of McKinlay's mills "can be seen at the machine shops of Messres. James M. Eason & Bro., in this city, where it has been erected by the inventor. We were present at a recent trial of its merits, and must say in all candor, that it appears to be just what is needed by those who grow rice for domestic use, or in small quantities for sale, at too great a distance from a rice mill to have it prepared for market, except at a cost which would consume all the profits. It can be put up any where, occupying, as it does, but little space, and can be worked by either horse, water or steam power."[167]

McKinlay designed his invention to be constructed "on a small scale for family use, and is employed for the purpose of cleaning rice—that is, to take off the outer shell [hull] and the dark cuticle [bran], which had to be removed to render the rice fit for food." It consisted of a pair of rockers that worked within two oblong chambers (see diagram 76). Rice was placed in the chambers. A lever attached to a connecting rod moved the rockers back and forth in the chambers. The flywheel, which was turned manually by the crank, vibrated the rod. Strips of india rubber or leather were fastened to the lower edge of the

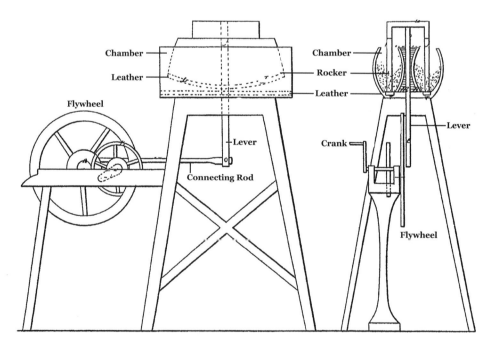

DIAGRAM 76. Patent diagram for McKinlay's portable rice mill, 1866. United States Patent Office.

rocker. The same kind of strips were put into grooves in the bottom of the chambers. The rocker action against the rubber strips of the chambers rubbed off the hulls and bran and at the same time polished the rice. The broken rice, hulls, bran, and polish passed off from the chambers through a wired opening in the bottom, and the prime rice was left in the chambers to be manually removed.

How many plantations used McKinlay's machine is unknown. Plantation photographs taken after the war show rice being milled with the mortar and pestle by freedmen. Because of their postwar financial distress, it was unlikely that many plantations would have invested in this machine just for milling domestic rice when the primitive mortar and pestle would have been sufficient.

The Brotherhood Improvement in Rice Pounding and Hulling Mills

On November 19, 1878, Fred Brotherhood of Charleston, South Carolina, was granted patent no. 210,002 for an "Improvement in Rice Pounding and Hulling Mills." Brotherhood's complicated patent application includes seven figures, on which only one is shown here, and the description of how his pounding machine operated is truncated (see diagram 77). Brotherhood did not state whether his machine was designed as a huller or to remove the bran. More than likely it was to remove the bran since millstones were almost always employed at this time to remove hulls.

Brotherhood's patent featured a unique method of raising a pestle and releasing it to drop by gravity into a mortar. Attached to the pestle was a board acted on by two cams or eccentric rolls, one on each side of the pestle board. A guide slot in the top of the frame or crosspiece held the pestle board in place as it rose and fell. Each cam was attached to the inside of a spur gear. The machine's power source turned the shaft, which was attached to spur gear A. As the shaft turned spur gear A clockwise, spur gear B turned counterclockwise because it meshed with spur gear A. As the two spur gears turned, the two cams turned and acted upon the lifting board on opposite sides throughout a portion of each revolution of the shafts. When the section of the cams acting on the lifting boards passed, the pestle board dropped down into the mortar by gravity.

Brotherhood claimed that his "peculiar organization and adaptation of devices . . . provide a light-running, portable, efficient mill for cleaning rice, which, both as regards the general feature of construction and mode of operation, and various minor parts or details, is unlike any rice-cleaning mill heretofore known, of which I have knowledge." The only documented use of Brotherhood's mill was in the Georgetown Rice Mill. Ingenious as his machine was, it probably found little success because it came at a time when financial restraints precluded investment in unproven machines.

The Lockfaw Rice Mill

On July 15, 1887, the *Wilmington Morning Star* reported on "A New Rice Mill" invented by one of its own citizens, John A. Lockfaw who was granted patent no. 365,191 on June 21, 1887, for a rice mill. In the patent Lockfaw stated that his machine was "a mill which

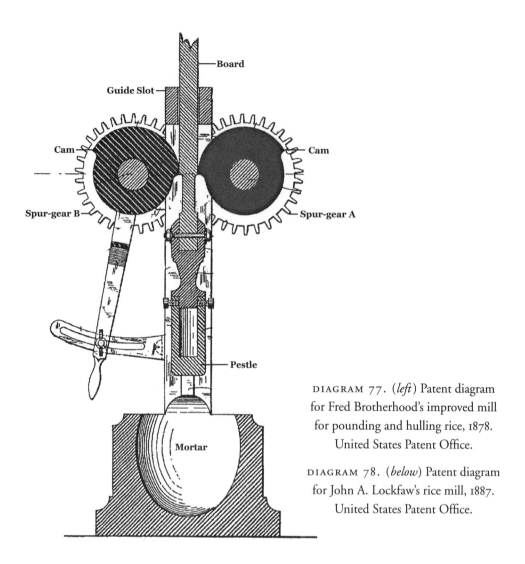

DIAGRAM 77. (*left*) Patent diagram for Fred Brotherhood's improved mill for pounding and hulling rice, 1878. United States Patent Office.

DIAGRAM 78. (*below*) Patent diagram for John A. Lockfaw's rice mill, 1887. United States Patent Office.

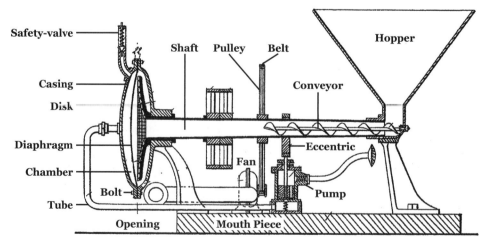

may be driven directly from the thrashing-machine and to which the rice may be fed in a continuous stream."

According to the *Star,* mills were so costly for individual plantations to operate that there were only four mills in North Carolina and that farmers had to send their rice great distances or sell it to mill agents. Lockfaw's mill was simple and cheap, the cost being within the means of any well-to-do farmer. The machine would "work a great revolution in rice milling, and make each farmer his own miller, cheapen the article and bring lowland and upland rice into general cultivation." The *Star* further reported that "if he only holds on and is able to retain its ownership, we predict for him a great fortune at an early age." *Scientific American* was so impressed with Lockfaw's machine it ran a feature article, including a diagram and detailed description of how it worked.[168]

In Lockfaw's mill (see diagram 78) rice from the thresher was directed into the hopper from where it passed into the hollow shaft and was forced down the shaft by a spirally flanged conveyor. The shaft was rotated by a belt from the power source passing over its pulley attached to the shaft. A disk located within a casing was attached to the far end of the shaft. The outer face of the disk was formed with ridges like those on the face of a file, but the ridges in this mill are much coarser and larger. Opposite the disk was a flexible canvas diaphragm held by bolts between the flanges of the two casing sections. The bolts also held together the two casing sections.

An air pump with a piston operated by an eccentric moved by the rotating shaft forced air through a tube into the chamber between the diaphragm and the casing. The rice passed down the shaft through the opening in the disk and into the space between the disk and diaphragm. Air pressure from behind pushed the diaphragm toward the disk. As the shaft rotated, the rice was caught between the roughened face of the disk and the diaphragm, and the hulls were stripped off. A safety valve let out excess pressure from the casing chamber so that undue pressure between the diaphragm and disk was avoided.

Lockfaw provided a fan to winnow the rice. The rice and hulls passed from the casing chamber through an opening where a blast of air from a mouthpiece connected to the fan blew away the hulls and left the rice for collection. The fan was set into motion by a belt attached to a pulley mounted on the shaft. As the shaft rotated, it powered the fan.

Despite what the *Star* and the *Scientific American* reported, it is difficult to see how Lockfaw's rice mill was a commercial success. His machine only hulled rice. His mill still had to have mortars and pestles to remove the bran and a brushing machine or a machine like Lachicotte and Bowman's to clean the rice. Lockfaw's rice mill may have been used briefly in the Wilmington area, but nothing was found in archival sources to indicate that it was used elsewhere in the Rice Kingdom. Nevertheless it was a most ingenious device, especially in its use of air pressure to push the diaphragm.

The Engelberg Huller Company, Syracuse, New York

In 1885 Evaristo Conrado Engelberg of São Paolo, Brazil, received a British patent for a rice-hulling machine. He received an American patent the following year, patent no.

341,324, dated May 4, 1886. He improved on this machine, and received United States patent no. 383,285, dated May 22, 1888. In 1889 the Engelberg Huller Company of Syracuse, New York, was formed to manufacture his huller for sale in North America. The huller was successful, and additional patents were granted for various improvements over the years. Called the "Engelberg steel huller," it was the last huller to see significant use in the Rice Kingdom. The Engelberg company also produced a disk huller or sheller for rice.

Engelberg continued to improve and modify the steel huller. Three of the improved Engelberg hullers have been preserved today on Lowcountry plantations—Kinloch Plantation in Georgetown County, Cockfield Plantation in Colleton County, and Turnbridge Plantation in Jasper County—where they were used until the end of the industry. Each of the three surviving hullers is a different model produced by Engelberg to improve on the 1888 model. The Engelberg huller is still in use today, adapted for use in nonindustrialized countries.

The Engelberg-type huller consisted of a fluted steel cylinder turning inside a horizontal casing. It was one of five systems used in the American rice industry after the Civil War. The other four were the disk sheller; a cone mill in which an emery-surfaced truncated cone revolved within a casing of steel-wire mesh; the rubber-band husker, in which grain passed between an endless belt of rubber and an iron-grooved roll; and the rubber husker, in which paddy rice passed between two horizontal rubber rollers, revolving in opposite directions and at different speeds.

As more and more planters after the Civil War struggled financially and with a decreased labor force to run their complex prewar mills, they turned to the Engelberg Huller and Polisher, advertised by the company thus: "The No. 1 Engelberg Huller and Polisher does away with the extravagant expensive pounders and produces in a single operation merchantable rice direct from the grain as received from the Thresher. It is suitable for Plantation or Mill use, mills of all kinds of rice when properly adjusted, reduces breakage to a minimum and not obtained by any other machine, while the yield is much greater."[169]

The Engelberg No. 1 Huller and Polisher (diagram 79A) did all functions of the prewar mills: it removed the hulls and bran in a single operation. A polisher could be attached to the huller if polished rice was desired. The Engelberg mill was less expensive than a prewar mill, required less maintenance and labor to operate, and ran on steam power. Although the large mills in the cities, where most planters after the war sent their paddy rice, still kept their mills that used stones and the mortar and pestle, at least some planters bought the Engelberg huller for use on their plantations. The main disadvantage of the Engelberg huller was that it did not separate the two byproducts, the hulls and the bran. As a result, despite the high nutritional value of the bran, the byproducts were unfit for animal consumption because the hulls caused irritations and bleedings in the intestine.

The essential component of the Engelberg huller was its ribbed-steel cylinder revolving inside a casing. The impact force in the steel huller was absent, which caused less

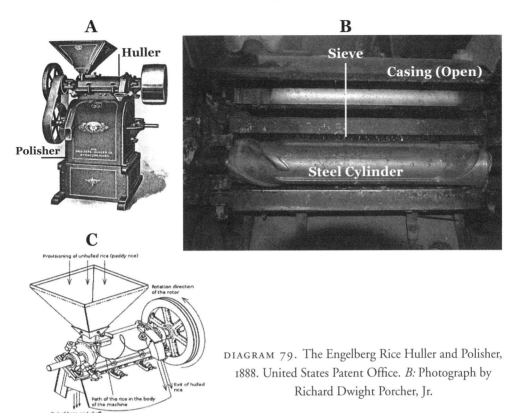

DIAGRAM 79. The Engelberg Rice Huller and Polisher, 1888. United States Patent Office. *B:* Photograph by Richard Dwight Porcher, Jr.

breakage. The cylinder and casing are shown in diagram 79B. Around the face of the cylinder were ribs, those at one end were spiral, and those at the other end were longitudinal. Rough rice was fed into the casing above the spiral ribs of the cylinder, and as the cylinder revolved, it acted as a screw, propelling the rice along the cylinder in a rotational movement between the straight ribs and casing (diagram 79C shows the path of the rice through the huller). An adjustable blade, held in position, prevented the rice from swirling around inside the casing. Great pressure and friction was exerted between the grains along the edge of the blade and the ribs of the cylinder, causing the hulls and bran to "rub" along the edge of the blade and the ribs of the cylinder, leading to their separation from the rice by scouring.

At the back of the casing was a sieve divided into two sections with different-sized openings. The contents of the casing were thrown against the sieve; the bran and germ passed through the smaller openings and the hulls through the larger openings. The bran and hulls were cut into small enough pieces to pass through the sieve. Rice was too large to pass through either opening.

While revolving around the cylinder, the rice was pushed toward the exit, driven by the rice still located in the feed hopper. Once it arrived at the exit, the rice escaped through a specially created spout. The portion of the release into the spout was placed especially high, which allowed the last separation between the rice and whatever byproducts had not yet evacuated by the sieve. The rice, because of its heavier weight, was projected into the exit by inertia. The lighter byproducts remained in the body of the huller and continued to turn until they exited through the sieve. The separation of the hulls and bran by this process eliminated the wind fan.[170]

If a very fine appearance was desired, Engelberg furnished a polishing machine, either a separate machine, or as part of one machine, the Rice Huller and Polisher (shown in diagram 79A). Instead of being discharged into a container, the rice from the casing was discharged through a spout into the polisher. A cylinder was covered first with sheepskin having the wool on, and over this was placed russet skin (skin with the wool off). The former was cut into strips from seven to nine inches wide, and nailed onto the cylinder by one edge, four inches apart; the other edge was left loose, overlapping each other like shingles. The russet skin was cut into strips about thirteen inches wide, and nailed to the cylinder in a similar manner. The rice was brushed between the skin and cylinder casing and thus polished.

The Habarnards Rice Huller

One other huller survives. At Cockfield Plantation there is a Habarnards Rice Huller, patented on July 13, 1897. The sole agent for the Habarnards huller was Philip Rahm of New Orleans. This huller is in poor condition, but there are enough parts to show how it operated. The Habarnards huller was similar to the Engelberg machine and featured a steel cylinder revolving inside a casing with two sieves of different-sized holes to remove the bran and hulls. Attached to the rear of the huller was a polisher that worked similarly to the Engelberg polisher. No other huller made by Habarnards was found.

The Engelberg Disk Huller

Only one disk huller is known from the Rice Kingdom. The Village Museum in McClellanville, South Carolina, purchased a Engelberg disk huller that was reported to have been obtained from the Santee Delta. Where it was used in the delta is unknown. Since it was only a huller, it might have been used for domestic rice. Because of its limited operation (removing only the hulls) and its introduction close to the end of the industry, this may have been the only disk huller ever used in the Rice Kingdom.[171]

The disk huller consisted of an adjustable feed hopper, a stationary upper disk, and a lower rotating disk or runner (see diagram 80A).[172] The driving shaft operated from below and was attached to the runner disk. The working faces of both disks were covered with an artificial stone dressing and special composition made up of emery or Carborundum, a special cement, and a solution of special salts—which was kept flat and rough. The rough surface functioned like the surface of the Peak millstones; it grasped the rice ends, forcing them to flip end over end and stripping the hulls. The disks were enclosed in a casing (diagram 80B). Paddy rice was fed into the hopper at the center of the upper stationary disk and carried outward by centrifugal force. The space between the disks

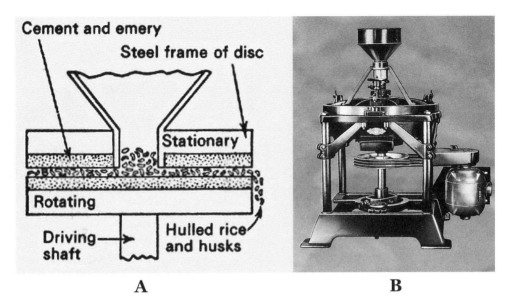

DIAGRAM 80. An Engelberg disk huller (left). Reproduced from D. H. Grist, *Rice* (6th ed., 1986). Disk huller from the catalogue of the Engelberg Huller Company, of Syracuse, New York (right).

was adjusted to allow the paddy to pass with the minimum damage to the rice. The disk huller removed only the hulls. A second operation was necessary to remove the bran and polish.

As time passed, buildings housing the marvelous machines that threshed and milled Carolina rice—and survived the ravages of the Civil War—were gradually torn down and their materials recycled, or they deteriorated from exposure to the elements. Accidental fire consumed many structures, including the threshing barn at Mansfield in the 1980s. Few owners thought enough about history to preserve these structures. Exceptions were the owners of Chicora Wood and Kinloch Plantations, where the threshing barns have been restored and preserved. Georgia and North Carolina have none. Fortunately fifteen chimneys in South Carolina and three in Georgia survive. The remains of Stoke Rice Mill and Middleburg Mill, although mostly gone, have been stabilized. Not one mill survives intact. Yet owners today, when informed of the significance of their rice artifacts, have taken a renewed interest in rice culture and set about preserving them.

7

City Mills and Large Plantation Toll Mills

Because of the high cost of maintaining a mill on a small plantation, over time milling shifted from small plantations to city mills or toll mills on more affluent large plantations. With their greater resources and larger and more elaborate steam-driven mills, large plantation toll mills served smaller plantations within barge range. With better means to transport rice (steamboats, for example), planters also began sending their rough rice to the city mills. The shift accelerated after the Civil War because many plantation mills were destroyed, and planters were unable to finance new mills. The day of the city mills had arrived, and they were the main source of milling Carolina rice until the end of the industry.

Documenting the machinery inside the mills and the steam-power source has been difficult. Few photographs or descriptions of the insides of the mills have survived, and documents of the mills' history have been lost or destroyed, making it hard to reconstruct their operation. Using extant archival materials and artifacts, however, we were able to piece together a reasonable description of the city mills and their operations.

The Hopper-Boy

When milling shifted to the city mills and large plantation toll mills, rough rice was shipped by river in open flats, sloops, or schooners, and the rice absorbed moisture before it reached the mills. Drying the rough rice before milling it was necessary. The city mills and plantation toll mills used an apparatus called the hopper-boy. When the hopper-boy was adapted to the rice industry is unknown, but by the time city mills and plantation toll mills were established, it was standard equipment. Small plantation mills probably would not have had to use a dryer because the rice went directly from the thresher to the mill.

The origin of the hopper-boy dates back to early flour mills. The flour, hot from the grinding action of the millstone, was passed to a large tublike enclosure called a "hopper,"

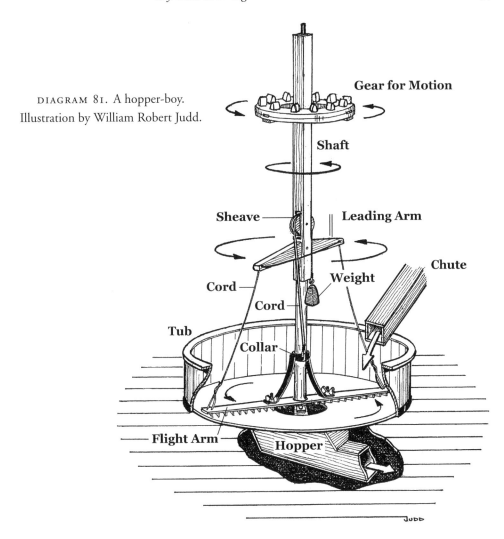

DIAGRAM 81. A hopper-boy. Illustration by William Robert Judd.

which was attended by a young boy whose task was to continually spread the hot incoming flour to let it cool and dry before it was bolted (sieved). To the millwright and miller, the hopper attended by a boy became the "hopper-boy." In 1785 Oliver Evans redesigned and mechanized the hopper-boy with a mechanical rakelike apparatus, doing away with the services of the young boys. Although Evans's device was unattended, and in some circles referred to as a dryer, to the "old guard" it was always the "hopper-boy."

Diagram 81 is based on Oliver Evans's diagram of the hopper-boy in his classic *The Young Mill-Wright and Miller's Guide* (1795). The tub enclosure was approximately ten feet in diameter with a centered opening above a self-emptying hopper beneath the floor. A wooden vertical shaft pivoted on a step gudgeon and plate supported below the enclosure, just above the hopper. Approximately four and one half feet of the shaft's lower end was made round to pass loosely through both an iron collar and a hole in the rakelike apparatus called a "flight arm," giving the arm liberty to rise and fall freely to suit any quantity of grain under it. The rakelike flight arm had adjustable, vertical wood slats called "flights." The flight arm was led around the bottom of the enclosure by a cord attached at each end and passing through a hole in each end of the leading arms attached to the vertical shaft. The vertical shaft was geared for slow motion—not above four revolutions per minute. (The power source for the shaft is not shown.) The forward edge of the flight arm was installed to hang loosely in order to make it rise over the grain. The arm's weight was nearly balanced by a weight suspended by a cord that passed over a sheave mounted in the shaft down to the collar/brace assembly supporting the flight arm. The weight controlled the arm's rising and falling motion. A plate on the bottom of the shaft held the flight arm from the floor of the enclosure.

When the hopper-boy was used in rice milling, rice entered the tub enclosure through a chute. The flights of the rotating flight arm spread the incoming rice in an outward direction around the enclosure so it would dry before it was swept toward the center, where it fell through an opening in the hopper beneath the enclosure. From the self-emptying hopper, the dried rice passed to a bin feeding an elevator that took the rice to a fan or cleaner on the mill's first floor.

The Early City Mills of Jonathan Lucas I

Little information exists in archival sources about the early mills built by Jonathan Lucas I and his family in Charleston. The only account was given by a Lucas family descendant, William Dollard Lucas.[1] In the early years of the nineteenth century, Jonathan Lucas I erected a tide rice mill on the Ashley River near the present site of the West Point Mill. Water power was furnished by a tide pond that backed up on present-day Spring Street. Years ago the millpond was filled in because of urban sprawl, and no trace or photograph of the mill remains. Water was let on the incoming tide, and at high tide, a gate was closed, creating a water head that turned a waterwheel. Lucas's mill made Charleston the center of the early city rice-milling business.

In 1817 Lucas built a steam rice mill at the foot of Mill Street. This mill is reported to have been the first mill operated by steam in South Carolina. Later Lucas and a partner named Norton built another steam mill on the Cooper River at Gadsden's Wharf.

Waverly Rice Mill

Waverly Rice Mill (figure 32) was a large plantation toll mill for planters of the surrounding tidewater plantations on the Great Pee Dee and Waccamaw Rivers in Georgetown County, South Carolina. The photograph of Waverly Mill depicts the mill during its peak activity at the turn of the century. Its physical size and operations rivaled the largest city mills. Waverly-milled rice won a gold medal in 1902 at the Charleston Exposition.

FIGURE 32. Waverly Rice Mill on the Waccamaw River in Georgetown County, South Carolina. Note the abandoned rice flat in the canal. Courtesy of Alberta Lachicotte Quattlebaum.

The first mill at Waverly was built prior to 1827. A plat produced in that year shows a pounding mill on a bluff above the Waccamaw River. Robert Allston, who came to control planting at Waverly Plantation in 1834 as the executor of his brother Joseph Allston's estate, rebuilt the mill in 1837, after he restored the plantation's finances. Robert Allston controlled Waverly until his two nephews, Joseph Blyth Allston (1833–1904) and William Allen Allston (1834–1878), who inherited Waverly at their father's death in 1834, came of age in 1857 and took control of the plantation. Joseph and William Allston divided the property, with Joseph obtaining the rice fields and rice mill. In May 1871 Joseph Allston sold Waverly Plantation to Philip Rossignol Lachicotte, who soon founded P. R. Lachicotte and Sons. Lachicotte and Sons modernized and enlarged the steam-driven rice mill and added a shipyard for construction of rice barges, a marine railway, a lumber mill, and a post office. Activity at Waverly was at its zenith from 1871 to 1911.[2] During this period Waverly Mill processed an average of two to three thousand bushels of rice a day.

Philip R. Lachicotte received patent no. 166,992 on April 17, 1875, for an "Improvement in Scouring and Polishing Rice." Lachicotte's patent employed very fine silica powder, made from burnt rice hulls, in connection with the ordinary scouring and polish machines that cleaned, whitened, and polished rice in less time and with less breakage than other polishing methods in use. Whether his process was used in his mill is unknown.

According to his granddaughter Alberta Morel Lachicotte Quattlebaum, after the mill closed in 1911, the family gradually sold off timbers, machinery, and bricks until nothing remained. Only the office was saved because it was moved to the Hammock Shops in Pawleys Island, where it stands today. Two sets of millstones were saved. Two runner stones are embedded in brick columns, one at the intersection of Waverly Road and Rice Mill Road (see color plate 18), the other at the intersection of Waverly Road and Lower Waverly Road. The two bed stones are at Brookgreen Gardens. According to Alberta Quattlebaum, the records for Waverly Mill in the 1871–1911 were destroyed by a family member who had been annoyed by sightseers rummaging through the building and invading her privacy.[3]

Although no photographs of the interior workings of the mill have survived and although the mill and mill records were destroyed, a photograph of the mill reveals some interesting features. A metal chimney anchored in a brick base was used instead of an all-brick chimney. Sometime after the Civil War, some larger city mills, regional mills such as Waverly Rice Mill, and larger plantations erected cylindrical metal chimneys to reduce construction and labor cost. England was the first industrial nation to introduce metal chimneys, and as it had with steam-engine innovation, America borrowed from England its metal-chimney technology, which evolved from maritime steam vessels. Limited space aboard steam vessels dictated that the furnace, boilers, and metal stack be combined into one unit.

With the introduction of steam-powered cranes, metal chimneys could be built in sections and erected on site. Like its brick chimney predecessor, the round metal chimney or stack also had an inner metal flue positioned in the center of the stack by stand-off rods spaced at intervals along the length of each section. Guy wires supported the chimney at multiple intervals along its length.

Metal chimneys were also superior to brick chimneys for other reasons: The enormous quantity of bricks for high chimneys increased the cost because the weight of a large brick chimney required a strong foundation and space; and a brick chimney's foundation could sink, resulting in cracks, which impaired function. Sudden changes in temperatures were also destructive to brick chimneys. Since most brick chimneys and flues in the Rice Kingdom were square, some efficiency was lost. The round metal flue expelled the ascending column of gases more efficiently. Unlike a brick chimney, which could sustain costly damage from a hurricane or deterioration from normal use, the metal chimney was easier to erect or replace in less time and at a minimal expense. A metal chimney could also be easily adapted to an existing boiler and furnace.

The flue coming from the engine house was above ground and entered at the bottom of the brick base. The manhole for cleaning the chimney is clearly visible in the photograph. An unusual feature is the elevated conveyor that Quattlebaum said carried the waste hulls from milling and dumped them into an adjacent rice field.[4] How the conveyor worked is unknown, but its simple pipe construction and length makes it unlikely that it housed a spiral conveyor. Water pressure may have forced the hulls through the conveyor.

Savannah Rice Mills

In 1829 a notice appeared in the *Savannah Georgian* that a new steam rice mill had been built on the lower end of the city by a Mr. McGin with the firm of Hall, Shapter & Tupper as a toll mill for the planters on the Savannah River.[5] The mill had a thirty-horsepower steam engine that drove twenty pestles and was calculated to pound and clean from seventy to one hundred bushels of rice per day. This original mill was probably replaced by a much larger mill. Other records indicate that Savannah had two other steam-powered rice mills located on the river wharfs, at a date later than 1829.

A series of nine depictions of one of the mills published in an 1884 issue of *Harper's Weekly* gives us a general idea of how this Savannah mill operated (see figure 31). Unfortunately a diagram of the steam engines that ran the mill was not included. Seven of the images reveal that the mill was automated, as were all the large city mills of the period.

Rough rice was brought to the mill by schooners or rice barges (flats) and mechanically conveyed into the upper story of the mill, where it was weighed, dried, and cleaned. It was then conveyed by a hopper and chute to the millstones (see figure 31 and figure 33). Rice was conveyed from the millstones to the chaff fans where the hulls were blown away, and the cleaned rice was then carried through a chute down to the mortar room on a lower floor. The mortar and pestles were motivated from below. The rice was then conveyed to a rolling screen (not shown), where the broken grains and bran were separated. Next the brush drum removed the polish, after which the rice was conveyed via a hopper and chute to barrels on the lowest floor of the mill. The rice was then loaded on ships for export.

The photograph of the millstones shows the four leveling rods that were employed to level the upper millstone as it became uneven from wear. Each rod passed through a hole in the millstone; the end of the rod rested on the rynd below. The hole through which the rod passed is shown in a similar millstone in color plate 18. The end of the rod was anchored in a nut secured on the underside. The nut above could be backed off and the rod turned to move past the securing nut, and since it rested on the rynd, its side of the millstone was raised. The nut was then tightened to keep the rod in place. The rod could be raised, lowering the respective side, by reversing the process.

William W. Gordon II described his experience overseeing rice processing on one of the Savannah mills: "I keep an account of the no. of sacks as the crew . . . bring in and empty them upon the floor of the mill, and an account of the no. bushels as the women of the mill measure them out and bear them into some . . . receptacle within.

FIGURE 33. Millstones at the Savannah Rice Mill. The enclosure and hopper were removed to allow the worker to conduct maintenance on the millstones. Collection of Richard Dwight Porcher, Jr.

The measures sing out the tally. . . . The thick cloud of yellow rice dust soften down the harsher features of the scene. . . . The whole, combined with the almost twilight in the mill—its machinery incessantly at work around us."[6] This is the only account of the operation of a city rice mill in the Rice Kingdom found in archival sources.

West Point Rice Mill

About 1840 Jonathan Lucas III built a steam-powered rice mill on the Ashley River in Charleston where the current West Point mill building now stands. The mill operated forty pestles and had a daily capacity of two hundred barrels of rice. The mill was housed in a four-story brick building. After Lucas died in 1853, the mill operated unprofitably, and in 1859 machinist William Lebby, acting as trustee for the West Point Mills Company, purchased the property along with eighty-nine slaves. This mill was destroyed by fire and replaced by a new mill in 1860–61. It operated until 1912.

The West Point Mill was operated by two massive English-made walking-beam steam engines of the condensing type. (Mill engines that were used on the plantations did not have enough horsepower to run a mill the size of the West Point.) These big engines could be used in unison or separately as power was needed. The engines not only milled rice twenty-four hours a day, they powered augers to remove rough rice from arriving vessels. The huge flywheels measured more than twenty feet in diameter. Since the engines were condensing types, freshwater had to be available. The Ashley River was salty at the mill site, so freshwater had to be obtained from elsewhere and stored at the mill. At the rear of the building were six large cylinders on top of the structure projecting from the main building (see figure 26). These cylinders held the freshwater for the engine. Where the freshwater for the cylinders was obtained and how the water was injected into the cylinders is unknown. After the mill closed, the engines were shipped to the Henry Ford Museum at Dearborn, Michigan.

The major portion of the rice was brought in bulk by boat. At the West Point Mill, a hoist at the water's edge lifted the rice out of the boats. From there it was conveyed along a screw conveyor on the dock floor to the hoist in the warehouse, where it was sent up to the large hoppers above the scales. After it was weighed, the rice was hoisted to another screw conveyor in the top of the warehouse through which it was directed to proper bins. From the bins it was carried by truck or conveyed to the mill as needed.

Rice next was cleaned in a rolling screen to remove trash. Then the rice was conveyed to a hopper above the millstones. The millstones operated in the same manner as in Jonathan Lucas I's Water Rice Machine. Several abandoned millstones can be seen leaning against the front of the mill in a photograph of the West Point Mill taken in the early 1900s (see figure 34). A wind fan blew away the hulls, and the rice was next conveyed to the mortars. Each mortar was about two feet in diameter and five feet deep.

The pestle ends were shod in Russian steel and lifted by spokes that rotated in the slots of the pestles.[7] The spokes were attached to the large shaft, both clearly seen in a photograph taken inside the mill (see figure 35). The pestle forced the rice in the mortar down through a loose ring and up the outside, causing intense friction between the rice which ground off the bran.

West Point Rice Mill, like Bennett's Rice Mill and Chisolm's Rice Mill, was a product of the flood tide of the rice industry of South Carolina. Such mills replaced most of the plantation mills, which were unable to process large quantities of rice.

Bennett's Rice Mill

Bennett's was the most architecturally elaborate of the city mills. The architect is unknown, but a hint of Italian design is seen in the windows. Bennett's Mill was built in Charleston by Governor Thomas Bennett (1781–1865) in 1845 as an addition to a sawmill built in 1830; he operated both mills until his death. Bennett kept the mills operating throughout the Civil War, even though they were under fire from Union forces and

FIGURE 34. Front view of West Point Rice Mill, in Charleston, South Carolina, early 1900s. Abandoned millstones are leaning on the front side of the mill. Courtesy of the Charleston Museum, Charleston, South Carolina.

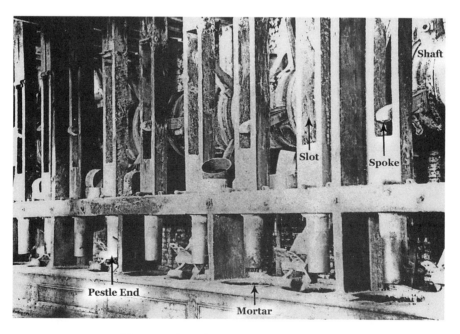

FIGURE 35. Mortars and pestles at West Point Rice Mill, early 1900s. Courtesy of the Charleston Museum, Charleston, South Carolina.

repeatedly struck. He was succeeded by his son M. W. L. Bennett, who operated the mills until his death in 1874. The mills then passed to C. S. Bennett & Co., which owned the mills until 1894, when they were acquired by A. B. Murray, who operated the mills until 1912, when rice ceased to be a major commercial crop.

The engine and boiler were housed in separate two-story brick structures with iron walls. The engine was a walking-beam, condensing engine typical of the period, requiring copious amounts of freshwater to condense the steam in the engine cylinder. Over the boiler house was a tank holding fifteen thousand gallons of water and over the engine house was a tank holding twenty thousand gallons of water. The water was piped to the engine through iron pipes. A storage house for rice was located across from the mill. Rice from schooners was hosted from the holds into the storage house and transported by a tramway in large-bodied railcars.

Today only the west wall of the rice mill is left standing. A 1959 hurricane blew down the other walls. No photographs of the machinery survive.

Daniel Hadley's Model of an 1850 Carolina Rice Mill

The Rice Museum in Georgetown, South Carolina, houses a model of a rice mill labeled "Carolina Rice Mill—1850." The model (color plate 30) was built in 1969 by Daniel I. Hadley of Wilmington, Delaware. Unfortunately no records were located, either at the Rice Museum or in Wilmington, of where Hadley obtained the information on which he based his model.[8] No extant mills in South Carolina were available for him to examine, and there are no existing diagrams or photographs of a mill's interior that Hadley might have used.

A 1969 letter in the archives of the Rice Museum gives information on Hadley's plans for his model. He said he intended to visit Mississippi and Louisiana, presumably to look at mills there.[9] He knew that, when the Southwest began growing commercial rice in 1886, it borrowed much of its early milling technology from the Rice Kingdom. With no local material on which to base his model, it is understandable that Hadley turned to the southwestern rice-growing region. Richard Porcher made two trips to this region, in 1984 and 2010, and found no material that might have been the source of Hadley's model. A model of an early rice mill in the Rice Museum in Crowley, Louisiana (now closed) is very different from the model in Georgetown. The sources for Hadley's model will presumably remain unknown.

Hadley's model represents how a large plantation rice mill or city mill operated. It is exceptional in its technical detail, but its accuracy is questionable in some regards. Yet material on the Savannah Mill and West Point Mill attests to the accuracy of much of it. Hadley's model adds additional information on how these large mills operated and demonstrates one main feature of city and large plantation mills: automation. Rice was placed in one end of the milling process and, virtually untouched, came out the other end as a finished market product and was fed into bags for shipment.

The use of a waterwheel to power the mill is the most obvious fault with Hadley's model. It would have been nearly impossible to power a large city or plantation mill with undershot water power, and there was no water source locally to drive an overshot waterwheel. All archival material on large mills of the 1850 period indicates that steam power was the only source used. The Georgetown rice mill and the big mills in Charleston and Savannah during this time period were all powered by steam. The first large rice mill in Crowley, Louisiana, was steam-powered, and no records of large water-powered mills in the Southwest were found.

The gearing in the mill, however, would have been the same whether the shaft was turned by water or steam power. Hadley's model shows a large shaft entering the mill at ground level. The series of gears that powered the mill all received their power from this large shaft, and the model accurately demonstrates how power was supplied to the various machines involved in the milling process.

Hadley omitted the polishing brush, which had been incorporated by Jonathan Lucas I in his Improved Water Rice Machine, described by John Drayton in 1802.[10] A brush is also shown in the depiction of the Savannah Rice Mill (figure 31). Why Hadley omitted this step is unknown. The brush was an important step in milling because it produced the white, polished rice esteemed by foreign markets.

Hadley's model is labeled with letters, which are visible in the photograph of the model. Terms have been added next to the letters and in the following description the letter corresponding to the term is presented in parentheses.

Step 1. By the mid-1800s, most planters shipped their rough rice from the plantations to the city mills by flats, sloops, or schooners, either in sacks or in bulk. Although Hadley's model shows rice being unloaded from a wagon (A), this was unlikely in the city mills, which were located adjacent to navigable rivers or in harbors. Wagons may have sometimes been used at large plantation toll mills. At the large city mills, rough rice was conveyed mechanically from the vessel to the weighing station and weighed on a scale (B) by a tallyman to determine how much planters were paid.

Step 2. Rice from the scale was temporally stored in a bin (C) until it could be moved to the cleaner (E).

Step 3. An elevator (D) carried the rice from the storage bin to the cleaner (E) on the upper floor. Whether rice was manually moved from the bin to the elevator or passed by gravity from the bottom of the bin into the elevator is not clear from the model.

Step 4. The cleaner was a wind fan, which operated in the same manner as all wind fans. The fan created a column of air. Rice passed by gravity down through a series of riddles and sieves, and the wind blew away the lighter tailings. A shaking motion of the sieves and riddles facilitated the passage of rice through the sieves and riddles, where it exited the wind fan for the next step in processing.

Step 5. Winnowed rice passed from the cleaner into a chute and moved by gravity to the second floor into a dryer (G) called a hopper-boy. It was difficult to process rice between millstones unless it was sufficiently dry. Dried rice exited the hopper-boy at the bottom of the enclosure into a chute that fed the rice to an elevator (H).

Step 6. The elevator carried the rice to the loft, where it passed into a screw conveyor (I), which passed the rice horizontally to a storage bin (J). The rice fell by gravity from the conveyor into the storage bin, where it was stored until it was sent to the millstones.

Step 7. Below the storage bin was a hopper. This hopper fed the rice into a chute (not shown), which in turn fed the rice into another hopper (L) positioned above the enclosure (M) that housed the millstones. A manually operated sliding gate built into the chute below the storage bin controlled the amount of rice flowing into the hopper above the enclosure.

Step 8. Rice passed from the hopper into the enclosure and then passed between the millstones (not shown), where the hulls were removed, leaving a mixture of hulls and rice with the bran layer attached.

Step 9. Rice and hulls passed from the millstones into a wind fan below. (The wind fan is hidden from view, but a label has been added to indicate the general location.) The wind fan blew away the hulls. The heavier brown rice fell by gravity into a hopper with two chutes (only one is shown in the model) below the wind fan.

Step 10. If brown rice was desired for market, a chute (PP) from the hopper below the wind fan was opened for the rice to bypass the mortars and pestles. The brown rice was conveyed to a separate area for bagging and shipment. Hadley mistakenly shows a chute (PP) emptying into a screw conveyor (Q). (The conveyor is hidden from view in the model and is located below the arrow.) Rice and bran from the mortars are also shown emptying into the same screw conveyor. Since brown rice would not have been processed the same way as the rice from the mortars, the brown rice must have been carried elsewhere by a different conveyor or other means not shown.

Step 11. The second chute (not shown) passed the brown rice from the wind fan into hoppers (O) above the mortars, which in turn passed the brown rice into the mortars. The pestles removed the bran layer from the rice.

Step 12. The cleaned rice and bran were manually emptied from the mortars into the screw conveyor (Q). The screw conveyor moved the rice and bran (direction indicated by arrow) laterally to an elevator that carried them up to a cleaner (R) on the second floor. (The elevator is not identified with a letter in Hadley's model; a label has been added.) The rice and bran passed from the elevator into a short chute and into the cleaner. The cleaner was a wind fan and blew away the lighter bran. The bran was disposed of, but the model does not show how.

Step 13. Some rice grains escaped hulling because they were shorter than the average rice. The hulled and unhulled rice were carried from the cleaner by an elevator (S) to the loft and fed into a separator (T). The separator had an inclined, vibrating, perforated tray. The smaller hulled rice and broken rice fell through a tray to a chute (U) that sent the rice to a grader (V). The larger unhulled rice was unable to pass through the tray and passed to another chute that carried it back to the hopper over the millstones. The chute is not labeled in Hadley's model. The letters "PC" (for paddy chute) are added to identify it. The millstones were adjusted to mill the unhulled rice.

Step 14. The grader was a rolling screen set on an incline with different-sized meshes. The smaller mesh was on the higher end. The revolving screen moved the rice from the higher end to the lower end. Below the rolling screen was a collecting bin with two compartments. Broken rice fell through the mesh at the upper end into the first compartment. The whole rice fell through the mesh at the lower end into the second compartment, the whole rice bin.

Step 15. Below each compartment of the collecting bin was a hopper and chute. Only the hopper under the whole rice bin is shown. From the hopper the rice passed into a bag (Z) ready for shipment. The model does not explain how the broken and small rice was treated. It was undoubtedly collected in some way and sold at a reduced price or given back to the planter for plantation use.

Georgetown Rice Mill

The Georgetown Rice Milling Company was set up in 1879 as a joint-stock company by planters and merchants attempting to work together in the difficult period of rice culture after the Civil War. The four-story brick building was located near the foot of Wood Street close to the Sampit River for ease of shipping rice to Charleston and New York. The mill was steam-operated; no record was found what sort of steam engine powered the mill.

The first president was Richard I'On Lowndes, and the first miller and superintendent was Oliver J. Butts. The mill proudly advertised that they used improved machinery in the mill, Brotherhood's patent, and touted that their mill had a higher "turn-out" than other mills, and a beautiful sample of milled rice. In Brotherhood's invention, the pestles fell only a certain distance during the entire pounding, and, whether the mortar was full or almost empty, as was governed by the shrinkage of the rice in the mortar, the mortar self-adjusted so it fell from the same distance above the shrinking rice. By the old style, the pestle was raised to the same point after each blow, and as the rice decreased in volume in the mortar, the weight of the pestle became greater by the increased distance of the fall, resulting in more breakage.[11]

The Georgetown mill advertised that it was the best market in the South for selling rough rice, and the mill's clean rice had markets in New York and Charleston. Steamers made two trips a week to Charleston.

No diagrams of the internal milling machinery has survived. The mill was steam operated and was fully automated like the mills in Charleston. The type of steam engine it used is unknown. The abandoned Georgetown mill was destroyed by fire in the 1940s.

8

The Golden Age of Rice, 1800–1860

The period from 1800 to 1860 was the Golden Age of Rice, and the last ten years before the Civil War was its zenith. The seeds of its future decline had already been cast. It was a one-crop industry based on slave labor and on a fragile hydraulic system that depended on a vast labor force to keep in order. Yet rice culture was still profitable up to the Civil War. Writing in 1972, historian Dale E. Swan concluded that the tidewater rice industry was in a healthy state at the end of the antebellum period; however, his study uncovered a paradox. While plantations producing in excess of one hundred thousand pounds of rice or more per year were doing well, plantations producing less than one hundred thousand pounds were in economic difficulty by 1859. The large-producing plantations, however, produced 96 percent of the total rice crop in 1859, the reason he concluded as a whole the industry was healthy. Swan suggested that rice plantations tended to be large and the industry highly concentrated because only through quantity production could profits be realized.[1]

Historian William Dusinberre has outlined how expansion and production in the rice industry increased from 1767 until the Civil War despite occasional setbacks: "American rice production was indeed lower in the early 1820s than the peak it reached thirty years earlier. . . . But by 1825 rice planters were recovering from Jefferson's embargo, the War of 1812, and the depression of the early 1820s. In the 1840s, rice production was higher than in the 1790s. And it continued to grow during the 1850s."[2] Dusinberre included production totals for clean, milled rice in three periods.

Tables 2 and 3 show a gradual increase in rice production during the decades before the Civil War. In general it ranged from 80 million pounds to 120 million pounds and generated receipts of $1.5 million to $2.5 million a year ($24 million to $48 million a year in 2009 dollars). The variation in production that occurred from region to region each year was mostly not under the control of the planters (see table 4). Hurricanes wreaked havoc on rice crops. After the hurricane of 1822, which struck Georgia and South

TABLE 2. American Rice Production, 1767–1860

Years	Average Annual Production (estimated millions of pounds of clean rice)	Years	Average Annual Production (estimated millions of pounds of clean rice)
1767–73	72	1835–39	85
1790–95	82	1840–44	94
1820–24	67	1845–49	109
1825–29	90	1850–55	108
1830–34	91	1856–60	114

Source: William Dusinberre, *Them Dark Days: Slavery in the American Rice Swamps* (Athens: University of Georgia Press, 2000), 389.

TABLE 3. American Rice Production, 1839–1860

Year	Pounds of Clean Rice (estimated millions of pounds)	Year	Pounds of Clean Rice (estimated millions of pounds)
1839	80.8	1850	107.8
1840	84.3	1851	110.9
1841	89.0	1852	107.5
1842	94.0	1853	106.4
1843	89.9	1854	74.4
1844	111.8	1855	108.7
1845	89.8	1856	104.5
1846	97.7	1857	113.5
1847	103.0	1858	122.0
1848	119.2	1859	124.4
1849	132.9	1860	106.6

Source: Dusinberre, *Them Dark Days*, 452.

TABLE 4. Lowcountry Rice Production, 1849 and 1859
(in estimated millions of pounds of clean rice)

	1849	1859	Change	% increase/decrease
New Hanover & Brunswick Counties, N.C.	2.5	5.6	3.1	124
Georgetown County, S.C.	28.9	34.4	5.5	19
Charleston, Colleton & Beaufort Counties, S.C.	66.8	44.8	-22.0	-33
6 coastal counties, Ga.	22.9	31.9	9.0	39
Total Lowcountry	121.1	116.7	-4.4	-4

Source: Adapted from Dusinberre, *Them Dark Days*, 390.[3]

Carolina, some planters contemplated diversifying crops and switching to Sea Island cotton because cotton fields were less likely to undergo major flooding during major storms and because cotton at that time had a higher market price. Sea Island cotton in 1823, 1824, and 1825 sold for 24.5, 24.6, and 54.3 cents per pound respectively ($5.18, $5.64, and $12.10 in 2009 dollars), an excellent return on investment.[4] Short-staple cotton, however, put pressure on Sea Island cotton prices, and planters returned to rice. Overseas markets were also fickle. Demand fluctuated yearly and influenced the price rice planters received for their crops. The hurricanes of 1850, 1851, 1852, and 1854 again forced the planters to consider alternatives to rice, but ultimately they returned to rice planting.

By the nineteenth century, rice production had expanded to areas on all Georgia's major rivers: the Savannah, Ogeechee, Altamaha, Satilla, and Saint Marys. Georgia planters increased production during the period from 1800 to 1860, reaching a peak in 1860, when they planted approximately thirty thousand acres of tidal rice. Nonetheless Georgia planted less acreage than South Carolina because Georgia had less tidal swamp conducive to tidal cultivation. By 1839 Georgia was producing more than 13 million pounds of rice, equal to 20 percent of the South Carolina rice crop, and by 1859 Georgia had quadrupled its output to 45 percent of the South Carolina crop.[5]

Large-scale rice production in Georgia began in the early nineteenth century, when Carolina planters expanded their operations across the Savannah River. John Potter of Charleston, who owned land on the Georgia side of the Savannah River, gave one son Colerain Plantation in 1817 and a younger son Tweedside Plantation in 1832. These two Georgia plantations were highly profitable. In 1845 Tweedside produced thirty thousand bushels of rice for a gross profit of $40,000 ($1,170,000 in 2009 money), an enormous sum for the day.[6]

At Hopeton Plantation on the Altamaha River, owner James Hamilton Couper planted his first rice crop in 1821, and it was so profitable that it replaced cotton and sugarcane as the plantation's main staple. From that year through 1826, his acreage increased dramatically.

In 1833 Charles Manigault of Charleston purchased Gowrie Plantation on Argyle Island in the Savannah River for $40,000 ($1,050,000 in 2009 money). A member of a distinguished Carolina family long involved in growing and marketing rice, Manigault started out planting rice at Silk Hope Plantation on the Cooper River in Berkeley County, South Carolina, and then expanded to Georgia, furthering his family's prominence as leading rice planters of the Lowcountry. Manigault continued to increase production and profits at Gowrie with a bumper crop of 578 barrels in 1838. In 1847 Manigault evaluated his earnings at Gowrie for the first fourteen years: "The last 3 years my Crop after paying all expenses [my profit was] about $12,000 [$323,000 in 2009 money]. But owing to my expendatives [*sic*] *on the place* the average annual Net Income spread over fourteen years is but $7,000 [$188,000 in 2009 money] per an[num]."[7] James Clifton estimated this was a return of 14 percent for fourteen years or 24 percent for the last three years. In 1855 Gowrie produced 23,800 bushels of rice, the largest crop ever. The period from 1855 to

1860 was the most productive of Gowrie's history. By 1860, Manigault was planting all 650 acres of tidal swamp in rice. Over this period Gowrie's profits amounted to a 10 percent return on capital. One factor in the profitability of Gowrie was its steam-powered threshing barn and water-driven mill installed after Manigault bought Gowrie. Both innovations allowed more time for slaves to perform other plantation tasks.

Hugh Fraser Grant operated plantations on the Altahama River from 1834 to 1861. Grant's net income from rice increased annually as he placed additional fields under cultivation with a greater labor force. In 1845 his yield from 200 acres was 10,168 bushels; in 1856 he planted 345 acres and produced 13,971 bushels. Unpublished census records for 1850 show that twenty-one planters in Georgia had annual yields from one hundred thousand to six hundred thousand pounds of rice; a few had yields of more than one million pounds.[8]

The Golden Age was characterized by factors that separated it from the pre-Revolutionary period: the final conversion of swamp to tidal fields, codification of the task system, the end of the legal slave trade in 1808, the mechanization of threshing and milling, and the expansion of manufacturing in the cities.

Tidal Rice-Field Expansion

The final conversion from inland rice culture to tidal culture occurred after the Revolution. Tidal swamp that could be converted to rice fields was limited to a narrow strip along the coast. Once these swamp lands were converted to fields, no further swamp lands were available. Sea Island cotton planters banked salt marshes to grow Sea Island cotton, but this was not an option for rice planters. Rice cannot tolerate salt, and it was not labor-wise or cost-wise to remove the salt from fields. The great expense involved in establishing a tidal plantation may account for the fact that ideal tidal swamp land was still available in the early 1800s. By the second decade of the Golden Age, however, most of the remaining tidal river swamp that could support tidal rice culture had been converted to rice fields. Whatever increase in crop production after this final conversion had to come from fields already constructed.

According to Lawrence S. Rowland, "the great growth and development of the Savannah River rice lands [in St. Peter's Parish on the South Carolina side] . . . began during the 1820s." He cited three reasons for the expansion: "the advances in the science of growing and processing rice; several years of exceptionally high prices . . . following the Napoleonic Wars; and the optimism and economic growth . . . of Savannah." According to Rowland, "the men who developed rice lands in St. Peter's Parish in the 1820s and 1830s were among the wealthiest and most influential men in two states," and "the years between 1830 and the Civil War were years of great enterprise and production in St. Peter's Parish. The engines of the Industrial Revolution, which were generally thought to have bypassed the agricultural old South, were very much in evidence on the Lower Savannah."[9]

As late as 1845, J. Motte Alston (1821–1909) began his new rice plantation at the intersection of Bull Creek and the Waccamaw River in Georgetown County. Alston supervised his overseer and slaves clearing the swamp and installing ditches and canals. Alston built a small cabin on the bank of Bull Creek and called his plantation Woodbourne. After he married, his slave Richmond built a twelve-room house, Sunnyside, for the Alston family. By 1858 Alston had made his fortune planting rice and sold Woodbourne for $45,000 ($1,210,000 in 2009 money), a plantation that had been worth only $5,000 ($127,000 in 2009 money) when his father gave it to him in 1841.[10] During the years he planted Woodbourne, Alston averaged a respectable profit each year.

In the Santee Delta, land was cleared through the first two decades of the 1800s, evidence that rice was still a profitable crop despite the Jefferson embargo, the War of 1812, and economic downturn of 1819. Henry Edward Lucas (1822–1900) purchased one half of Crow Island, a tract of eight hundred acres of swamp along Big Duck Creek in the Santee Delta. Slaves banked off four hundred acres to make rice fields, and Lucas built a threshing mill, a large barn, heavily-framed slave houses to resist storms (see color plate 32), and a small dwelling house. Lucas recorded no date for when he began his work, but since he was born in 1822, it must have been around 1845.

The Labor Force

By the turn of the nineteen century, the task system for rice culture was well entrenched. The foreign slave trade was outlawed in 1787 and reopened from 1803 until 1808, when the U.S. Constitution declared the trade closed. Henceforth, the slave population grew almost entirely by natural increase. In spite of the closing of the slave trade in 1808, rice plantations functioned uninterrupted in regard to labor. Since further construction of new fields (a major use of plantation labor) ended around 1820, natural increase and the task system kept the labor force adequate in the years before the Civil War.

The number of slaves on tidal plantations in the Rice Kingdom actually fell from fifty-eight thousand in 1850 to forty-eight thousand in 1860.[11] One reason for the decline was a shift of slaves from rice production to Sea Island cotton fields. With the advent of mechanical threshers and advances in rice milling, especially the introduction of steam power around 1817, larger crops could be processed with less labor. Planters who owned cotton plantations could shift slaves to Sea Island cotton plantations and still operate their rice plantations profitably.

Mechanization and Manufacturing Expansion

The many steam threshing barns and mills built in the first half of the 1800s are evidence of an expanding and profitable rice industry. A steam mill in 1833 cost $7,500 ($181,000 in 2009 money), a large sum for the time; a threshing barn in 1850 cost $8,000 (approximately $227,000 in 2009 money). Our field surveys have documented ruins of around fifty steam-driven threshing barns and mills in the tidewater area of South Carolina alone, most dating after 1840, with an estimated cost of $11 million based on 2009 values. Georgia and North Carolina planters also built steam-driven threshing barns and mills. (No estimates on the number are available.) A Butts thresher alone was a considerable expense for a rice plantation. Plantations spent additional funds on repair and replacement of warn-out parts and other plantation expenses. Even keeping water-driven systems in operation was expensive.[12]

In the two decades before the Civil War, Charleston industry expanded and reached its peak in 1856, with similar expansion in Savannah. Charleston industry was varied and included iron foundries, rice mills, sawmills, cotton mills, turpentine distilleries, shipyards, and brickyards—all in support of local agriculture, rice being a mainstay. Machinery for city rice mills was an important aspect of manufacturing. The foundry of Eason and Dotterer, for example, built a two-hundred-horsepower steam engine for Chisolm's Rice Mill. In 1848 Charleston had six rice mills with an aggregate of 155 pestles to mill rice sent from the plantations. Five were run by steam power. The largest rice mill, the West Point Mill, cleaned two hundred barrels a day in 1860. Georgetown had a dozen rice mills with an output of processed rice in 1860 valued at $1,110,000 ($26,600,000 in 2009 money). Mill engines, boilers, and threshers were all produced for plantation threshing barns and mills. Savannah and Georgetown were also centers of manufacturing, although not on the scale of Charleston.[13] All these mills depended for their livelihoods on the cotton and rice plantations.

The increase in production of rice was owed in part to the increased efficiency in new machinery used in the market production of rice. It remains for future historians to explore fully the influence of these new machines on market economy, crop production, slave labor, and plantation economy—and their role in the Atlantic world.

By 1860 virtually every highland that bordered on a tidewater river in the Rice Kingdom was adorned with a plantation mansion with an unobstructed view of the river and its rice fields, the vista kept clear by slaves. Avenues of live oaks, the saplings dug in the dead of winter from surrounding forests and planted, were still small trees, nothing like the behemoths one sees today. Formal gardens graced the landscape around the plantation house, adorned with exotic plants the planters ordered from overseas. One or two slave streets were set off from the main house. Slaves tended small subsistence gardens next to their cabins after their tasks were done. In the plantation yard slaves would have been busy making trunks, barrels, or other goods necessary to keep the plantation operation going. At harvest time, flats would be transporting sheaves from the fields through the canals to the threshing yard or barn. On a large plantation, a steam pounding mill would be cleaning the rice, the sounds of the pestles pounding the mortars reverberating through the fields and rivers. A schooner would be docked at the edge of the mill, waiting to take the clean rice to city markets, with enormous profits for the planter. If it were the social season, the planters themselves would be in Savannah or Charleston in elaborate houses, the plantation operations left to overseers and drivers. In summertime, some planters would head to the mountains or Lowcountry pineland villages, or go "to the salt" of the barrier islands.

This was life in 1860 in the Rice Kingdom: a mixture of servitude and privilege, manual drudgery and technology. A manufactured landscape maintained with the energy of slaves, plantation life proceeded as it had since rice culture resumed after the Revolution.

On April 12, 1861, the cannons fired on Fort Sumter in Charleston Harbor. Planters and slaves in the Cooper River area, where rice culture had begun so many years before, heard the explosions in the distance. Did planters know that the cannons sounded the beginning of the end of their privileged life? Did slaves, hearing the same cannons, know it was the beginning of their emancipation which took over a century to reach fruition? The answer would come four years later.

9

The Last Days of Rice Planting

The rice aristocracy of the tidewater Rice Kingdom—which prospered on the toil and resourcefulness of bound labor, on great outlays of capital, and on the ingenuity and resourcefulness of a handful of planters—represented one of the greatest concentrations of wealth and social privilege in the antebellum South. The ascendancy of this elite began to wane in the carnage and ashes of the Civil War. War and Reconstruction created almost insurmountable land ownership, labor, and capital difficulties, which led to production and profit declines—obstacles the planters ultimately could not overcome. There is no single date when rice culture ended in the Rice Kingdom; planting ceased in different areas and at different plantations at various times over half a century, starting with the Union invasion of the Beaufort area in November 1861. While the Civil War disrupted the production of rice and crippled the southern economy, however, it did not destroy rice production. Production stumbled along for another fifty years or so, but under different conditions. More financially secure plantations were able to continue production longer than those that were less financially sound.

The Civil War, did, however, set in motion a series of events that caused added difficulties for the planters trying to produce successful commercial crops and contributed to the eventual end of rice culture. Rice production declined drastically during the war. The northern naval blockade cut off the Lowcountry from its overseas export market. A portion of the labor force was taken off the rice plantations and set to work building fortifications or performing other tasks for the Confederacy. Many plantation owners and their sons and overseers went to fight or served the war effort in other capacities. The drain of white men into the Confederate army left many rice plantations under the supervision of wives or elderly and infirm men, and slaves increasingly challenged authority. The Confederate government called on southern forges and foundries to make guns rather than plows and machinery for agriculture. Agricultural implements and machines became expensive, and most planters had to get by with the tools and machines they already owned. Defective or broken implements and machines placed a strain on getting a crop to market. New parts for steam engines were difficult to find because foundries and machine shops were producing materials for the war effort. Rice plantations, along with all southern farms, lost many horses and mules to impressment by the Confederate army or capture by Union forces. Impressment of food for Confederate troops also place a drain on plantations.

Before the war, slaves often ran away and escaped capture for a while by hiding out in the Lowcountry swamps and forming maroon colonies. During the war it was much easier to escape detection and form such colonies because there was less manpower to track down fugitives. Fugitives who escaped during the war caused additional drain on a plantation's operation. Some slaves refused to work and devoted their time to tending their domestic crops.

The Union occupation of Port Royal and adjacent areas on November 7, 1861, gave the Union not only a supply base for its blockade of the Carolina coast but also a base from which Union forces could harass Lowcountry rice plantations. Charles Heyward of Rose Hill on the Combahee River commented in his diary in 1862: "In March, 15 of my negroes including 3 women and 1 child left the plantation and went over to the enemy on the Islands."[1] Heyward moved the remaining slaves from his Combahee River plantations inland to Goodwill Plantation on the Wateree River and gave up planting on the Combahee until after the war.

Wherever Union forces approached plantations, many slaves abandoned them at the first opportunity and set out in search of freedom, following the Union troops. Many of those who left were young and skilled laborers; there was no way to replace these workers after the war. Slaves desired freedom itself, not just freedom from abusive masters or overseers. Slaves of kindly and benevolent masters sought freedom as eagerly as those of cruel ones. Uppermost in the planters' fears were slave insurrections, followed closely by apprehensions about Union raids of pillage and destruction.

Planters were shocked when their "trusted servants" deserted them. Louis Manigault wrote, "The war has taught us the perfect impossibility of placing the least confidence in the Negro. In too many instances those we esteemed the most have been the first to desert us. It has now been proven also that those planters who were the most indulgent to their Negroes when we were at peace, have since the commencement of the war encountered the greatest trouble in the management of this species of property."[2]

Heyward and Manigault were not the only planters who found it hard to believe that slaves would desert as soon as the opportunity presented. Edmund Ruffin, an agricultural reformer who touted the use of lime to restore old fields, found it difficult to believe that his slaves were not content. Upon learning from his son that his slaves from his farm outside Richmond were running away to the Federals as they pushed closer to Richmond in May of 1862, Ruffin lamented, "No where were they better cared for, or better managed & treated, according to their condition of slavery?"[3]

The destruction that accompanied the Civil War along the Georgia coast is well documented. The Union navy recognized the importance of St. Simons and Brunswick, where there was a perfect harbor. The occupation of Brunswick and St. Mary's, the capture of Darien, and raids from Sherman's army, did considerable damage to the infrastructure of Georgia plantations. Some rice planters destroyed their own materials and infrastructure so they would not fall into the hands of the enemy. Summer homes of planters along the coast were burned. For example Maybank, Woodville, and Social Bluff on Colonel's Island in Liberty County, Georgia, were burned by freed people who had gathered on nearby St. Catherine's Island during the closing days of the war and had come across the sound by boat. According to historian Erskine Clarke, "the fires they set at Maybank and the other island homes had been kindled by long smoldering rage and apparently by the hope to possess abandoned land." Although the loss of these summer houses had no direct impact on rice culture, it caused financial distress as well as robbing planters of places to escape the unhealthy summer months on the rice plantations.[4]

The Rice Kingdom was a main target of northern forces throughout the war because of the enormous value of the tidewater region to the Confederacy. When Sherman's army marched through the tidelands of Georgia into South Carolina, his soldiers destroyed much of the infrastructure, as did many of the freed slaves in reprisal for their enslavement and to ensure that they would never have to plant the fields again. Many Savannah River plantations were not able to replant after the war. Mills and barns were burned, and equipment was destroyed. On plantations not in the path of Sherman's army, slaves disrupted the plantation routine by doing poor work, encouraging others to do the same, and spreading rumors that caused disruption of work. On plantations abandoned during the war, canals became blocked with mud and weeds, trunks deteriorated, and fields left unattended quickly became choked with weeds. Efforts to reclaim fields after the war were costly and contributed to the eventual demise of cultivation.

When Sherman's army marched into South Carolina on February 1, 1865, anarchy and rebellion reigned in the army's path. In South Carolina, more than anywhere else in the Confederacy, slave resentment burst forth in violence and destruction. Plantation houses were ransacked, and goods and food were stolen.

On March fifth and sixth in 1865, a month and few days before Lee's surrender of the Army of Northern Virginia to Grant, Union soldiers and the United States Navy sailed up Georgetown's rivers and bays and announced the end of slavery. Soldiers and bands of freedmen roamed the countryside looting plantation homes and buildings, carrying away livestock, food and families' valuables, and burning buildings. Georgetown was the center of rice production in South Carolina, and damage to the infrastructure was a major blow to planters when they tried to restart growing after the war.

When all is considered, there were eight main factors that ended rice cultivation on most plantations around 1911 although they did not affect plantations or river systems equally: neglect of the rice fields; destruction of the infrastructure by Union soldiers and slaves as well as by natural disasters such as hurricanes; lack of an adequate labor force; severe shortage of capital; competition in the European market with low-cost rice from Asia;

competition from rice production in the southwestern states, which began in 1886; and alteration of the freshwater/saltwater interface.

In an effort to restore their plantation operations and some semblance of the lifestyle they enjoyed before the war, planters from Charleston, South Carolina, to south Georgia, first had to reclaim their lands. Their plantations had become part of the Sherman Reservation created by Special Order 15, issued January 16, 1865. Reaching the coast after his march through Georgia, General Sherman had ordered the entire area of the coast—the Sea Islands and thirty miles inland from Charleston to Jacksonville—set aside as a reservation of land to be distributed to freedmen to own and farm. After the Confederate surrender and Lincoln's assassination, President Johnson pardoned most rice planters and restored their property. Once they reclaimed their lands, there were still many problems to overcome, but at least they had land to plant.

Returning to their rice fields after the war, the planters found that four years of neglect left many fields with broken banks, ditches and canals chocked with weeds, and much of the infrastructure destroyed. Even keeping the milling and threshing machinery in working order posed a problem. Edward Barnwell Heyward wrote to Catherine Maria Clinch from Combahee in 1867 that "Cousin James is quite disappointed with his Threshing Mill, and is very mad about it. The machine works badly and does very little credit to the Maker, as there was no stint of money, and a first rate machine was bargained for."[5] The war had destroyed much of the manufacturing industry in the cities too, and planters had neither money nor qualified repairmen to attend to the machines on the plantations. Unable to afford repairs to water-powered and steam-powered threshing barns and mills, planters reverted to manual threshing and milling. They were able to plant into the early 1900s, but at a much reduced production level. Surviving photographs show workers threshing with the flail and milling with the mortar and pestle well into the 1900s.

Many planters that returned home after the war found their plantation in utter ruin. The fall of the Confederacy left General M. C. Butler of South Carolina "with one leg gone, a wife, and three children to support, seventy slaves emancipated, a debt of $15,000, and in his pocket, $1.75 in cash."[6] The war took another toll on rice planting: many planters' sons did not survive the war. The loss of these young man, who had been expected one day to run family plantations, was deeply felt.

When Louis Manigault returned to Gowrie and East Hermitage in March 1867, he found damage that was probably typical throughout the Rice Kingdom, certainly to plantations in the path of Sherman's army. At Gowrie Sherman's forces had destroyed Manigault's house, the servant house, the barn and its contents, and the pounding mill; at East Hermitage the threshing mill had been burnt. Louis described Gowrie in his plantation journal:

> Standing near the ruins of my former dwelling I contemplated the spot. Where once stood this Country House could alone now be seen a few scattering brick, &

the tall chimney to denote the spot. No remnant of my Kitchen, Fine Stable, both built just previous to the War, remained. . . .

The great loss in the Settlement however was the huge Rice Pounding Mill burnt by the Yankees and from which I am told the flames extended to the other Buildings. . . . The change in the appearance of Gowrie Settlement is, I may say, from a Village to a Wilderness. The Settlement is a barren waste and presents a most abandoned and forlorn appearance.[7]

One option was not available as planters pondered how to resume planting again: it would have been difficult, if not impossible, to plant the thoroughly altered landscape with any crop other than rice. No other crops would have been successful in the tidal fields. If they were to profit from the fields again, rice was their only viable option. As with the adjacent Sea Island cotton planters, planting rice was their heritage and their passion. It was their life before the war, and it would be their life after the war, for that was all the former rice aristocracy knew. For them growing rice was an honorable and noble life and separated them from the yeomen of the upcountry. They believed that the world would still demand their Carolina Gold and that somehow the next crop would start them back on the road to recovery and restore some part of their prewar wealth and some semblance of their antebellum grandeur. They were convinced that their privileged lifestyle—big house, abundant land for hunting and nearby rivers for fishing, field slaves and house servants, overseas sojourns, and social life in the cities—was just a few good crops away. So they planted rice against all odds, and they were successful to a point.

Planters faced two major problems if they were to resume planting: adequate financial resources and suitable work agreements with freedmen to acquire a reliable and quality labor force. Confederate money was now worthless; wealth in human chattel had ended with emancipation. Having burned much of their stored rice so it would not fall into Yankee hands, planters deprived themselves of capital after the war. Rice production was labor intensive, and there was little capital to hire replacements for slave labor. Carpetbaggers were piling debts on the South; interest rates rose; and federal taxes mounted to repay the war debt.

In the Georgetown area, planters borrowed money at 2½ to 3 percent per month, a high interest rate that almost guaranteed failure. Successive crop failures drove planters deeper into debt, and some were forced to cease planting because of their tenuous financial condition. Those who wanted to plant again were forced to borrow again. Similar situations faced planters throughout the Rice Kingdom. Even after Reconstruction they always faced uncertain financial conditions. Factors in Charleston and other cities, who financed and sold most of the rice, were fearful of the possible effects of unsettled labor and political conditions and were reluctant to advance money to rice planters.

Unlike the Sea Island cotton planters who sold off some of their land to raise capital and planted reduced acreage, it was difficult to divide or sell a rice field without disrupting the means to control the flow of water to the fields. Only limited cash could be raised by selling the highlands since they were not as valuable as the rice fields.

Some planters were so destitute after the war that they could not hire workers. At Chicora Wood on the Great Pee Dee in Georgetown County, the widow of Robert Allston, Adèle Petigru Allston, was asked by her daughter why she had to let a servant called Daddy Aleck go. She replied: "Child, you don't understand. Aleck really wants to stay now, but I have no right to keep him. He is a valuable groom and hostler, can manage and drive any horses, and he can easily get a good place in Georgetown, whereas I could not only not pay him, but I could not possibly feed him."[8] Similar scenarios played out on many plantations after the war, as skilled freedman were free to leave and find good employment elsewhere. On the Combahee River, Edward Barnwell Heyward, who with his sister inherited four rice plantations, proposed in July 1866 to abandon rice growing and turn to the more profitable cotton: "Our capital is too much reduced for us to attempt the Rice again for long time if ever," he reported.[9]

By 1879 there had been a significant recovery of production on the hardier plantations, especially in South Carolina and North Carolina. This recovery was helped by infusion of northern capital into the southern banking system and a tariff imposed on foreign rice (about 100 percent), which allowed southern planters to sell their rice for four to four and one-half cents per pound less than imported rice.[10] Even with the labor problem and the primitive methods of market preparation that many planters had to employ after the war, the tariff made it possible to produce rice profitability. Planters also realized they had to modernize their methods of cultivation by using animals and machines and changing the growing regimen to reduce the need for certain kinds of field work. They began to employ animal-driven plows instead of hoes, mule-driven planting drills, and harrows. One aspect of cultivation they could not change: labor to maintain the hydraulic infrastructure of the tidal fields, and this inability to change ultimately proved fatal to the industry. Freedmen were adamant that they would not stand in cold water during the winter to clean ditches and repair dikes.

Securing an adequate labor force was a daunting problem. Some former slaves remained on plantations and others found their way back to begin rice growing after hostilities ceased. The plantation was the only home they knew even if it had been a place of bondage. Despite intense racial and social tension in the aftermath of slavery, most freed people were not vengeful enough to inflict violence on former masters or mistresses, though some cruel overseers were not as fortunate. On Butler's Island and St. Simons Island in Georgia many freed people returned after the war, including some who had been sold off the plantation in 1859 by Pierce Butler (1810–1867). His daughter, Frances Butler Leigh (1838–1910), said "Nearly all who have lived through the terrible suffering of these past four years have come back, as well as many who were sold seven years ago." She considered this loyalty an endorsement of how they had been treated by the Butlers.[11] She quickly found, however, that they were not eager to return to the rice fields and signed contracts only after long negotiations. It is uncertain how frequently this scenario played out elsewhere.

When the generation of prewar skilled workers died or became too old to work in the fields, however, their sons and daughters had already left the plantations to work in the cities for better jobs and higher wages. The new phosphate industry in South Carolina drew freedmen from the plantations. John Girardeau Legaré of Darien, Georgia, who planted General's and Champney's Islands along the Altamaha River, recorded in his journal on October 18, 1898: "I find it impossible to get hands. The lumber people are paying common laborers from $1.25 to $2.50 pr. day & rations and the negroes will not therefore work for me for 75 cents pr. day and so not much has been accomplished so far."[12] This pattern was repeated throughout the Rice Kingdom after the war, so the skill of the labor force deteriorated in the late 1800s, and without a skilled labor force, the quality of cultivation suffered. The generation of workers who grew up after the war did not understand the intricacies of rice culture, nor did they have the desire or patience to learn. The United States Commissioner of Agriculture observed that "the power to compel the laborers to go into the rice swamps was utterly broken."[13] Duncan Clinch Heyward lamented on the labor problem: "No crop, and especially rice, can be a success unless it is thoroughly and properly cultivated, particularly when the work is done principally be hand Labor. In our rice fields the hand hoe was always the most important implement, for, with the soil packed as the result of irrigation, it was essential during the season of dry growth to stir the ground thoroughly."[14]

Planters who could not find freedmen to do wintertime field maintenance tried other labor sources. Frances Butler Leigh, for example, leased an island on her estate "to an energetic young planter" who "brought down thirty Chinamen to work it."[15] Other planters hired gangs of Irishmen to ditch and tend banks. Planters found out, however, that acclimated freedmen were the only laborers who could work with impunity in the rice fields during the appalling conditions of the summer months. Irish and Chinese workers did not solve the labor shortage.

In the Georgetown area, the first planting season after the war revealed the new problems that planters would encounter with labor. Freedmen were hired for cash wages, but the planters often found that following payday on Saturday, many workers failed to show up for work the following Monday.[16] After years of a slavery system that had forced them to work under appalling conditions, freedmen rebelled at the initial working conditions planters tried to impose.

Many freedmen choose not to work on the plantations, for the land was "God's pantry" to the former slaves. The woods were full of game, and the ocean and tidal creeks teamed with fish, oysters, and crabs. Frances Butler Leigh wrote: "There are about a dozen on Butler's Island who do no work, consequently get no wages and no food, and I see no difference whatever in their condition and those who get twelve dollars a month and full rations. They all raise a little corn and sweet potatoes, and with their facilities for catching fish and oysters, and shooting wild game, they have as much to eat as they want, and now are quite satisfied with that, not yet having learned to want things that money alone can give."[17]

After two hundred years of slavery, it is understandable why some freedmen were content to live off the land and not share the planter's value of money. The primary goal for many freedmen was a small plot of land to raise crops to sell in local markets and a house to live in. The loss of these freedmen's skill in the rice fields reduced the quality of the labor force, straining an already tenuous revival of rice culture.

Some planters became their own worst enemies by refusing to acknowledge the new order. One thing was certain. Emancipation eliminated coercion as a method of labor management. Some planters actually went so far as to not inform their former slaves of emancipation. Ultimately, however, planters realized that they needed a stable labor force and that they needed freedmen to continue planting. They had to make concessions. Most planters came to understand that they were no longer masters but employers. At the same time freedmen realized they had some power in the new labor agreements because rice planting could not proceed without their labor. If one planter's rules and regulations seemed too onerous, freedmen could usually find a nearby planter desperate for labor who would give a better contract. Using this power, freedmen were able to gain some control over contracts.

Freedmen's Bureau officials and planters did agree on one thing: the necessity for binding contracts and the need for some mode of compelling laborers to honor them. Military authorities often supervised the making of contracts between the planters and freedmen, and ordered signers to honor the contracts or leave the plantation in ten days. Freed people who did not honor the contracts were forced off the plantations; sometimes they burned barns and houses in reprisal, adding to the problem of financing a plantation. Military authorities sided with the planters and restored some sense of normalcy to the labor force, allowing planting to proceed. Planters, however, were warned that if they decided not to plant crops on their land, they would still be responsible for providing for their former slaves and possibly other freed people on their land.

No matter what other problems the planters faced in reviving rice culture, the key to success was getting an adequate supply of good labor, which involved continual negotiation with freedmen, who wanted as much control on their own lives as possible. Many freedmen voluntarily remained on the plantations and worked the rice fields under contract systems because during and after Reconstruction other employment was scarce. Various types of contracts were tried. Under one type, the laborers would get one-half of the year's crop after a fifth had been deducted to cover plantation expenses. The planter would furnish all necessary tools and planting equipment. The workers agreed to supply all labor and furnish their own provisions.[18]

Some planters initiated a "two-day" tenant system similar to an arrangement implemented by the Sea Island cotton planters.[19] The planter provided the African American laborer's family a house, the right to gather wood for fuel, and from five to seven acres of land. In return, the tenant agreed to work for the planter two days a week (usually Monday and Tuesday); the rest of the week the family could work their own fields. The tenant laborer was expected to work "two tasks," or one-half acre, per day. If more work was required for the crops, the planter would find more tenants or pay his own tenant a wage for the extra work. This system had one major appeal to black laborers: land. Freedmen were not interested in large tracts of land or even market agriculture; small acreages of land were sufficient to give them the autonomy they had not known as slaves. Any sort of labor that provided land where they could grow crops and raise animals for home use or to sell locally provided a limited income that could be used to buy more land. Planters who realized this were able to have a more stable labor force.

On Butler's Island in Georgia, the plantation developed by Major Pierce Butler (1744–1822) and operated after the war by his grandson Pierce Butler, the former slaves who had either stayed on or returned to the island agreed to work on a crop-sharing basis. Owner and workers each received one-half the crop. As Malcolm Bell points out, however, Butler "was discouraged to find the willingness to work was measured. Almost half the force left the fields by one o'clock and all were out by three." As the Butlers feared, their first rice crops failed, and says Bell, "the inability to share earnings with the blacks further strained the tenuous relationship between Pierce Butler and those who worked his lands."[20] The Butlers, like some other planters, changed from crop sharing to wages. This suited the workers but strained an already tenuous financial situation.

On Gowrie Plantation on the Savannah River, Louis Manigault rented 390 acres of rice land to Gen. George P. Harrison in 1867. Harrison, in turn, entered into a contract with freedmen to plant rice. The 390 acres were divided into divisions of 78 acres each, and a freedman experienced in rice culture was chosen as foreman to run the division until the crop was ready for market. The foreman selected his own hands. A contract was signed between each foreman and Harrison in Savannah at the Freedmen's Bureau, where the contracts were drawn up according to established law. Little or no contact occurred between Harrison and the workers. Harrison furnished land, trunk lumber, mules, ploughs, plantation tools, and one-half the seed rice. The foreman furnished labor and the other half of the seed rice. The workers furnished their own provisions. At the end of the year, after all plantation expenses were paid, Harrison retained one-half the net profits, and the workers divided the other half. The system pleased Manigault because it was the kind of "gang system" that he saw as the most effective labor unit for the rice fields.[21]

No matter the type of work system, problems were common on the plantations after the war. Freed people proved unwilling to work under the harsh regimen that a tidewater rice plantation required, and there was no way to alleviate or reduce the harsh conditions required to cultivate rice in the tidelands. Before the war natural energy had been shaped and channeled by slaves' energy to produce crops. Nowhere was the energy of slaves and the flow of natural energy more linked than in the tidal rice fields of the Rice Kingdom— and nowhere were the working conditions more harsh. Without an adequate postwar labor force and the newer machines available for cultivation, planters such as Duncan Clinch Heyward and his father before him knew that the industry would soon end.

In the Georgetown area, a few planters tried to consolidate a number of plantations and operated them under the direction and financial control of joint-stock companies. Planter Philip R. Lachicotte used a corporate structure of Lachicotte & Sons on the Waccamaw River, James LaBruce formed the Guendalos Company to plant the Pee Dee River plantations, while Samuel Mortimer Ward formed the S. M. Ward Company to plant on the Santee River. These companies combined planting with milling and worked together to reduce costs.[22] The Georgetown area and Santee River planters, however, suffered the same fate as the Combahee River planters, and hurricanes ended their rice growing around 1906. Elizabeth Waties Allston Pringle wrote in 1906: "But it almost seems as though I was meant to give it up. The rice-planting, which for years gave me the exhilaration of making a good income for myself, is a thing of the past now—the banks trunks have been washed away, and there is no money to replace them."[23]

Peter Coclanis has argued that starting just after the Civil War, competition from the Far East created another problem for the Rice Kingdom: "The transfer of low-cost eastern rice to the West, facilitated by the advent of the steamship and the opening of the Suez Canal [in 1869], gradually displaced traditional supply areas such as the South Carolina Lowcountry."[24] Burmese rice exports to Europe alone averaged 682,210 tons between 1881 and 1890, a critical period for planters trying to revive the Lowcountry rice industry.

By 1880 the cost of producing an acre of rice had risen dramatically. Production of an acre on Causton Plantation on the Savannah River was estimated at $27 ($585 in 2009 money). At Gowrie Plantation it was $26 ($563 in 2009) and at Proctor Plantation $30 ($649 in 2009). As James Clifton has pointed out, "With rice selling at approximately $1.00 per bushel in the rough and the average yield per acre being in the vicinity of thirty bushels, it [was] obvious that the great profits of the antebellum period were no longer possible." Polishing rice and selling it at about four cents per pound would bring perhaps ten dollars more per acre, but this gain was more than offset by the cost of milling and other factors.[25]

Hardly had recovery in the Rice Kingdom begun, when competition came in 1886 from rice growers in the southwestern United States. Prior to 1886 rice had long been grown in Louisiana on the floodplains of the lower Mississippi River below New Orleans. But it was grown on a small scale for domestic use and was only minor competition for the Rice Kingdom. The Civil War, however, stimulated the Louisiana rice industry. Along the Mississippi floodplain, planters returning to the area after the war found their sugar plantations desolated, equipment destroyed or dilapidated, and themselves without adequate resources to finance reconstruction of sugar planting. As an alternative, the impoverished sugar fields were split up and converted to cheaper rice cultivation relying on rainwater. The industry expanded and began to share the market with the Rice Kingdom, although the Lowcountry maintained its competitive advantage because of the superior quality of its Carolina Gold. By 1880 rice production in Louisiana had reached twenty-three million pounds, or about one fifth of the country's total output, but after 1880 limitation of suitable acreage and inability to adopt modern production methods slowed the rate. This Louisiana competition along the river was minor, however, compared to the rice growing that followed in the prairie lands of southwest Louisiana.

Jabez B. Watkins, a financier and promoter, set up several corporations under the Watkins Syndicate and purchased 1.5 million acres of land in Louisiana from the Vermillion River to the Sabine River and inland thirty to sixty miles. Two-thirds of the land was marsh where Watkins planned an extensive reclamation project to convert the land to rice fields. Begun in 1884 this project was abandoned because the syndicate's prairie lands proved ideal for rice cultivation at considerable less cost than the marsh lands.

Attracted by promotional literature touting the farming opportunities in the prairie of southwest Louisiana, Midwestern farmers flocked there and observed that Cajuns grew rice in small plots using rainwater. Prairie lands in southwest Louisiana had been considered fit for nothing but cattle grazing, but they discovered these lands had a silt-loam topsoil peculiarly adapted to rice cultivation and a clay pan below the topsoil that could hold water on the fields. Large tracts of level prairie could be treated in units of much greater size than in the Rice Kingdom. Abandoning the crops the promoters had expected them to cultivate, the midwesterners quickly adapted the machines they used for wheat culture to rice. Mechanical reapers with binder attachments (see figure 36), gang plows, broadcast seeders and drills, disk harrows, and steam threshers used for wheat were adapted to rice cultivation.

Rice culture using modern machines had begun on the prairie lands in Calcasieu Parish in southwest Louisiana and ultimately spread throughout the prairie lands. Arthur H. Cole described the state of planting in the Southwest around 1886:

> The fields were now prepared for the crop by the gang-plow and disc harrow, and the seed was planted by the drill, or less frequently by the broadcast seeder. The fields then were flooded once for the germination of the seed, and after a few days drained off. When the plants reached a height of six or eight inches, the water was again allowed to overflow the fields, being retained there by surrounding dikes, and it was maintained at the desired height of four to six inches until the grain was ripe. Thereupon the fields were again drained, and as soon as the ground was dry, the crop was cut and tied into bundles by the twine-binder. Usually the bundles were shocked and allowed to remain in the fields for a time; and subsequently, by use if the harvester [actually, the thresher] "set" in the fields, the grain was separated from the stalks.[26]

In 1888 David Abbott used a small steam engine and an endless bucket chain to raise water from a bayou to his fields. More farmers turned to irrigation, and extensive canals crisscrossed the prairie lands to supply water to fields. As canal irrigation was developing, a water strata close below the surface was detected, allowing the sinking of artesian wells

FIGURE 36. Harvesting with a twine binder in Crowley, Louisiana, early 1900s. Reproduced by permission of Arcadia Parish Library, Freeland Archives, Crowley, Louisiana.

for irrigation using pumps. Ten years from its beginning, the prairie lands of Louisiana had been transformed into a highly mechanized and modern rice-growing region. A culture that had been carried on since time immemorial in a peculiarly labor-intensive manner in Asia and which for two hundred years had been practiced in a like manner in the Rice Kingdom, had been modernized in a few years in Louisiana. Land previously worth twenty-five to fifty cents per acre brought fifty dollars per acre during the early boom years. The beginning of the modern rice industry in the United States was established on Louisiana prairie, separated in culture, time, location, and technology from the tidelands of the Rice Kingdom and without bound labor.[27]

From its conception, the prairie rice culture exhibited a nonsouthern profile. Southwest fields were not subject to the vagaries of tidal water, which sometimes was unreliable. More important, the heritage of slavery was not the basis for the southwest labor system. The Rice Kingdom, which clung to its proud past of tidal cultivation, could not produce rice at a price to compete with the new industry.

In the late 1800s Seaman Asahel Knapp with the U.S. Department of Agriculture introduced Japanese Kiushu rice seed into Louisiana, which gave a 25 percent greater yield over other varieties. The Kiushu rice had 50 percent less breakage during milling. By 1889 Louisiana produced more rice than any other rice-producing state, and by 1899 it produced 70 percent of the total American crop.[28]

Rice production spread from Louisiana west across the Sabine River, and in 1889 Texas grew its first commercial crop. Early Texas rice farmers had many problems with the water supply and land. Texas soils were more tenacious and clayey and hard to work. Through hard work and innovation, however, rice became a viable crop in the Texas Gulf Coast by 1900.

Arkansas rice production began in 1902, when a farmer in Lonoke raised an acre of rice in Southwest Arkansas's Grand Prairie region. The first attempt was so successful that he expanded production and encouraged his neighbors to grow rice as well. Arkansas had land similar to that of southeast Louisiana, and commercial production began seriously in 1905. The first rice mill was erected in Stuttgart in 1907. Aided by a new land boom during which people came from all over to buy cheap land, Arkansas "rice fever" swept the area.

Developments in milling centered in New Orleans soon gave added impetus to the expanding rice industry in the Southwest. Improvement in brushing, separating broken grain from the whole, and polishing with an artificial coating of glucose and talc came in rapid succession. These improvements were followed by a new invention called a "rice scourer," which replaced mortars and pestles to remove the bran. Next shelling stones made of carborundum replaced the old shelling stones. Before long, southwestern states produced rice at less cost than Georgia and the Carolinas, undercutting the price of Carolina Gold rice. The consumption of rice was limited in the United States, and these states quickly flooded the market with home-grown rice. For several years the price of rice fell below what it cost to produce, making it harder for the old rice-growing states to compete.

Even after prolonged drying, the swamp soils of most fields that grew Carolina Gold were too soft to support heavy machines, and harvesting always had to be done by hand, making it more costly for growers in the Rice Kingdom to produce a crop for market. Modern threshing machines that were housed in barns were generally too expensive for most planters, although several Carolina planters purchased them. The southwestern states made use of newer machines to cultivate and harvest rice, reducing their cost of producing rice crops for market. The fields in the Rice Kingdom were small, about fifteen to twenty acres, making it difficult for the new machines to turn around. The quarter drains in the fields also posed a problem for machines. Roswell Trimble and his partners purchased two Deering reapers and binders to experiment with near Charleston. Their experiment failed because "the necessary presence of the smaller, parallel ditches [quarter drains] . . . prevented the expeditious harvesting with machinery and the economy of its operation."[29] The planters understood the need to modernize; however, most did not have the money to buy newer machines or their fields were not suited for modern machines.

The southwestern states did not suffer from hurricanes and freshets, and far greater acreage of land was planted using powered irrigation that replaced the less certain tidal water supply of the Atlantic states. The vast acreage of prairie lands (supplied by pumps from streams and wells) could be converted to rice fields because their elevation enabled the planter to drain the fields easily. Some growers in these states labeled their rice "Carolina Gold Rice" to capture the market demand for Carolina Gold. In short development of modern production and milling industries in the Southwest contributed collectively to the end of rice growing in the Carolinas and Georgia. A revolution had occurred in the American rice industry: the upland farmers of the Southwest had replaced the tidewater planters.

The production figures in table 5 document the decline of rice production in the Rice Kingdom and its replacement by the Southwest as the major rice producer.

TABLE 5. Rice Production in the United States, 1859–1919
(per million pounds of clean rice)

	1859	1869	1879	1889	1899	1909	1919
Total production	187.2	73.6	110.1	128.6	250.3	658.4	1065.2
Percentage of total by state or region:							
South Carolina	63.6	43.9	47.3	23.6	18.9	2.5	0.4
Georgia	28.0	30.2	23.0	11.3	4.5	0.7	0.2
North Carolina	4.1	2.8	5.1	4.6	3.2	0.0	0.0
Southwest	3.4	21.7	21.2	58.9	71.9	96.7	99.1

Source: Adapted from Peter A. Coclanis, "Bitter Harvest: The South Carolina Low Country in Historical Perspective," *Journal of Economic History* 45 (June 1985): 258.

Around 1890, some Rice Kingdom planters turned to the abundant longleaf-pine and slash-pine forests that graced the uplands of their plantations as a secondary source of income, hoping that what they could make on selling the timber, combined with what they could make on rice, would save their plantations. Tar burning had long been practiced in the Atlantic states as part of the naval-stores industry. Controlled burning of pine logs in kilns produced tar used to coat the rigging of sailing ships. Burning tar produced pitch, which was used to caulk wooden ships. Tar burning was a short-term solution because wooden ships were being replaced by steam-powered steel ships. Turpentining was another source of income from the pine forests. Crude turpentine extracted from the living pine trees was distilled to produce spirits of turpentine and rosin used for a multitude of purposes.[30]

Planters became absentee landowners and turned to a variety of pursuits for income away from the plantation. They leased their lands for hunting and turned to time-honored occupations that they and their ancestors had eschewed for the planter's life: law, medicine, banking, mercantile, or public office. A few even turned to growing and selling manually threshed seed rice.[31] Others turned to growing cotton on the highlands as a secondary source of income. Some were able to stave off abandoning rice planting for a while, but in the end it was a futile pursuit.

Around 1892, the United States government constructed jetties in the south channel of the Savannah River, which altered the saltwater-freshwater interface. The jetties altered the normal longshore flow of currents and sediment deposition, and the river deepened so that it was no longer possible to control the flow of water over the rice fields of Argyle Plantation, which lay upriver.[32] Other river projects changed the saltwater-freshwater interface along the coast, each contributing to the demise of the industry.

A series of hurricanes pushed walls of salt water up rivers, breaking the banks, flooding the fields, and destroying the crops. Damage was not restricted to the plantations. City infrastructure was damaged. Wharves were washed away and boats sunk in the harbors. The hurricane of August 27–28, 1893, the center of which hit Beaufort, South Carolina, destroyed all the crops between Savannah and Charleston. Anne Simons Deas, who was living in Summerville, South Carolina, wrote in her diary on Friday, September 8, 1893, about damage to Comingtee Plantation on the Western Branch of the Cooper River: "Heard from Eva (Porcher), she gave further particulars of the storm. Two trunks blasted out, cuts in the banks, mill-chimney down, carrier-shed washed away, half the engine-house blown off. Some of the small trees between the dairy and the house down and some of the oaks on the place."[33] On September 29, 1898, the Lowcountry again felt the force of a hurricane, and although it was not a severe storm, damage was still inflicted on a struggling rice economy.

The last crop of rice planted on the Lower Cape Fear region of North Carolina was harvested in 1909. Cape Fear rice growing ended for the same reasons it did in South Carolina. A series of disastrous hurricanes, especially the hurricanes of 1893, 1894, 1898, and 1906 all contributed to its demise. The hurricane of August 27, 1893, broke rice-field banks and flooded the fields along the Lower Cape Fear, the principal area of cultivation. Then, on October 13 of the same year, another devastating hurricane hit, and the storm surge was "the highest ever known" in the area, "stretching in an unbroken sea across the rice fields as far as the eye could reach."[34]

By 1910 the Combahee and Edisto were the only rivers on which rice was planted to any extent in South Carolina. (Georgetown County rice growing ceased around 1906.) In 1910 yet another hurricane brought devastation. Approaching the coast, it turned in a northward direction parallel to the coast and piled up great tides south of Charleston. Waves of salt water were pushed up the rivers, breaking the banks, blowing out trunks, and flooding the fields with salt water. This time the planters knew that they could never protect the fields from hurricanes. Constructing riverbanks high enough and strong enough to stop the rising waters of a hurricane was impractical.

A few Carolina planters borrowed money after the hurricane of 1910 to repair the broken banks and make the fields ready for planting the following year, hoping to pay off their 1910 debts with the next crop. It was not to be. A 1911 hurricane destroyed this crop as well, and most planters could not financially afford to borrow money again.

Some planters reluctantly acknowledged the inevitable. After the hurricane of 1911, Duncan Clinch Heyward lamented that when "I saw the ocean actually coming up

Meeting Street [in Charleston] . . . I knew . . . that the death-knell of rice planting in South Carolina was sounded."[35] In 1915 Charles Spalding Wylly, a descendent of antebellum planters, wrote a poignant statement about the end of rice culture in Georgia. The rice industry had "entirely disappeared, and the old plantation lands had already reverted into swamp, or are fast passing into a jungle of marsh, wood and water; great sums of money have been lost by the former owners in the effort to re-establish their [former] prosperity and productiveness, which in every instance known to me have ended in financial disaster . . . [and] at present every rice place on the Savannah, the Ogeechee and the Altamaha is practically abandoned, and are no longer an asset in their owner's ledgers, but rather a charge and incumbrance."[36] Heyward's and Wylly's feelings about the end of rice culture was undoubtedly shared by many other planters.

Although most rice planters ceased planting around 1911, rice growing in South Carolina continued on until 1927. In 1860 there were at least 250 planters; in 1913 there were about two dozen. A few hardy souls continued to defy the odds and plant. Most of the planting was done along the Combahee River or its tributaries: John P. Gregorie at Brewton planted until 1910; Duncan Clinch Heyward at Myrtle Grove until 1913; Oliver Middleton Read at Hobonny until 1921; Cheves brothers at Newport [Nieuport] until 1923; and Theodore DuBose Ravenel at Laurel Springs and Rose Hill until 1927. They survived in part because World War I brought a surge in commodity prices. They planted larger acreages while higher prices prevailed. Because of their expertise in rice culture, they recognized the Combahee River had unique qualities at the time to produce a profitable crop. After the war prices again fell. New problems now faced the remaining few planters. One by one the city mills shut down, leaving planters few options for milling their rice. The West Point Rice Mill in Charleston was the last to mill rice and shut down for good in 1924. Theodore Ravenel continued to plant Laurel Springs until 1927, and to him goes the honor of being the last major commercial rice planter of his lineage.[37]

After the Civil War planters could not and would not accept a new world order that had passed them by. The environmental challenges the tidewater planters met and conquered before the Civil War were no longer under their control. Bound by a plantation mystique they inherited, they were unable to cope with a new world market. For two hundred and twenty-five years rice had reigned supreme. Human chattel for labor and overseas markets warping the Lowcountry economy and rendering alternative economic venues difficult, the Rice Kingdom economy collapsed, and rice production ceased. Even the most determined planters were finally forced to cease planting, closing forever one of the most colorful chapters in the history of American agriculture, but one built on the work of enslaved Africans. The old order was swept away by natural events and technology. The antebellum age of human exploitation and agricultural magnificence was forever past. This book documents its legacy.

Epilogue

Richard Dwight Porcher, Jr.

The wings of the Cessna 152 bit into the cold air as it lifted off from East Cooper Airport in Mount Pleasant, South Carolina. When the air is cool, the lift is so strong it feels like a giant rubber band attached to the top of the plane and pulling it upward. Late fall is the best time to fly in the Lowcountry if you want spectacular, clear vistas, exactly what I wanted that day in 1994 for a flight over the Cooper River. I had passed my flight exam the previous summer and was taking my first fall solo flight. A cold front had just passed through and cleared the air of haze. At three thousand feet, visibility was thirty miles; the abandoned rice fields made fascinating patterns. Although I had studied the Cooper River fields from aerial photographs and field surveys, seeing them firsthand from the air gave me a new perspective and heightened my interest in the history of rice culture. No one seeing the fields from the air could not be curious about their history. I was and still am.

Sixteen years later, in the winter of 2010, I repeated the flight. It became a voyage into the past. It reminded me that today's Cooper River landscape—and inland swamps and other rivers and deltas as well—is a creation of that past, a past that will be part of the Lowcountry forever. Rice culture left relict features on the landscape, and the plantation enterprise, based on slave labor and large landholdings, left lasting impressions on the minds of twenty-first-century Americans.

My flight plan was to first follow the path of John Beaufain Irving, who in 1842 wrote *A Day on Cooper River*. After flying over the Cooper River, I planned to fly to the Santee Delta. The first point of interest in my "Day over Cooper River" was the "T," where the Western Branch and Eastern Branch join to form the Cooper River, which empties into Charleston Harbor. From the "T" the history of the different periods and methods of rice culture can be seen.

To the west flows Back River and Goose Creek, tributaries of the Cooper River where the Goose Creek Men settled and experimented with rice in the last decade of the 1600s.

They succeeded in establishing rice as a market crop, and rice culture spread north and south to the area that became the Rice Kingdom. To the east lie Nicholson and Turkey Creeks, tributaries of the Eastern Branch, where there are remains of extensive inland-swamp fields, long ago abandoned when tidal cultivation commenced. The inland fields and their reservoirs have been reclaimed by hardwood or swamp forests that harbor wading birds and a diverse flora of wildflowers, shrubs, and trees.

Straight ahead is Lake Moultrie, which in 1942 flooded the lands and plantations of my forefathers, the Ravenels and Porchers of Middle St. John's Parish. I lament again construction of Lake Moultrie (and its sister Lake Marion), an environmental and cultural abomination without equal in the state (and some argue in the South). The loss of Pooshee and Cedar Spring plantations owned by my family since the 1700s was especially painful. Many of my ancestors still lie under the lake in the land of their forefathers, where my ashes one day will be scattered over the site of Black Oak Church. Carolina rice, nurtured by rain-fed reservoirs and artesian springs, was a staple crop in Middle St. John's Parish until the Revolution, when it was replaced by Santee long cotton. The genteel life my Huguenot and English ancestors thought would last forever ended with the American Civil War; the lake erased the last physical trace of their plantations.

Below are remnants of an artificial ordering of the landscape created by bound labor. The Cooper River is flanked on both sides by abandoned tidal rice fields that replaced reservoir-fed fields beginning in the mid-1700s. The tidal fields are remnants of an agricultural system unique to America, and they represent the zenith in rice production. Tidewater rice cultivation in the Rice Kingdom was limited geographically to cleared freshwater tidal swamps along a narrow strip of the freshwater rivers where there was at least a three-foot difference in tidal amplitude for flooding and draining a rice crop. Swamps that took nature eons to produce were converted to rice fields in a matter of years. Working under a task system in a unique ecological environment, slave labor satisfied the unusual demands of the industry dependent on a remarkable degree of efficiency between planters and slaves—unlike any other agriculture endeavor in the South.

As I fly over the fields, the past becomes part of the present. Fishermen, hunters, and boaters enjoy the abandoned fields where Carolina rice once grew, having no idea of their history, no idea that where they find pleasure, bondservants toiled with no hope of freedom or material reward. Perhaps only one in a thousand who pass by the fields knows of their history. Today the rice fields are empty of their human chattel. Where slaves once toiled in the August heat, cutting rice with sickles to a rhythm and song like that of their ancestors along the wetlands of the Niger River, are fields teaming with wildlife and a plethora of marsh wildflowers. Slaves passed their entire lives in the fields and the villages on the adjacent uplands, many perhaps never going to a nearby town. How many lie in unmarked graves across the landscape, known to no one today, not even to their descendants? Their burying sites lie hidden in a wall of vegetation, a testimony to how completely and quickly nature reclaims her own. How easily or unwittingly people have desecrated the burial sites to build a home or business.

In these rice fields, slave labor produced immense crops of Carolina rice that made many planters wealthy. Slaves, who never benefited from that wealth, produced more rice the next year, bringing yet more wealth to their masters, who created more fields for them to cultivate. Here they labored without freedom in fields of rice far from their home and their culture. Rice was a cruel and harsh master for slaves. How many thousands of their lives were sacrificed in order to furnish the world with Carolina rice? And the planters with their mansions? How do we judge slaves' contribution to today's Lowcountry? The idea that Africans contributed nothing except labor to rice culture has been firmly discredited by scholars. The legacy of rice culture is a mixture of African and American European achievement and ingenuity.

My mind comes back to the present as I notice a popcorn tree growing on a rice field bank below. Popcorn tree is especially prominent in the fall and winter because of its white berries, even from an airplane. Planters imported popcorn tree, Cherokee rose, and other plants from across the seas to grace plantation gardens. Many escaped and became part of the naturalized flora of the Lowcountry. Cherokee rose, white mulberry, chinaberry, and others were incorporated into slave folk remedies. For the first century rice was cultivated, medical resources on plantations were at best rudimentary, and Africans turned to the native and introduced plants growing around the plantations as a medicine source, just as they had in Africa. Scattered around the ruins of slave villages today are these very plants that were incorporated into a slave pharmacopeia.

I reflect on my West African trip in 2005 to Goree Island off the coast of Dakar, Senegal. On the island is the slave house with the "Door of No Return," where Africans passed through to the ships that brought them to the rice fields of Carolina and other destinations in the New World. They were taken away from the only home they knew, toward darkness. They had no hope of a gorgeous destiny awaiting the end of their voyage. Only slavery!

Gazing down from the air at the abandoned fields, questions come to mind: What if that door never existed and slavery never existed? How different would the view be on this flight over the Cooper River? What would the Cooper River look like today? A black-water river flanked with primordial swamp forest? None of the other crops the Lords Proprietors planned for the colonists to grow (for example, grapes for wine, white mulberry for silk worms, olive trees, and currants) could have been profitable in the swamp land or could have justified the cost of constructing the banks and installing water-control structures. I know of no other crops available in the colonists' time that would have justified transforming tidal swamp into fields. White colonists maintained that they themselves were unfit for laboring in the Lowcountry humid swamps, so it is unlikely that, without African slaves, they would have converted the swamps to agriculture. Would the plantation houses of Middleburg and Mulberry Castle be here today if the Goose Creek Men had not been successful in developing rice as an export crop using slave labor? Would the colonists have found another use for the swamp land bordering the rivers that did not require clearing and banking? The timber might have been cut at least once, and secondary forest might have quickly reclaimed the land, so we might be looking at a mature, secondary swamp forest flanking the river, but there would be no banks or abandoned rice fields.

As I fly north along the Western Branch, the first point of interest is Stoke Rice Mill, which sits on a bluff adjacent to the river and provided easy access for flats carrying rice to and from the mill. Stoke Rice Mill was on Comingtee Plantation, named after a Captain Coming who settled land adjacent to the confluence of the Cooper River's eastern and western branches—hence, the name of his plantation. Only the brick wall of the mill's first floor remains; the second floor and loft, made of wood, have deteriorated. The millrace is close to being intact. Stoke Mill was typical of water-driven mills throughout the Rice Coast.

Anne Simons Deas, in her reflections of the Ball family and Comingtee, mentioned the "busy pestles kept rising and falling." What a sensuous experience it must have been to see a rice mill in operation, and to hear the rhythmical sounds of the pestles as they fell in sequence into the mortars. Or to see and hear the sounds of flails threshing rice. No one alive today has seen a Carolina rice mill in operation or seen golden grains of Carolina rice swaying in a field waiting for harvest. What a sight it must have been to see a twenty-foot-high waterwheel in operation. My life's passage is lessened by not having these experiences.

My plane next passes over Mepkin Plantation. Today Mepkin is a community of Roman Catholic Trappist monks. A canal that slaves dug from the Cooper River to the water-powered mill Jonathan Lucas I built for Henry Laurens is clearly visible from the air at low tide. The Mepkin mill, built a year or two after Lucas built a water mill on Millbrook on the Santee River in 1792, was an improved, fully automated tide mill with rolling screens, elevators, packers, and other features. I discovered the ruins of the mill in 2008. Nothing remains except part of the foundation of the mill building and part of the millrace. The mill sat on raised land between a rice field and the canal. I experienced the scene: water was let into the rice field from the river with the flood tide. A trunk gate at the head of the millrace held back the water until the field was fully flooded. Then the gate was opened at ebb tide, and the water flowed from the field through the millrace, turning the waterwheel before exiting into the canal that carried the water to the river. The ruins are a mecca for rice-milling historians because they are the last traces of Lucas's first automated water-powered mill. I hope that one day archeologists will conduct a detailed survey of the site. On every visit to the site, I feel a spiritual connection to Lucas and admire his inventive genius.

I leave the Western Branch and fly to the Eastern Branch toward Middleburg Plantation. The original water-powered mill at Middleburg was built in 1801 by Jonathan Lucas I as a toll mill. The mill offered milling service to small rice plantations in the Cooper River area whose owners were unable to afford their own mills. Later it was converted to a steam-powered mill. The rice chimney of Middleburg comes into view, mute and smokeless in a clear sky, a lonely sentinel, a reminder that change is inevitable, a relic of the threshing and milling technology that made processing large crops of Carolina rice possible, bringing enormous profits to the planter elite. I imagine hearing a slave on a rice flat bringing rough rice to the toll mill at Middleburg, calling to the steerer that he sees the chimney identifying their destination. Each plantation with a steam mill boasted a chimney with a distinct pattern for identification. Rivers were the main means of transportation during the 1700s and early 1800s, and once chimneys became part of the landscape, they were easily visible from the river. Only two chimneys stand today on the Cooper River. The Middleburg chimney no longer guides flats; yet it stands as a silent witness to the past.

Leaving the Eastern Branch, I fly up Quenby Creek on the way to the Santee Delta. From above, the outer banks of the rice fields built by slaves two hundred years ago are clearly visible along the creek. Most outer banks on the Cooper River and its tributaries are breached, as are most other outer banks in the Rice Kingdom. After rice growing ceased, most owners had no reason to keep the banks intact since no other type of crop could be grown profitably in the former fields. Without constant upkeep, the banks could not withstand erosion, burrowing from animals, and other natural forces. The owners abandoned the fields to the forces of nature, and swamp forest encroached on the fields. Clearly visible in some Quenby fields are breached outer banks. Saplings of red maple and swamp gum, tree species that will dominate the future swamp forest that is developing in the abandoned rice fields, rise above species they will replace (see color plate 31). Along any river of the former Rice Kingdom, abandoned fields with breached banks dominant the landscape.

Leaving the Cooper River area, I fly northward to the Santee Delta formed by the North Santee and South Santee Rivers. The same observations and questions raised about the Santee Delta can be applied to the other deltas where rice was grown (Cape Fear, Winyah Bay, Altamaha, Savannah, and Ogeechee). First, imagine the Santee Delta as a primeval sixteen-thousand-acre bald-cypress and tupelo-gum swamp forest nurtured by the waters of the Santee River and tributaries that originated in North Carolina. The river was the ecological lifeblood of the delta—and later the rice fields—bringing nutrients from the upcountry to fertilize the fields. Now visualize the entire delta as one vast rice field ready for harvest, golden grains bending to and fro in the ocean breeze. How can one measure the energy African slaves exerted to create the rice fields out of swamp forest or to construct and operate the threshing barns and mills and their own slave villages. What would the Santee Delta look like today if there had been no slaves, no rice culture?

The Santee Delta is a treasure trove for archeologists and social historians studying slave life. Once the rice industry ended, the delta was abandoned by permanent residents. Except for brick thieves, who confined their activity above ground, the lands around the villages were left undisturbed. A poignant reminder of a long-lost time, brick chimneys from the abandoned slave cabins rise from the ground, overgrown with vegetation. Wooden structures deteriorated, but artifacts of daily life are still there, relatively undisturbed in the soil, a vast store of knowledge of slave life waiting to be explored.

Along the North Santee River, the former rice fields of Dr. Philip Tidyman are visible. John Drayton's sketch of the mill Jonathan Lucas I built for Tidyman is the only diagram of a tidal water mill that has survived; it is of inestimable value to historians studying the process of rice milling. Nothing remains of the mill today except a few scattered bricks. Lucas's being ship-wrecked off the Santee River and settling nearby was fortunate for rice culture. His use of millstones and his automated mill made processing large crops of rice more efficient and therefore less laborious and time consuming. Rice culture probably would never have reached the advanced state it did without his mill. There would have been only one world-class crop in the Lowcountry—Sea Island cotton—and the land in the Rice Kingdom would be vastly different today.

Crow Island on the North Santee comes into view next. The bank constructed around the slave village to keep out tidal water is plainly visible (see color plate 32). In the early 1970s I visited the site several times and found the remains of nine slave-house chimneys. A return trip in 2008 revealed that most of the bricks had been pirated, and the locations where the slave houses stood were no longer apparent. Henry Edward Lucas (1822–1900) recounted that he purchased one half of Crow Island, a tract of eight hundred acres of rice land lying in the North Santee River. He put four hundred acres under bank and built a threshing barn, a large barn, slave houses heavily framed to resist storms, and a small dwelling house. Unfortunately I could not locate the site of the threshing barn.

On one visit to Crow Island, I sat on the remains of a slave-cabin chimney and tried to visualize the daily routine of a slave family (see color plate 32, insert). There is, or course, no way I can comprehend what it was like to have been a slave, but I can ask questions of myself, and through this book, the reader. What did they discuss sitting in front of the fire in their cabins at night? Or as they worked in the fields? Did they aspire to a better life for their offspring, and if so, how did they envision it happening? Did they even have the faintest notion of what such a life might be, or were they too numbed from working in the "deadly fields" to even envision such thoughts? What did they eat for dinner, or breakfast before beginning the daily tasks that probably never varied? Were they able to grow plants in the village for medicines or plant a garden? Gardens were one means that slaves developed a sense of place and self-worth. Did any of them go to the mainland, or did they spend their entire lives in the delta?

Slaves remained on the plantations during the summer, partially protected from malaria by their inherited sickle-cell trait. How many sick children did a mother rock to sleep in front of a fire? How many children died in this village? Were they buried in the village? Totally lost from history! I saw no signs of a burial ground. Did any of these slaves try to escape their bondage? Living so close to the ocean and certainly seeing masts of passing ships, some must have thought of rowing across the bar in hope of catching a ship to freedom even knowing the harsh punishment that would result if they were unsuccessful. Even people four or five generations removed from Africa still had an unbridled desire for freedom. There is no recorded history of Crow Island village or the other slave villages in the delta, so these questions remain unanswered. Their history and the history of their village vanished when they died. History closed its eyes forever on Crow Island and on many other slave villages.

Turning west, I see the Santee Coastal Preserve. At the height of the rice industry, many plantations lay within what are now the boundaries of this preserve. After the end of rice culture, twenty gentlemen from Philadelphia bought twenty thousand acres of former rice plantations and operated the land as a private hunting preserve called the Santee Gun Club. The club maintained the outer banks of the fields and managed them for waterfowl hunting. In 1985 they deeded the land to South Carolina and kept the hunting rights for twenty years. In 2005 the land became invested to the State as the Santee Coastal Preserve under management of the Department of Natural Resources. Having spent my entire personal and professional life in the Carolina Lowcountry and having trekked protected lands for years on field trips with Citadel students and nature groups, the value of this gift and others like it are well known to me. Although descendants of the original owners of Lowcountry plantations may have resented losing of their land to "outsiders," it was fortunate for all South Carolinians that northerners purchased the former rice plantations. These northerners loved the fields and forests and the Lowcountry way of life so much that many preserved the land for future South Carolinians. Rather than see the land serving a chosen few as golf courses and high-end developments, they took steps to keep the land forever in its natural state. South Carolinians are the fortunate beneficiaries.

Turning north and flying up the South Santee, I see a unique piece of American history and architecture—a storm tower. The Moreland Hunt Club converted a former storm tower into a clubhouse (see color plate 33), preserving the priceless structure. The brick wall of the storm tower is intact; the club replaced the roof, steps, and door. Slaves living in the delta in banked villages or on small parcels of high ground had no advance warning of hurricanes in time to move to the mainland. On September 27–28, 1822, a hurricane's eye came ashore near the delta. Nowhere in the delta is the land more than five feet above mean sea level. Many slaves were lost to the rising water and wind. Afterward, planters built storm towers for the delta slaves as refuge against a hurricane's wrath. The towers were built of brick, twenty to thirty feet in diameter with conical roofs, and their floors were ten feet above ground level. (In 1989 Hurricane Hugo, with its nineteen-foot surge, would have put water nine feet above the tower floor.) How many towers were built in the delta is unrecorded, but there are remains of two others. What did the slaves experience during a hurricane? Waves pounding against the walls and the wind shrieking outside must have caused pure terror for those inside, especially for those who had never experienced a hurricane. The storm tower is a reminder that so much of Carolina rice culture was unique in America—threshing barns and mills, trunk gates, and storm towers occurred nowhere else.

I fly by vacant bluffs along the rivers that were once adorned by stately mansions with mills and slave villages nearby, structures gone forever except for scattered bricks and foundation and wall remnants. Mill sites and slave villages are now urban developments.

Some plantations exist only in old court documents, census records, old family papers, history books, or articles; only a few left their own written records. Nor did the slaves leave written records. Much has been lost even since we started documenting artifacts.

The planters who lived in the Lowcountry were once the wealthiest class in British North America. Based on the enormous profits from rice, they lived like their European ancestors in houses surrounded by formal gardens adorned with plants from across the world; camellia, azalea, mimosa and many others owe their presence in the Lowcountry to planters. The streets of Charlestown, and later the streets of Beaufort, George Towne, Savannah, Wilmington, and Darien, were lined with majestic mansions where planter families spent the fall and winter social season. During the summer, they sojourned in pineland villages or coastal settlements away from the malarial rice fields and swamps. The planters became leading statesmen of the Lowcountry, their state, and the whole South because of their education and wealth, which was made possible largely by slave labor. Many were ardent secessionists, believing slavery was their right and necessary for their plantation system; the resulting war that their political and social positions fostered contributed to the demise of their privileged life—and began a new, richer life for freedmen although it took more than a century to achieve full citizenship in the society and landscape they helped create.

The history that happened in these rice fields and on the adjacent uplands exacted profound ramifications—yesterday, today, and tomorrow. Real people, both free and enslaved, some with names we know, most with names we will never know, toiled, lived and died here. Along the tidal rivers of the Atlantic Coast, planters and slaves created a unique agricultural industry and played important roles in a momentous process of cultural, economic, and ecological change unlike anywhere else in the New World. Although most of the physical remains of the industry are gone, the magnificent accomplishments of the slaves is not diminished by the disappearance; likewise the industry that the planters oversaw and created is not diminished by the end of rice culture. The lives of everyone living in the former Rice Kingdom were affected by what the slaves and planters accomplished. I hope this book will pique the reader's curiosity about the history of rice culture, a legacy that is nowhere and everywhere at the same time.

Departing the delta and heading back to East Cooper Airport, I reflect on the main goal this book hopes to accomplish: preservation of the legacy of rice culture, no matter how one views its consequences—and more specifically of market preparation. As a fellow author and friend, Alberta Quattlebaum, once said to me, we write about Lowcountry history and the legacy of rice culture in particular "so it won't be forgotten!"

Appendix 1

Original Field Research Plantations and Sites

The following list includes plantations and sites the authors surveyed to document ruins of threshing barns, mills, and floodgates. The major river system associated with each plantation or site is included as a means of location.

Georgia

McIntosh County

Butler's Island, Altamaha River

South Carolina

Beaufort County

Hobonny Plantation, Combahee River
Nemours Plantation, Combahee River
Nieuport Plantation (now part of Nemours Plantation), Combahee River
Rose Hill Plantation, Combahee River

Berkeley County

Bluff Plantation, Western Branch of the Cooper River
Cedar Hill Plantation, Eastern Branch of the Cooper River
Comingtee Plantation, Western Branch of the Cooper River
Dean Hall Plantation, Western Branch of the Cooper River
Limerick Plantation, Huger Creek–Gough Creek
Mepkin Abbey, Western Branch of the Cooper River
Middleburg Plantation, Eastern Branch of the Cooper River
Oakland Hunt Club, Santee River
Stoke Rice Mill, Western Branch of the Cooper River
Wappaoolah Plantation

Charleston County

Barnhill Plantation, North Edisto River
Bennett's Rice Mill, Cooper River

Caw Caw County Park, Tea Farm Creek
Charles Pinckney National Historic Site, Horlbeck Creek
Chisolm's Rice Mill, Ashley River
Drayton Hall Plantation, Ashley River
Jehossee Island, South Edisto River
Magnolia Gardens, Ashley River
Murphy Island, South Santee River, Santee Delta
Santee Coastal Reserve, South Santee River
Lucasville, Shem Creek
Tibwin Plantation, Tibwin Creek
West Point Rice Mill, Ashley River
Willtown Plantation, North Edisto River

Colleton County

Cockfield Plantation, Combahee River
Hope Plantation, South Edisto River
Lavington Plantation, Ashepoo River
Social Hall Plantation, Ashepoo River

Dorchester County

Middleton Place, Ashley River

Georgetown County

Black Out Plantation, North Santee River
Chicora Wood Plantation, Great Pee Dee River
Crow Island, Big Duck Creek, Santee Delta
Estherville Plantation, Winyah Bay
Fairfield Plantation (now part of Arcadia), Waccamaw River
Hobcaw Barony, Winyah Bay
Kinloch Plantation, North Santee River
Nightingale Plantation, Great Pee Dee River
Laurel Hill Plantation (now part of Brookgreen Gardens), Waccamaw River
Mansfield Plantation, Black River
Rice Hope Plantation, North Santee River
Rochelle Plantation, North Santee River
Waverly Mill Plantation, Waccamaw River
Weehaw Plantation, Black River

Jasper County

Poplar Grove Plantation, Back River
Turnbridge Plantation, Wright River

Appendix 2

Museums with Originally Manufactured Products and/or Machinery

Cape Fear Museum, Wilmington, North Carolina
Charleston Museum, Charleston, South Carolina
Crowley City Hall, Crowley, Louisiana
Greenfield Village and the Henry Ford Museum, Dearborn, Michigan
Rice Museum, Crowley, Louisiana
Rice Museum, Georgetown, South Carolina
Roundhouse Museum, Savannah, Georgia
South Carolina State Museum, Columbia, South Carolina
Village Museum, McClellanville, South Carolina

Appendix 3

United States Patents for Rice-Processing Machinery, 1829–1887

Fred Brotherhood, Charleston, South Carolina. "Improvement in Rice Pounding and Hulling Mills." November 19, 1878. Letters Patent no. 210,002.

Fred Brotherhood, Charleston, South Carolina, Assignor of One-Half to James Brotherhood, of Stratford, Canada. "Rice-Pounding Machinery." December 6, 1881. Letters Patent no. 250,340.

Jehiel Butts, Charleston, South Carolina. "Rice Thrashing Mill." May 23, 1848. Letters Patent no. 5,600.

O. J. Butts, Georgetown, South Carolina. "Machine for Brushing Rice." October 27, 1857. Letters Patent no. 18,496.

William C. Chapman, Charleston, South Carolina. "Improvement in Rice-Screens." June 14, 1870. Letters Patent no. 104,112.

Calvin Emmons, New York. "Improvement in the Threshing Machine." July 27, 1829. Letters Patent no. 5584X.

William Emmons, New York. "Improvement in the Threshing Machine." February 7, 1831. Letters Patent no. 6366X.

E. C. Engelberg, Piracicaba, Brazil. "Rice-Hulling Machine." May 22, 1888. Letters Patent no. 383,285.

Philip R. Lachicotte, Waccamaw, Georgetown County, South Carolina. "Improvement in Scouring and Polishing Rice." April 17, 1875. Letters Patent no, 166,992.

Philip R. Lachicotte and T. B. Bowman, Charleston, South Carolina. "Improvement in Machines for Cleaning Rice." July 28, 1857. Letters Patent no. 17,882.

J. A. Lockfaw, Wilmington, North Carolina. "Rice-Mill." June 21, 1887. Letters Patent no. 365,191.

William Mathewes, Charleston District, South Carolina. "Improvement for Threshing Rice & Other Small Grain." August 27, 1835. Letters Patent no. 9059x.

Peter McKinlay, Charleston, South Carolina. "Improvement in Mills for Cleaning Rice." April 1, 1851. Letters Patent no. 8,010.

———. "Rotary Rice Cleaner." March 30, 1852. Letters Patent no. 8,841.

———. "Rice-Huller." March 30, 1852. Letters Patent no. 8,841.
———. "Machine for Cleaning Rice." November 6, 1866. Letters Patent no. 59,432.
John F. Taylor, Charleston, South Carolina. "Machine for Cleaning Rice." June 23, 1857. Letters Patent no. 17,646.

Notes

Introduction

1. We use the name "Rice Kingdom" for this Atlantic region. The term "Gold Coast," a reference to Carolina Gold rice, also appears in the literature to designate this rice-growing area. However, Carolina Gold was introduced after the American Revolution, so "Gold Coast" cannot be used to refer to the region before this date. Two other terms, "Rice Coast" and "South Atlantic rice industry," are used occasionally.

2. In the general discussion of rice, the term "Carolina Gold rice" is not used. Researchers have documented that a variety of rice called "white rice" was the initial commercial rice grown in South Carolina, Georgia, and North Carolina. Although several other varieties were introduced, the white rice became the first commercial crop. The variety known as Carolina Gold was introduced after the Revolution. Since this study of the milling steps starts with the beginning of rice culture, the term "Carolina rice" is used to cover both crops and any other varieties that were prepared for market.

CHAPTER 1. A Brief History of Rice

1. Pangaea (Greek for "all earth") was the vast continent comprising all the continental crust of the earth. It is postulated to have existed before the earth's crust broke up into Gondwanaland and Laurasia 200 million years ago. Gondwanaland comprised present-day Arabia, Africa, South America, Antarctica, Australia, and the peninsula of India. Laurasia comprised present-day North America, Greenland, Europe, and most of Asia north of the Himalayas. Gondwanaland began to break up about 167 million years ago when East Gondwana—Antarctica, Madagascar, India, and Australia—began to separate from Africa. Beginning about 130 million years ago, South America began to drift slowly westward from Africa and the South Atlantic Ocean opened up. East Gondwana then began to separate about 120 million years ago, when India began to move northward and ultimately collided with Asia, forming the Himalayan Mountains.

2. Te-Tzu Chang, "The Origin, Evolution, Cultivation, Dissemination, and Diversification of Asian and African Rices," *Euphytica* 25 (1976): 426–29.

3. The most important tool for examining the origin and evolution of rice is a powerful biochemical method known as isozyme analysis. Isozymes can be regarded as neutral genetic indicators or markers of evolutionary relationships and are believed by many workers to be better indicators of phylogenetic relationships than the traditional morphological comparisons. This method has been available only for the last twenty years and has the potential for greatly expanding our knowledge of the origins of cultivated plants.

4. G. Second, "Evolutionary Relationships in the *Sativa* group of *Oryza* Based on Isozyme Data," *Genetics, Selection, Evolution* 17 (1985): 89–114; G. Second, "Isozymes and Phylogenetic

Relationships in *Oryza*," in *Rice Genetics, Proceedings of the International Rice Genetics Symposium, 27–31, May 1985* (Manila: International Rice Research Institute, 1986), 27–39.

5. Te-Tzu Chang, "Rice," in *Evolution of Crop Plants,* edited by N. W. Simmonds (London: Longman, 1976), 102.

6. David Catling, *Rice in Deep Water,* International Rice Research Institute (London: Macmillan, 1992), 109–10.

7. Judith A. Carney, *Black Rice* (Cambridge: Harvard University Press, 2001), 38–41.

8. Ibid., 40–43.

9. Catling, *Rice in Deep Water,* 109.

10. Chang, "Rice," 101; J. W. Purseglove, *Tropical Crops: Monocotyledons* (New York: Wiley, 1972), 166–67; Leslie S. Cobley, *The Botany of Tropical Crops* (London: Longman, 1977), 26–27.

11. Andrew M. Watson, *Agricultural Innovation in the Early Islamic World* (Cambridge: Cambridge University Press, 1993), 17.

12. Ibid.

13. Ibid.

14. Ibid., 18.

CHAPTER 2. The Origins and Introductions of Rice Seeds

1. The state of South Carolina was originally part of the Province of Carolina. In Charles II's 1665 charter to the eight Lords Proprietors, the boundary of the province stretched from north 36° 30' (the southern border of Virginia) to north 29° (about fifteen miles south of present-day Daytona Beach). From this land was carved the states of North Carolina, South Carolina, and Georgia as well as the northern part of Florida. North Carolina was separated in 1735, but the exact boundary was in dispute until around 1764. In 1787 the boundary between Georgia and South Carolina was established as the Savannah, Tougaloo, and Chattooga Rivers. This book uses "Carolina" instead of "South Carolina" when the discussion centers on the time before the state of South Carolina was established.

2. S. Max Edelson, *Plantation Enterprise in Colonial South Carolina* (Cambridge, Mass.: Harvard University Press, 2006), 59. For another analysis of the "black rice hypothesis," see David Eltis, Philip Morgan, and David Richardson, "Agency and Diaspora in the Atlantic History: Reassessing the African Contribution to Rice Cultivation in the Americas," *American History Review* 112 (December 2007): 1329–58. These authors agree that Africans were the primary cultivators of rice and that some introduced African customs of sowing, threshing, milling, and winnowing the crop to the New World. However, they argued, "there is no compelling evidence that African slaves transferred whole agricultural systems to the New World; nor were they primary players in creating and maintaining rice regimes in the Americas," as espoused by the "black rice hypothesis." They argue that rice planters in the Atlantic region sought little support from slaves from the rice-growing regions of Africa during the crucial formative period prior to 1750, when the foundations of the early rice economy was laid. Europeans also knew much about drainage and banking wetlands, and they did not have to depend on African knowledge for these skills. Like Edelson, Eltis and his coauthors say the rice industry was a "hybrid" synthetic rather than solely European or African in character.

3. Edelson, *Plantation Enterprise,* 62–64.

4. A. S. Salley, Jr., secretary of the Historical Commission of South Carolina, wrote an extensive article on the introduction of rice into Carolina: "The Introduction of Rice into South

Carolina," *Bulletin of the Historical Commission of South Carolina,* no. 6 (Columbia: State Company, 1919). Salley cited a series of primary documents that are referred to in this chapter.

5. Ibid., 3.

6. William J. Rivers, *A Sketch of the History of South Carolina* (Charleston: McCarter, 1856), 382.

7. José Miguel Gallardo, "The Spaniards and the English Settlement in Charles Town," *South Carolina Historical and Genealogical Magazine* 37 (April, July 1936): 93, 94, 98, 99.

8. Salley, "The Introduction of Rice," 4.

9. Ibid., 11.

10. One of the enduring myths about the origin of rice into Carolina is David Ramsay's crediting its introduction to Landgrave Thomas Smith. Ramsay's account has been thoroughly discredited.

11. *Gentleman's Magazine* 36 (June 1766): 278–79. The article was reprinted in the *Southern Agriculturist* 4 (January 1831): 7–9. Salley, who included the quotation in his article, said that it was reprinted inaccurately, but he does not say how.

12. John Stewart, "Letters from John Stewart to William Dunlop," *South Carolina Historical Magazine* 32 (January 1931): 22; hereafter cited as Stewart, "Letters," pt. 1. Stewart was visiting Governor James Colleton at his seat at Wadboo Barony, located along Wadboo Creek, a main tributary of the Cooper River in Berkeley County, South Carolina.

13. Stewart, "Letters from John Stewart to William Dunlop," *South Carolina Historical Magazine* 32, no. 2 (April 1931): 85; hereafter cited as Stewart, "Letters," pt. 2. It is unclear to whom Stewart was referring with his use of "he."

14. Salley, "The Introduction of Rice," 4.

15. Ibid., 5.

16. "Inventory of Barnard Schencking," November, 24, 1692, p. 60, Wills and Miscellaneous Records, Charleston County, 1692–1693, typescript copy, South Carolina Department of Archives and History, Columbia, South Carolina.

17. Bill of lading, November 9, 1695, Wells and Miscellaneous Records, 1694–1704, p. 214, South Carolina Department of Archives and History.

18. Salley, "The Introduction of Rice," 5, 6, and 7.

19. Alexander Moore, "Daniel Axtell's Account Book and the Economy of Early South Carolina," *South Carolina Historical Magazine* 95 (October 1994): 299.

20. John Lawson, *A New Voyage to Carolina,* 1709, new edition, edited by Hugh Talmadge Lefler (Chapel Hill: University of North Carolina Press, 1967), 81.

21. Robert F. W. Allston is Robert Francis Withers Allston (1800–1864), rice planter and author of many articles on rice planting. He resided at Chicora Wood Plantation on the Pee Dee River in Georgetown County. Since he is the only Robert Allston mentioned in this book, he will hereafter be called by this short version of his name.

22. Robert Allston, "Rice," *Commercial Review of the South and West* 1 (1846): 326.

23. William Salmon, *Botanologia: The English Herbal* (London: H. Rhodes and J. Taylor, 1710).

24. John Norris, "Profitable Advice for Rich and Poor," 1712, republished in *Selling a New World: Two Colonial South Carolina Promotional Pamphlets,* edited by Jack P. Greene (Columbia: University of South Carolina Press, 1989), 95.

25. Salley, "The Introduction of Rice," 9.

26. Invoice of Richard Splatt, Charles Towne, January 16, 1726, South Caroliniana Library, University of South Carolina, Columbia. Richard Splatt to William Crisp, January 17, 1726, written on back of the invoice.

27. Duncan Clinch Heyward, *Seed from Madagascar* (Chapel Hill: University of North Carolina Press, 1937), 4–5.

28. Anna M. McClung and Robert Fjellstrom are scientists at the United States Department of Agriculture–ARS, Rice Research Unit, Beaumont, Texas. Their article, "Using Molecular Genetics as a Tool to Identify and Refine 'Carolina Gold,'" was presented at the Carolina Gold Rice Symposium, sponsored by the Carolina Gold Rice Foundation, Charleston, South Carolina, in August of 2005.

29. Allston, "Rice," 326.

30. Frederick A. Porcher, "Historical and Social Sketch of Craven County, South Carolina," in *A Contribution to the History of the Huguenots of South Carolina* (N.p.: Republished for private circulation by T. Gaillard Thomas, 1887), 73, 76.

31. *Charleston City Gazette,* February 21, 1793, Charleston Library Society.

32. After having been abandoned for almost two hundred years, many of the banks of inland-swamp fields have washed away in places where they crossed a swamp.

33. The Committee, "Report," *American Farmer* 5 (September 1823): 24.

34. Henry Laurens ledger, Charleston, S.C., Special Collections, Robert Scott Small Library, College of Charleston.

35. Dumas Malone, *Jefferson and the Rights of Man,* vol. 2 of *Jefferson and His Time* (Boston: Little, Brown, 1951), 126.

36. Thomas Pinckney, "Letter from T. Pinckney to Jas. Gregorie, Chairman, &c.," *Southern Agriculturist* 2 (April 1829): 148–49.

37. Carolina Gold rice lives today commercially, thanks to Richard Schulze, M.D., an ophthalmologist who resides on Hoover Plantation, in Tillman, South Carolina. In 1985, on the three-hundredth anniversary of the introduction of rice into the Lowcountry, Schulze obtained some original Carolina Gold seed that had been accessioned at the U. S. Department of Agriculture in Beaumont, Texas. He received two small bags of seed totaling fourteen pounds and planted the seed on another of his plantations, Turnbridge on the Wright River in Jasper County. From the first crop in 1986, sixty-four pounds of Carolina Gold were harvested. From this small beginning, Carolina Gold now grows commercially on a large scale at Plumfield Plantation on the Pee Dee River in Darlington County, South Carolina.

38. Anna McClung to Richard Dwight Porcher, Jr., October 21, 1998.

39. William Mayrant, "On the Cultivation of Bearded Rice," *Southern Agriculturist* 2 (February 1829): 75.

40. A Black-River Planter (pseud.), "Observations on the Bearded Rice," *Southern Agriculturist* 3 (November 1830): 578–79.

41. William Washington, "On the Advantage of Planting the Bearded Rice," *Southern Agriculturist* 2 (April 1829): 19–20.

42. A Marsh-Planter (pseud.), "Account of the Culture and Product of a Field of Bearded Rice," *Southern Agriculturist* 3 (June 1830): 292–93.

43. Allston, "Rice," 326.

44. Joshua John Ward, "Letter from Col. Ward, on the Big Grain Rice," *Proceedings of the Agricultural Convention and of the State Agricultural Society of South Carolina* (Columbia: Summer & Carroll, 1846), 56–57.

45. William M. Lawton, *An Essay on Rice and Its Culture, Read before the Agricultural Congress, Convened at Alabama, December 5, 1871.* (Charleston, S.C.: Printed by Walker, Evans & Cogswell, 1871): 7.

46. Robert Allston to Edmund Burke, November 22, 1847, Allston Family Papers, South Caroliniana Library.

CHAPTER 3. The Culture of Carolina Rice

1. Stewart, "Letters," pt. 2: 110.

2. Thomas Nairne, "A Letter from South Carolina," 1719, in *Selling a New World: Two Colonial South Carolina Promotional Pamphlets,* edited by Jack P. Greene (Columbia: University of South Carolina Press, 1989), 40.

3. Norris, "Profitable Advice," 98.

4. Mark Catesby, *The Natural History of Carolina, Florida, and the Bahama Islands,* 2nd British edition, 2 vols. (London: Printed for C. Marsh, T. Wilcox, B. Stichall, 1754; originally published 1731–43), 1:iii–iv.

5. After the initial contingent of Barbadian planters came in 1670, more came and settled in Goose Creek. Some of the Goose Creek Men were Governor John Yeamans, Maurice Mathews, James Moore, Thomas Smith, Thomas Smith, Jr., Peter St. Julien, George Chicken, Robert Daniel, Benjamin Schenckingh, John Newe, Arthur Middleton, Ralph Izard, Robert Gibbes, and Benjamin Mazÿck.

6. The term "provision ground system" comes from Michael J. Heitzler, *Goose Creek: A Definitive History,* vol. 1: *Planters, Politicians and Patriots* (Charleston: History Press, 2005), 66. Heitzler did not give a source for the term.

7. B. W. Higman, *Slave Population and Economy in Jamaica, 1807–1834* (Cambridge: Cambridge University Press, 1976), 23–24; 220.

8. Stewart, "Letters," pt. 2: 86.

9. Ultisols are highly weathered, extremely old soils. Abundant in the southeastern United States, they are low in nutrients, acidic, and have a clay accumulation in the *B* horizon.

10. Philip D. Morgan, "Work and Culture: The Task System and the World of Lowcountry Blacks, 1700 to 1880," *William and Mary Quarterly* 39 (October 1982): 566.

11. J. Motte Alston, *Rice Planter and Sportsman: The Recollections of J. Motte Alston,* edited by Arney R. Childs (Columbia: The University of South Carolina Press, 1953), 46.

12. James Ritchie Sparkman to Benjamin Allston, 10 March 1858; quoted in J. H. Easterby, ed., the *South Carolina Rice Plantation as Revealed in the Papers of Robert F. W. Allston* (Chicago: University of Chicago Press, 1945), 346.

13. James C. Darby, "On Planting and Managing a Rice Crop," *Southern Agriculturist* 2 (June 1829): 249.

14. Sparkman to Allston, March 10, 1858, in Easterby, ed., *South Carolina Rice Plantation,* 346.

15. Frances Anne Kemble, *Journal of a Residence on a Georgian Plantation in 1838–1839,* edited by John A. Scott (New York: Knopf, 1961), 114. Frances Kemble was married to Pierce Butler, owner at the time she visited Butler's Island

16. Ulrich B. Phillips, *Plantation and Frontier Documents: 1649–1863* (Cleveland: A. H. Clark, 1909), 81.

17. James M. Clifton, "Jehossee Island: The Antebellum South's Largest Rice Plantation," *Agricultural History* 58 (January 1985): 63.

18. Alston, *Rice Planter and Sportsman,* 47.

19. Clifton, "Jehossee," 61.

20. James M. Clifton, ed., *Life and Labor on Argyle Island: Letters and Documents of a Savannah River Rice Plantation, 1833–1867* (Savannah: Beehive Press, 1978), 74.

21. Geoffrey Day, *Tide Mills in England and Wales* (Woodbridge, Suffolk, U.K.: Published by the Friends of Woodbridge Tide Mill, 1994), 3.

22. Carney, *Black Rice,* 27–28.

23. Walter Blith, *The English Improver Improved; or, the Survey of Husbandry Surveyed* (London: Printed for John Wright, 1653), 47.

24. For a detailed description of seventeenth-century British agriculture and water-control measures, see Richard I. Groening, Jr., "The Rice Landscape in South Carolina: Valuation, Technology, and Historical Periodization," master's thesis, University of South Carolina, 1998, 1–30.

25. Judith A. Carney, "Landscapes of Technology Transfer: Rice Cultivation and African Continuities," *Technology and Culture* 37 (1996): 5–35.

26. John Mortimer, *The Whole Art of Husbandry,* 2nd ed. (London, 1708), 32.

27. Stewart, "Letters," pt. 1: 16.

28. Catesby, *Natural History,* xvii.

29. Ibid., ii.

30. Ibid., iii–iv.

31. Thomas Cooper and David J. McCord, eds., *Statutes at Large of South Carolina,* 10 vols. (Columbia, SC: A. S. Johnston, 1836–41), 3:609–10.

32. Peter Manigault to Thomas Gadsden, May 14, 1766, Peter Manigault Letterbook, series 11-493, South Carolina Historical Society, Charleston.

33. Josiah Smith to George Austin, April 22, 1774, Josiah Smith Letterbook, Southern Historical Collection, University of North Carolina, Chapel Hill.

34. James Glen, *A Description of South Carolina* (London: Printed for R. and J. Dodsley, 1761), 26.

35. William Gerard De Brahm, *De Brahm's Report on the General Survey of the Southern District of North America,* edited by Louis De Vorsey, Jr. (Columbia: University of South Carolina Press, 1971), 78. *De Brahm's Report* was not made public until De Vorsey edited it for publication in 1971.

36. David Ramsay, *History of South Carolina, from Its First Settlement in 1670 to the Year 1808,* 2 vols., (Newberry: W. J. Duffie, 1858), 2:fn 116.

37. James L. Petigru to Robert F. W. Allston, August 25, 1843, in Easterby, ed., *South Carolina Rice Plantation,* 91–92.

38. An American (pseud.), *American Husbandry,* 2 vols. (London: Printed for J. Bew, 1775), 1:392.

39. William B. Lees, "The Historical Development of Limerick Plantation, a Tidewater Rice Plantation of Berkeley County, South Carolina, 1683–1945," *South Carolina Historical Magazine* 82 (January 1981): 53.

40. Mathurin Guerin Gibbs Plantation Register, entry for November 27, 1845, call no. 34/701/01, South Carolina Historical Society.

41. There is no clearly delineated area where reservoir culture occurred, in particular inland-swamp culture, which formed the greater acreage of reservoir culture. No field studies by scientists or geographers have mapped the abandoned inland-swamp fields or reservoirs. Aerial surveys are not reliable because inland-swamp fields cannot be readily detected in aerial photographs or aerial surveys. Inland-swamp fields have been abandoned so long that secondary and tertiary forests,

which developed after the fields were abandoned, make it difficult to distinguish the fields from upland forests. Inland-swamp fields were made in a variety of habitats, unlike tidal fields, which were all on tidal river swamps and are still easy to recognize. No study has even determined a rough estimate of the acreage that was relegated to inland-swamp culture. Georgia and North Carolina also had considerable inland-swamp fields, but not as many as South Carolina. Only through time-consuming field surveys in combination with plantation plats will the full extent of inland-swamp culture be ascertained.

42. Mabel L. Webber, ed., "Col. Senf's Account of the Santee Canal," *South Carolina Historical and Genealogical Magazine* 28 (January 1927): 18.

43. Edmund Ravenel, "The Limestone Springs of St. John's Berkeley, and Their Probable Availability for Increasing the Quantity of Fresh Water in Cooper River," *Proceedings of the Elliott Society of Natural History* (September 1860): 28–32.

44. Mary R. Bullard, *Cumberland Island: A History* (Athens: University of Georgia Press, 2003), 71–72.

45. The Quaternary period runs from 2 million years before the present to the present. It is divided into the Pleistocene epoch, which ran from 2 million years before the present to 10,000 years before the present, and the Holocene epoch (sometimes referred to as the recent epoch) which runs from 10,000 years ago to the present.

46. Hugh Meredith, *An Account of the Cape Fear Country, 1731*, edited by Earl Gregg Swem (Perth Amboy, N.J.: Charles F. Heartman, 1922), 20–21.

47. William L. Saunders, ed., *The Colonial Records of North Carolina*, 10 vols. (Raleigh: Printed by P. M. Hale, 1886–90), 3:168.

48. James M. Clifton, "A Half-Century of a Georgia Rice Plantation," *North Carolina Historical Review* 47 (October 1970): 391.

49. John G. W. De Brahm, *History of the Province of Georgia* (Wormsloe, Ga., 1849), 17.

50. Charles C. Jones, *The History of Georgia* (Boston: Houghton, Mifflin, 1883), 494.

51. Julia Floyd Smith, *Slavery and Rice Culture in Low Country Georgia, 1750–1860* (Knoxville: University of Tennessee Press, 1985), 30–34.

52. Lawrence S. Rowland, "Alone on the River: The Rise and Fall of the Savannah River Rice Plantations of St. Peter's Parish, South Carolina," *South Carolina Historical Magazine* 88 (July 1987): 125–31.

53. William A. Courtenay, "Early Crops and Commerce," in *Year Book of the City of Charleston, 1883* (Charleston, n.d.), 398. Courtenay's account was accepted by Edward McCrady, Jr., in his *The History of South Carolina under the Royal Government, 1719–1776* (New York: Macmillan, 1899), 387–88, and by Ulrich B. Phillips, *Life and Labor in the Old South* (Boston: Little, Brown 1929), 116. These authorities have been the popular sources for the Estherville introduction.

54. "Journal of an Officer's [Lord Adam Gordon's] Travels in America and the West Indies, 1764–1765," in *Travels in the American Colonies*, edited by Newton D. Mereness (New York: Macmillan, 1916), 400.

55. Ibid., 396.

56. Henry Laurens to Joseph Clay, September 2, 1777, in *Papers of Henry Laurens*, 11:482.

57. James M. Clifton, "Golden Grains of White Rice: Rice Planting on the Lower Cape Fear," *North Carolina Historical Review* 4 (October 1973): 394.

58. Henry Laurens to Edward Bridgen, September 23, 1784, photocopy in the Papers of Henry Laurens, South Caroliniana Library.

59. George Baillie to John McIntosh, September 7, 1783, John McIntosh, Jr., Papers, Georgia Historical Society, Savannah, Georgia.

60. David J. McCord, ed., *Statutes at Large of South Carolina,* vol. 7 (Columbia: A. S. Johnston, 1840): 475–588.

61. John B. Irving, *A Day on Cooper River,* 2nd ed., edited by Louisa Cheves Stoney (Columbia: R. L. Bryan, 1932), 152.

62. Clifton, ed., *Life and Labor,* 25.

63. Ibid., 190.

64. David Doar, *Rice Planting in the South Carolina Low Country.* Contributions from the Charleston Museum, no. 8 (Charleston: Charleston Museum, 1936), 12.

65. Richard Carew of Antony, *The Survey of Cornwall,* 1691, reprinted edited by F. E. Halliday (London: Melrose, 1953), 110.

66. David S. Shields, "George Ogilvie's *Carolina or the Planter 1776*: An Introduction & Edition," *Southern Literary Journal,* special issue (1986): 54.

67. Doar, *Rice and Rice Planting,* 11.

68. Nathaniel Heyward Manuscript, 1802, Southern Historical Collection.

69. Clifton, ed., *Life and Labor,* 174.

70. Ibid., 278.

71. Ibid., 361.

72. Ibid., 165.

73. A. M. Forster, "Rice Culture in South Carolina," *Proceedings of the Elliott Society* 2 (December 1860): 41.

74. Allston, "Rice," 335.

75. Theodore D. Ravenel, "The Last Days of Rice Planting," in Doar, *Rice and Rice Planting,* 49–50.

76. Allston, "Rice," 339.

77. This outline of flow culture is an attempt to synthesize the variety of practices we discovered in primary and secondary sources into what we believe is an accurate depiction of the standard flow culture that prevailed throughout the rice industry.

78. William Butler, "Observations on the Culture of Rice," circa 1786, call no. 43/0742, South Carolina Historical Society.

79. John Drayton, *A View of South-Carolina, as Respects Her Natural and Civil Concerns* (Charleston: W. P. Young, 1802), 120. The fortieth governor of South Carolina, John Drayton (1766–1822) was born on Magnolia Plantation in St. Andrews Parish, near Charleston. He was admitted to the Charleston bar in 1788 and embarked on a political career in 1792 with his election to the South Carolina House of Representatives. He subsequently served as lieutenant governor (1798–1800), governor (1800–1802, 1808–10), state senator (1805–8), and U.S. District Court judge (1812–22). One of his most important acts as governor was the establishment of South Carolina College (later the University of South Carolina). In addition to *A View of South-Carolina,* Drayton's writings also include *Carolinian Florist* and *Memoirs of the American Revolution from its Commencement to the Year 1776,* an extensive biographical sketch of his father, William Henry Drayton, with a compilation of his Revolutionary War papers.

80. Thomas Pinckney, "Thomas Pinckney's Letter on the Water Culture of Rice," in *Report of the Committee Appointed by the South Carolina Agricultural Society* (Charleston: Printed by Archibald E. Miller, 1823), 16–24.

81. Clifton, ed., *Life and Labor,* 82.

82. Ibid., 114.

83. Elizabeth Waties Allston Pringle (pseud. Patience Pennington), *A Woman Rice Planter* (New York: Macmillan, 1913), 36.

84. Clifton, ed., *Life and Labor,* 123.

85. Ibid., 97.

86. Editor, "Management of a Southern Plantation," *De Bow's Review* 22 (January 1857): 41.

87. The term "floodgate" is used in written sources to refer both to the main water-control structures at the heads of the large canals and to the tide trunks that controlled the flow of tidal water into and out of individual rice fields. This book uses "floodgate" for the main water-control structures at the heads of the main canals, and uses "tide trunk" for the structures that control the flow of water into and out of individual fields.

88. Louise P. Ford and Marion J. Pelleu, "The 'Carolina Gold' Crop: Rice," unpublished manuscript in Richard Porcher's collection.

89. Doar, *Rice and Rice Planting,* 10.

90. Ibid., 10–11.

91. The James Hamilton Couper Plantation Records (1818–1854) in the Southern Historical Collection at Chapel Hill consist of four volumes. The first two volumes are detailed plantation accounts; volume three gives a complete summary of the crops at Hopeton from 1814 to 1841; and volume four, "Agricultural Notes," is a compilation of Couper's observations on the crop production and market processing of crops, bills of scantling for construction of trunks, and description of a rice mill.

92. Alexander Hewatt, *An Historical Account of the Rise and Progress of the Colonies of South Carolina and Georgia,* 2 vols. (London: Printed for Alexander Donaldson, 1779), 159.

93. Kemble, *Journal,* 114.

94. Ibid., 69–70.

95. Smith, *Slavery and Rice Culture,* 209.

96. Joseph I. Waring, *A History of Medicine in South Carolina, 1825–1900* (Charleston: The South Carolina Medical Association, 1967), 41.

97. William Dusinberre, *Them Dark Days: Slavery in the American Rice Swamps* (Athens: University of Georgia Press, 2000), 49.

98. The bobolink (*Dolichonyx oryzivorus*) is also called the "reed bird" and "rice bird." Fewer birds are more aptly named. Its specific epithet, *oryzivorus,* means "rice eating."

99. An Observer (pseud.), "Protecting Crops from Birds," *Southern Agriculturist* 6 (August 1833): 424–27.

100. Doar, *Rice and Rice Planting,* 23–24.

101. Clifton, ed., *Life and Labor,* 9.

102. Amory Austin, *Rice: Its Cultivation, Production, and Distribution in the United States and Foreign Countries,* U. S. Department of Agriculture, Miscellaneous Series, report no. 6 (Washington, D.C.: U.S. Government Printing Office, 1893), 32.

103. Ibid., 31–32.

104. Doar, *Rice and Rice Planting,* 12.

105. Ibid., 12–13.

106. Roswell King, Jr., "Queries on the Culture of Rice," *Southern Agriculturist* 1 (September 1828): 409–10.

107. Joe Kelley and Dan Tufford, "Plant Succession in Cooper River, South Carolina, Tidal Former Rice Fields: Ecological and Human Use Implications," in *The Golden Seed: Writings on the History and Culture of Carolina Gold Rice,* edited by David S. Shields (Charleston: Douglas W. Bostick, 2010), 111–28.

CHAPTER 4. Harvesting

1. Easterby, ed., *South Carolina Rice Plantation,* 164.

2. Robert F. W. Allston, "Best Mode of Curing and Milling Rice," in *Reports Submitted to the Winyah and All-Saints Agricultural Society on the 20th April, 1848* (Charleston: Printed by Burges and James, 1848), 6.

3. Albert Virgil House, ed., *Planter Management and Capitalism in Ante-Bellum Georgia: The Journal of Hugh Fraser Grant, Ricegrower* (New York: Columbia University Press, 1954), 36.

4. Charles Munnerlyn, "Queries on the Culture of Rice; by William Washington, with Answers, by Charles Munnerlyn," *Southern Agriculturist* 1 (May 1828): 220.

5. Thomas J. Molony, Jr., "Rice Cutting Machines," *Rural Carolinian* 2 (1871): 183.

6. Ibid., 183–84.

7. C. C. Pinckney, "On the Advantage of Using Animal Power and Machinery in the Culture of Rice," *Southern Agriculturist* 10 (April 1837): 171.

8. Although turning under the stubble provided excellent manure, some planters burned the stubble instead. Turning it under also buried the seeds of the volunteer red rice, which caused a problem the following year. Planters avoided this problem by allowing the red rice to sprout first and then turning under the stubble.

9. The Committee, "Report on Cradling Rice," *Southern Agriculturist* 5 (October 1845): 388–89.

10. Anne K. Gregorie and Flora B. Surles, "Minute Book, South Carolina Agricultural Society: 1825–1860," W.P.A. Project 65-33-118 (1936).

11. Pinckney, "Advantage," 172.

12. Thomas Spalding, "On the Culture, Harvesting and Threshing of Rice, and on the Rust in Cotton," *Southern Agriculturist* 8 (April 1835): 171–72.

13. Pocket notebook, 1852, Louis Manigault Manuscripts, Duke University Library, Durham, North Carolina.

14. Austin, "Rice," 23. We found no detailed information about curing practices on Carolina plantations. A letter from Skinner to Manigault quoted in *Life and Labor on Argyle Island* mentions a "Temperature Stick" which undoubtedly was used in the same manner as the stake.

CHAPTER 5. Threshing

1. When they were interviewed in the 1970s, several persons who worked in rice fields during the early 1920s used the term "thrashing," which is an older spelling and pronunciation of "threshing."

2. Ramsay, *History of South Carolina,* 2:117.

3. On December 15, 1836, all patents then held in the U.S. Patent Office were lost in a fire. The staff was faced with the mammoth task of restoring records and patent models, basing the restoration on the memory of a single patent examiner and a book that a draftsman named William Steiger had earlier removed from the Patent Office.

In the forty-six years prior to the fire, the United States government had issued about ten thousand patents. Most of these could never be reinstated, but Congress acted to restore those records that could be reconstructed from private files and to reproduce models that were deemed critical. Patent holders were instructed to resubmit their diagrams and descriptions. Some did; others did not. Patents whose records were not restored were cancelled. Most of the 2,845 restored patents were eventually given numbers beginning with "X". All new patents granted after July 1836 were numbered as a new series (without the X), beginning with a new "Patent No. 1," granted to John Ruggles.

4. Lewis Cecil Gray, *History of Agriculture in the Southern United States to 1860,* 2 vols. (Gloucester, Mass.: Peter Smith, 1958), 1:281.

5. Charles Manigault, "Plantation Journal," in Clifton, ed., *Life and Labor,* 8–9.

6. Heyward, *Seed from Madagascar,* 21.

7. James Bagwell, *Rice Gold: James Hamilton Couper and Plantation Life on the Georgia Coast* (Macon: Mercer University Press, 2000), 93.

8. Charles Manigault to Jesse T. Cooper, July 12, 1848, in Clifton, ed., *Life and Labor,* 63.

9. Doar, *Rice and Rice Planting,* 16.

10. Clifton, ed., *Life and Labor,* 207.

11. J. Bryan, "On the Culture of Rice," *Southern Agriculturist* 5 (October 1832): 530.

12. *Charleston Morning Post, and Daily Advertiser,* Thursday, March 2, 1786.

13. Max L. Hill III, of the family that owns Middleburg, documented that the structure was a seed barn, not a slave jail as some accounts state.

14. South Carolina Commons House of Assembly, *Journal of the Commons House of Assembly, November 12, 1754–September 23, 1755,* edited by Terry W. Lipscomb (Columbia: Published for the South Carolina Department of Archives and History by the University of South Carolina Press, 1986), 34. This Kogar is Joseph Kogar, Sr., not his son Joseph Kogar, Jr., who built the Kogar House on Wiregrass Road in Dorchester County.

15. Charles Drayton Diaries, 1784–1820, Drayton Hall Historic Site, National Trust for Historic Preservation, Charleston, South Carolina. Charles Drayton, M.D. (1743–1820) was the uncle of Governor John Drayton, author of *A View of South-Carolina.*

16. R. Douglas Hurt, *American Farm Tools* (Manhattan, Kans.: Sunflower University Press, 1982), 69.

17. An American, *American Husbandry,* 1:393.

18. Thomas Jefferson, *Jefferson's Memorandum Books: Accounts, with Legal Records and Miscellany, 1767–1826,* edited by James A. Bear, Jr. and Lucia C. Stanton, vol. 2 (Princeton, N.J.: Princeton University Press, 1997), 907.

19. Jefferson, *The Papers of Thomas Jefferson,* vol. 26: May 11–August 31, 1793, edited by John Catanzariti, Eugene R. Sheridan, J. Jefferson Looney, Elizabeth Peters Blazejewski, and Jean-Yves M. Le Saux (Princeton: Princeton University Press, 1995), 62.

20. Spalding, "On the Culture," 169.

21. N. and D. Sellers to H. Laurens, July 23, 1802. Pinckney Lowndes Papers in the Ravenel Collection 11/332A/24, South Carolina Historical Society.

22. Ibid.

23. Clifton, ed., *Life and Labor,* 50.

24. The advertisement appears in an unidentified paper or journal in the Pinckney Lowndes Papers in the Ravenel Collection, call no. 11/332A/22, South Carolina Historical Society.

25. Gregory and Surles, "Minute Book," 17.

26. A Friend to Improvement (pseud.), "On Rice Threshing Machines," *Southern Agriculturist* 2 (September 1829): 404–5.

27. W. (pseud.), "On Rice Threshers," *Southern Agriculturist* 7 (November 1834): 581.

28. These four quotations are among six that appeared in the February 19, 1831, edition of the *Georgian*.

29. Robert F. W. Allston (1800–1864) was governor of South Carolina (1856–58) just before the outbreak of the Civil War and a firm proponent of southern rights. He was also one of the wealthiest planters in the Georgetown area of the Lowcountry, the richest rice-producing region in the United States. He owned four plantations: Chicora Wood, along the Pee Dee River, and three others on the Waccamaw River. In 1860 he owned 630 slaves. He was a prolific writer on agricultural topics, especially rice, of which he was an acknowledged leader in planting and marketing.

30. Allston, "Rice," 341–42.

31. Ibid., 340.

32. W., "On Rice Threshers," 581.

33. Daniel Blake to [illegible], 1832, Daniel Blake Financial Records, 1830–32, MS 72, Georgia Historical Society.

34. W., "On Rice Threshers:" 581–82.

35. The original drawing is located in the South Caroliniana Library.

36. Charles Manigault to James Hayes, January 1, 1847, in Clifton, ed., *Life and Labor*, 45.

37. Conversion of values to 2009 has been based on Lawrence H. Officer and Samuel H. Williamson, "Purchasing Power of Money in the United States from 1774 to 2010," Measuring Worth, 2009, on-line at http://www.measuringworth.com/ppowerus/result.php (accessed May 16, 2013).

38. Calvin Emmons, "Dear Sir," *American Farmer* 1 (February 5, 1840): 293.

39. This patent was restored after the fire.

40. It was too difficult to remove the original letters from fig. 2, but new labels have been added.

41. Charles Manigault to James Haynes, January 1, 1847, in Clifton, ed., *Life and Labor*, 44–45. Manigault said this threshing machine was "the first of its great dimensions which has ever been made—my being the first to order such a one." Manigault did not imply that he was the first to use the Butts thresher, only that he was the first to order one built on a larger design that the original. Evidently manufacturers were able to construct the Butts thresher on a larger plan that enabled increased threshing capacity.

42. Charles Manigault to R. Habersham & Son, from Paris, February 1, 1847, Charles Manigault Letterbook, South Carolina Historical Society.

43. Ibid.

44. Charles Manigault to James Haynes, August 15, 1846, ibid., 37.

45. James Haynes to Charles Manigault, April 22, 1847, ibid., 55.

46. An extensive study of the threshing barn at Chicora Wood concluded that, toward the end of its active use, it was only a threshing barn. Whether it once housed milling machinery is undetermined. No physical artifacts relating to milling machinery were found in the barn, and no archival references indicate that it was ever a mill.

47. Allston, "Rice," 340.

48. Thurston & Johns to C. Petigru Allston, August 6, 1869, Allston Family Papers, folder no. 20, South Caroliniana Library. C. Petigru is Charles Petigru Allston (1848–1922), second son of Robert F. W. Allston.

49. The opening in the wall was boarded up, and the chute no longer exists.

50. Pringle, *A Woman Rice Planter,* 116. P. D. Wragg and Vareen are the names of two rice fields; on rice plantations, each field was given a name.

51. Solon Robinson, "Governor Aiken's Extensive Rice Estate," *DeBow's Review* 1 (August 1850): 201–3.

52. Charles Manigault described how the sheaves were fed into the Butts thresher in a letter to James Haynes, in Clifton, ed., *Life and Labor,* 37. Manigault wrote that "Mr. Butts (the maker) says that it must be fed by the two-feeding hands with a Constant supply of sheaves—the head of the one touching the but[t] of the foremost sheaf just as it is about to disappear in the thresher."

53. Anne Simons Deas, "Two Years of Plantation Life, 1850 and 1880," unpublished manuscript, 1910, Anne Simons Deas Papers, accession 2118, p. 13, South Caroliniana Library. Anne Deas (1845–1928) wrote a fictionalized account of her childhood at Buck Hall, which she called Ashland; her description of the threshing barn was written just prior to the Civil War. Her cousin Col. Alston Deas, who inherited her papers, made notations that, when Deas referred to Ashland, it was Buck Hall on the Western Branch of the Cooper River. Historian Cecile Ann Guerry from Moncks Corner, once part-owner of Buck Hall, has done research, both in archives and in the field on Buck Hall, and confirms that Ashland is Buck Hall.

54. See Robert F. W. Allston to Benjamin Allston, June 25, 1858, in Easterby, ed., *South Carolina Rice Plantation,* 144. According to Allston, "I had left with Jas. Adger and Co. an order to Mr Tighe of Richmond for a larger grain Separator (Solomon's) to be used as a chaff-fan."

55. A sloop is a single-masted, fore-and-aft-rigged sailing boat with a single headsail set from the forestay; a schooner is a sailing ship with two or more fore-and-aft-rigged masts. The sloop in diagram 32 was typical of the early 1800s.

56. Clifton, ed., *Life and Labor,* 139.

57. Heyward, *Seed from Madagascar,* 41–42.

58. Clifton, ed., *Life and Labor,* 162.

59. The hopper in the loft, which directed the rice to the two chutes, is the most important remaining artifact in the storage barn. According to the present owners of Chicora Wood, who oversaw some of the restorations, the hopper is original.

60. Memorandum of agreement between Anthony Weston and Henry A. Middleton, November 25, 1851, H. A. Middleton Papers, call no. 1168.02.05.02, folder 12/161/26, South Carolina Historical Society.

61. A. H. Seabrook, "Reminiscences of A. H. Seabrook," Tampa, Florida, 1932. A copy of this paper, which exists only in the family's collection, was given to Richard Porcher.

62. James Haynes to Charles Manigault, April 22, 1847, in Clifton, ed., *Life and Labor,* 55.

63. Clifton, ed., *Life and Labor,* 271.

64. Editor, "Management of a Southern Plantation," 42.

65. B. T. Sellers to Mr. [Williams] Middleton, September 29, 1860, Williams Middleton Papers, folder no. 3, South Caroliniana Library.

66. Ibid.

67. Doar, *Rice and Rice Planting,* 21.

68. The Invincible was manufactured by Kingsland and Ferguson of St. Louis and distributed by the Charleston firm of Cameron and Barkley. An Invincible is housed in the threshing barn at Kinloch Plantation in Georgetown County and is in good condition but not operable. Another Invincible, at Cockfield Plantation in Colleton County, is in poor condition. There are no records of many other postwar threshers operating in the Rice Kingdom.

69. S. W. Barker, "Implements and Machinery Adapted to Rice Culture," in *Address Delivered before the Agricultural Society of South Carolina at Their Seventy-sixth Annual Meeting, January 12, 1871* (Charleston: Walker, Evans & Cogswell, 1872), 38. S. W. Barker is Dr. Sanford William Barker (1807–1891), a physician and botanist who married Christiana Constant Broughton of South Mulberry Plantation on the Western Branch of the Cooper River in St. John's Parish, Berkeley.

CHAPTER 6. Milling

1. Thomas Cooper and David J. McCord, eds., *Statutes at Large of South Carolina,* 10 vols. [covering 1682–1838] (Columbia: A. S. Johnston, 1838–41), 3:599, 4:30, and 5:69.

2. Henry Laurens, *Papers of Henry Laurens,* vol. 9: April 19, 1773–December 12, 1774, edited by George C. Rogers, Jr., and David J. Chesnutt (Columbia: Published for the South Carolina Historical Society by the University of South Carolina Press, 1981), 52.

3. Gregorie and Surles, "Minute Book," 49.

4. Nothing in archival sources indicates that milling described in this book was significantly different for the various strains of commercial rice grown in the Rice Kingdom, which include the three major varieties—white, Carolina Gold, and long-grain Carolina Gold. The mortar and pestle—and the subsequent machines that powered a series of mortar and pestles for milling—all depended on friction between the grains to clean the rice; different strains did not require different machines. Even when millstones were first used to mill rice, the stones could be adjusted for the different lengths or rice, either between strains, or for the different lengths of rice within the same strain. Both were a minor adjustment.

5. The term "rough rice" is used for unmilled rice because "paddy rice" is a term also used to refer to rice grown in flooded fields, called "paddies."

6. Lawson, *New Voyage,* 216.

7. Rev. William Ellis, *History of Madagascar* (London: Fisher, 1838), 205.

8. Lydia Parrish, *Slave Songs of the Georgia Sea Islands* (1942; reprint, Athens: University of Georgia Press, 1992), 227.

9. Peter Manigault to John Owen, February 20, 1794, Peter Manigault Letterbook, South Caroliniana Library.

10. The *Oxford English Dictionary* defines *engine* as a "machine with moving parts that converts power into motion." In this sense Guerard's mill qualifies as an engine. Today an "engine" is generally understood to be a steam engine or an internal-combustion engine.

11. Cooper and McCord, eds., *Statutes at Large of South Carolina,* 2:63.

12. A copy of the handwritten document is at the South Carolina Historical Society in the Lucas Family Papers 1792–1936, folder 11/270/1. The unsigned letter is not dated and the recipient is unknown. However, a sentence at the beginning of one paragraph states: "I John grandson of Jon Lucas 1st assisted dressing the Mill Stones for Egypt." More than likely this John Lucas is Jonathan Lucas III (1800–1848), son of Jonathan Lucas, Jr. (1775–1832), who went to England in 1823 and died there in 1832. Jonathan Lucas III may have gone by the name "John." However, this

grandson may also have been John Hume Lucas (1822–1853) or Jonathan Lucas (1823–1881), both sons of William Lucas (1789–1878), brother of Jonathan Lucas, Jr.

13. *Papers of Henry Laurens,* vol. 8: October 10, 1771–April 19, 1773, edited by George C. Rogers, Jr., and David J. Chesnutt (Columbia: Published for the South Carolina Historical Society by the University of South Carolina Press, 1980), 204.

14. Ibid., 224.

15. Laurens, *Papers,* 9:6.

16. Ibid., 52.

17. Ibid., 454, 648.

18. Ibid., 648–49. According to footnote 3, the record of Deans's shipment of his model to the United States is "Memorial of Robert Deans of Fludyer Street, Westminster," November 1776, AO 13/127, Public Record Office, London.

19. Catesby, *Natural History,* xvii.

20. Glen, *A Description of South Carolina,* 202.

21. Henry Laurens to Peter LePoole, August 14, 1772, in Laurens, *Papers,* 8:409.

22. J. F. D. Smyth, *A Tour of the United States of America,* 2 vols. (Dublin: T. Henshall, 1784), 1:67–68.

23. Glen, *A Description of South Carolina,* 202.

24. Allston, "Rice," 342.

25. Joseph W. Barnwell, ed., "Diary of Timothy Ford, 1785–1786," *South Carolina Historical and Genealogical Magazine* 13 (October 1912): 183–84. Timothy Ford (1762–1830), who received his education at Princeton, came to Charleston in 1785 and was admitted to the South Carolina bar in 1786. He recorded his observations on the mill while visiting Dr. John Beamer Waring at his plantation Pine Hill, near Beaufort.

26. Luigi Castiglioni, *Viaggio: Travels in the United States of North America 1785–87,* translated and edited by Antonio Pace (Syracuse: Syracuse University Press, 1983).

27. Charles Drayton Diaries, Drayton Hall Historic Site.

28. R. A. Wilkinson, "Production of Rice in Louisiana," *DeBow's Review* 6 (July 1848): 55.

29. Ibid., 57.

30. Nairne, "Letter from South Carolina," 40.

31. *Georgia Gazette,* Wednesday, October 26, 1768.

32. Elkanah Watson, *Men and Times of the Revolution,* edited by Winslow C. Watson (New York, 1856), 52.

33. Ulrich B. Phillips, ed., "Some Letters of Joseph Habersham," *Georgia Historical Quarterly* 10 (June 1926): 161.

34. Drayton, *A View of South-Carolina,* 117–18.

35. John Lucas document, South Carolina Historical Society.

36. *South Carolina Gazette,* Saturday, July 21–Saturday, July 28, 1733, Charleston Library Society.

37. Ibid., Saturday, May 11–Saturday, May 18, 1734.

38. *South Carolina Gazette,* May 18, 1734.

39. Ibid., September 14, 1734.

40. John Lucas document, South Carolina Historical Society.

41. Charles Drayton Diaries, Drayton Hall Historic Site.

42. John Lucas may have based his 1850 description and sketch of the mill Charles Drayton wrote about on someone else's recollection or from notes in the Lucas family papers.

43. Charles Drayton Diaries, Drayton Hall Historic Site.

44. Drayton, *A View of South-Carolina,* 121.

45. Glen, *A Description of South Carolina,* 202.

46. An American (pseud.), *American Husbandry* (London, 1775), 1: 393.

47. Hewatt, *An Historical Account,* 1: 159.

48. Allston, "Rice," 348.

49. James M. Clifton, "The Ante-Bellum Rice Planter as Revealed in the Letterbook of Charles Manigault, 1846–1848," part 2, *South Carolina Historical Magazine* 74 (October 1973): 303.

50. "Pounding" was a common term used for the action of the pestle moving down into the rice in the mortar—hence a pounding mill.

51. House Journal, 37, pt. 2 (November 3, 1767–November 19, 1768), 490–91, 496, South Carolina Department of Archives and History, Columbia, South Carolina.

52. *South Carolina Gazette,* February 1–8, 1768.

53. The bill of scantling called for fifty "arms," more often called "spokes," the structure that lifted a pestle. Theoretically there should have been forty-eight spokes instead of fifty since there were twelve pestles, each pestle requiring four spokes. However, it is understandable that mill workers would order extra spokes in case of breakage.

54. Day, *Tide Mills in England,* 3.

55. Moore, "Axtell's Account Book," 292–96.

56. Ibid., 299.

57. E. Merton Coulter, ed., *The Journal of William Stephens, 1743–1745* (Athens: University of Georgia Press, 1959), 60.

58. George Fenwick Jones and Renate Wilson, trans. and eds., *Detailed Reports on the Salzburger Emigrants Who Settled in America,* vol. 18: 1744–1745 (Camden, Maine: Picton Press, 1995), 213–14.

59. William Bartram, *Travels Through North & South Carolina, Georgia, East & West Florida, the Cherokee Country, the Extensive Territories of the Muscogulges, or Creek Confederacy, and the Country of the Choctaws,* 1791 (Reprint, New York: Penguin, 1988), 37.

60. Alston, *Rice Planter and Sportsman,* 42.

61. Charles Drayton Diaries, Drayton Hall Historic Site. The rush was probably *Juncus roemerianus* Scheele, a common plant of the high salt marsh along the Atlantic coast.

62. Henry Laurens, *Papers of Henry Laurens,* vol. 16: September 1, 1782–December 17, 1792, edited by David R. Chesnutt and C. James Taylor (Columbia: Published for the South Carolina Historical Society by the University of South Carolina Press, 2003.), 627–28.

63. The mill diagram is one of two related documents in the Ball Family Papers at the South Caroliniana Library. The first document is a July 15, 1793, letter from Jonathan Lucas I addressed to "Dear Sir" (Elias Ball [1752–1810]). It is a bill of scantling for some part of Ball's mill work. Lucas wrote on the letter: "This Will Be followed By a Drawing and Bill of Scantling For your Barn and mill house." The second document is "A Bill of Scantling For Elias Ball Esq," which is written on the diagram of a mortar and pestle mill.

64. Jonathan Lucas, Jr., to Jonathan Lucas I, December 12, 1795, Lucas Family Papers, South Carolina Historical Society.

65. Drayton, *A View of South-Carolina,* 123.

66. Clifton, ed., *Life and Labor,* 358.

67. James Haynes to Charles Manigault, October 25, 1847, in Clifton, ed., *Life and Labor*, 59–60.

68. Charles Manigault, "Visit to 'Gowrie' and 'East Hermitage' Plantations, Savannah River, 22d March 1867," in Clifton, ed., *Life and Labor*, 358.

69. *Georgia Gazette*, November 24, 1763.

70. *State Gazette of South Carolina*, November 19, 1787.

71. *Charleston City Gazette*, November 4, 1799.

72. Historians question whether the mill for Bowman was actually the first mill Lucas built. A February 20, 1793, letter to Lucas from Plowden Weston states: "I was the first person to engage you in building a mill." Bowman, however, is usually credited with owning the first mill built by Lucas.

73. Lucas did not restrict his mill construction to water-powered mills. His advertisement in the March 4, 1796, issue of the *Charleston City Gazette* reads: "FOR SALE: The MACHINERY and iron work complete for a MILL for grinding rice, to be worked by horses or oxen, constructed by Mr. Lucas. For terms apply to Mr. LUCAS, or the Printers. February 23." Lucas also built windmills, including the wind-powered sawmills he constructed in Charleston Harbor. In 1792 he built a wind-powered sawmill for the same Bowman for whom he had built the Water Rice Machine. This windmill's cast-iron shaft, which was cast in England, was recovered in 1992 and is on display at the Village Museum in McClellanville. Lucas also proposed to build a wind-powered rice mill for John Hume, but after the windmill shaft arrived from England, Hume decided on a water mill.

74. On the bottom left side of the engraving is "B. H. Latrobe Esq. Del.," indicating that Latrobe was responsible for the initial diagram of the mill. The actual engraving was done by J. Akin of Philadelphia.

75. Alexander Lucas Lofton, *The Lucases of Haddrells Point*, book 1: 1785–1835 (Mount Pleasant, S.C: Privately published by the author, 1998), 54.

76. Clifton, ed., *Life and Labor*, 125.

77. Allston, "Rice," 348.

78. Ibid.

79. Ibid., 343fn.

80. Doar, *Rice and Rice Planting*, 21.

81. During the 1970s this mortar bottom was in the yard at Middleburg Plantation. After Middleburg was sold, the mortar bottom disappeared.

82. Austin, *Rice*, 27–28.

83. Rice Millers Association, unpublished document in a collection of rice facts by unknown authors, Southwestern Louisiana University, Lafayette, Louisiana.

84. S. A. Knapp, *Rice Culture*, U. S. Department of Agriculture, Farmers' Bulletin 417 (Washington, D.C, U.S. Government Printing Office, 1910), 22. Seaman Asahel Knapp (1831–1911) was a physician, college instructor, and administrator who took up farming late in life. By 1902 he was working with the federal government to promote good agricultural practices in the South, including rice culture.

85. Beth Bailey McLean, ed., *The Story of Rice: The World's Most Popular Food* (New Orleans: Rice Millers Association, 1934), 13–14.

86. Allston, "Rice," 347–48.

87. Charles Drayton Diaries, Drayton Hall Historic Site.

88. George Jobey, "Millstones and Millstone Quarries in Northumberland," *Archaeologia Aeliana* 5th ser., 14 (1986): 61.

89. Charles E. Rowland, "Queries on the Culture of Rice," *Southern Agriculturist:* 353.

90. In *A View of South-Carolina,* 122, John Drayton mentioned a pendulum screen by Lewis Dupre that was brought into use in 1798, but since Drayton did not state that it removed the unhulled rice, the authors are not certain it was the same screen Dupre patented in 1807.

91. Several letters to and from the Lucas family in Charleston refer to the family's obtaining their millstones from England. For example Thomas Naylor wrote to William Lucas September 11, 1820, from Olterton, near Knutsford: "My good friend William do let me entreat you to take into your hands what mill stones are remaining unsold, the quantity shipped to him were in the whole 28 pair" (Lofton, *The Lucases of Haddrells Point,* 193).

Field studies located two runner millstones made from a clay-colored rock of unknown material and source. One was recovered from Hobcaw Barony in Georgetown County and is displayed at the Rice Museum in Georgetown (see color plate 18D). The second lies in a marsh near Daniel Heyward's mill site at Hazzard's Creek in Jasper County, South Carolina. Duncan Heyward reported they were "brought from Scotland" (Heyward, *Seed from Madagascar,* 49). Thus not all Lowcountry millstones were Peak millstones.

92. D. Gordon Tucker, "Millstone Making in the Peak District of Derbyshire: The Quarries and the Technology," *Industrial Archeology Review* 8 (Autumn 1985):42–58.

93. *Charleston City Gazette,* May 5, 1804.

94. Charles Howell and Allan Keller, *The Mill at Philipsburg Manor, Upper Mills, and a Brief History of Milling* (New York: Sleepy Hollow Restorations, 1977), 70.

95. D. Gordon Tucker, "Millstones North and South of the Scottish Border," *Industrial Archaeology Review* 6 (Autumn 1982): 186.

96. Robert Scott Davis, Jr., "As Good as the French: The Rise and Decline of the Georgia Buhr Stone Industry," *Georgia Historical Quarterly* 77 (Fall 1993): 560–668.

97. D. Gordon Tucker, "Millstone Making in Scotland," *Society of Antiquaries of Scotland* 114 (1984): 544.

98. Greville Bathe and Dorothy Bathe, *Oliver Evans: A Chronicle of Early American Engineering* (New York: Arno Press, 1972), 11.

99. The size of the belts and buckets was determined by measurements from pieces of belts and buckets found in abandoned mills.

100. Jonathan Lucas, Jr. (1775–1832), was born in England, to of Jonathan Lucas I and his first wife, Mary Cook.

101. Only the foundation of the mill and part of the brick millrace remain of the Mepkin mill. The location of the mill ruins was made known to officials at Mepkin Abbey and the staff of the monastery will insure its preservation.

102. American inventor Oliver Evans (1755–1819) designed his first invention, a machine for making card teeth for carding wool, in 1777. Going into business with his brothers, he made several improvements in the flour-milling industry. His most important invention was an automated grist mill, which used bulk-material handling devices, including bucket elevators and conveyor belts. Evans described this invention in *The Young Mill-Wright and Miller's Guide* (1795). His second most important invention was an improved high-pressure steam engine.

103. Wheat and rice are so different that the process for milling wheat cannot be used for milling rice. When wheat is harvested, the nonflowering parts, the paleas and glumes, are left behind

and do not adhere to the fruit. In wheat milling, the object is to grind the starch-containing endosperm into flour. In rice milling, however, the object is to keep the starch-containing endosperm whole. Wheat seeds were processed whole between millstones and later between steel rollers. The seeds are submitted to breaking, grinding, and rolling. Then the products are sifted to remove the bran and germ, leaving the white ground endosperm, the wheat flour. If whole wheat flour is desired, no sifting is done. Rice, on the other hand, retains the hard paleas and glumes, called the hulls, after harvesting and threshing. First the hulls have to be removed and discarded. At first hulls were removed with a mortar and pestle and then separated from the fruit. In 1787 millstones were used first to remove the hulls and then millers used mortars and pestles to complete the milling process, taking care not break the rice. Although the milling process for wheat and rice were different, rice mills employed other machinery used in wheat mills, such as elevators designed by Oliver Evans.

104. Drayton, *A View of South-Carolina,* 123–24.

105. Allston, "Rice," 348–49.

106. The Editor, "Observations on Strong & Moody's Patent Rice Mill," *Southern Cabinet* 1 (December 1840): 736–40. The editor followed his praise for the Lucas's mill with a description of Strong and Moody's patent mill, where "large wire teeth, similar to those used for cards and fixed on the face of sections of wooden cylinders, would clean the rice more efficiently that the mortars and pestles in the Lucas mill." As best as we can determine, Strong and Moody's cleaning cards were never employed in mills in the South. Eight were established in South Boston in 1834, and eight were sent to Great Britain.

107. Allston, "Rice," 349–51. None of these mills was preserved.

108. William Dollard Lucas, *A Lucas Memorandum,* edited by David Henry Lucas (N.p.: Privately published, n.d.), 23. William Lucas stated that the quoted material was from the records of Thomas S. Lucas of Atlanta, Georgia.

109. James Hamilton Couper, "Agricultural Notes," vol. 4, p. 55, James Hamilton Couper Plantation Records (1818–54), Southern Historical Collection.

110. The brick chimney and structure are along the west side of U.S. 17 south of Darien on Butler's Island.

111. John D. Legaré, "Account of an Agricultural Excursion made into the South of Georgia in the Winter of 1832," *Southern Agriculturist* 6 (March 1833): 142–43.

112. John Ravenel (1793–1862) was the son of Daniel Ravenel III, of Wantoot Plantation in St. John's Parish, Berkeley County. Ravenel and Samuel N. Stevens established the Charleston mercantile house Ravenel & Stevens, which became Ravenel, Stevens & Company when John's brother William joined the firm, and Ravenel & Company after John retired in 1850 and William and his two nephews continued the business. Part of their business was to trade rice to European ports.

113. J.G. (pseud.), "On Rice Mills," *Southern Agriculturist* 2 (November 1829): 505.

114. Clifton, ed., *Life and Labor,* 10.

115. Gregorie and Surles, "Minute Book," 44.

116. Ibid., 44.

117. Ibid., 42.

118. This description of the operation of the Newcomen engine was adapted from Carroll W. Pursell, Jr., *Early Stationary Steam Engines in America* (Washington, D.C.: Smithsonian Institution Press, 1969), 4.

119. Louis C. Hunter, *A History of Industrial Power in the United States, 1780–1930,* vol. 2: *Steam Power* (Charlottesville: University Press of Virginia, 1985), 138.

120. Pursell, *Early Stationary Steam Engines,* 45.

121. Ibid., 116.

122. Ernest M. Lander, Jr., "Charleston: Manufacturing Center of the Old South," *Journal of Southern History* 26 (August 1960): 330–51.

123. Lofton, *The Lucases of Haddrells Point,* 192. William Lucas (1789–1878) was the son of Jonathan Lucas I and brother of Jonathan Lucas, Jr.

124. Pursell, *Early Stationary Steam Engines,* 73.

125. Allston, "Rice," 344.

126. Middleton Papers, box 7, folder 3, Middleton Place Manuscript Collection, 1783–1899.

127. Middleton Papers, box 8, folder 2, Middleton Place Manuscript Collection, 1783–1899.

128. The accession description reads: "Horizontal boxbed steam engine. Side crank, slide valve, throttling governor. Boxbed designed so that the flywheel can be either right or left hand. Engine is equipped with two boiler feed pumps which are vertical, driven off of the crosshead. Governor is watt type, belt-driven and located above crosshead. Bore 8.25 inches, stroke 30 inches. Max governed speed 66 r.p.m. Horsepower 20 estimated. Built around 1840."

129. This overview of plain cylindrical boilers was taken from Hunter, *History of Industrial Power,* 2: 327.

130. Lofton, *The Lucases of Haddrells Point,* 192.

131. These dimensions are based on measurements of the existing boilers and from several notes in plantation journals.

132. The measurements of the boilers found in the field are in keeping with those recorded in archival sources. For example, Cameron McDermid Mustard of Charleston contracted with Mr. Munnerlyn for two boilers, each thirty inches in diameter and thirty feet long. This contract may be found in the H. A. Middleton Papers, no. 1178.02.05.02, folder 12–161–25, South Carolina Historical Society.

133. This description is based on one cast-iron cover assembly found in the threshing barn at Chicora Wood, the only one found in years of field work.

134. When fuel burns, hot gases are given off. The gases contain the heat that converts water to steam.

135. Robert Habersham & Son to Charles Manigault, January 30, 1852, Manigault Family Papers, box 11/277/41, South Carolina Historical Society.

136. Clifton, ed., *Life and Labor,* 124.

137. Hunter, *History of Industrial Power,* 2: 341.

138. Charles Manigault to Louis Manigault, October 11, 1856, in Clifton, ed., *Life and Labor,* 228.

139. Ibid., 212.

140. John Farey, *A Treatise on the Steam Engine, Historical, Practical, and Descriptive* (London: Printed for Longman, Rees, Orme, Brown & Green, 1827), 99–101.

141. Steam-power technology was introduced into the United States shortly after 1800. It is possible that shipyards and foundries were the first to use steam power. The Charleston Foundry of William Lebby was building steam engines for the Lucas family in 1832. Possibly, through an earlier association with Lebby (prior to 1817), Jonathan Lucas became versed in steam technology as a new source of power to operate rice machinery.

142. J. Hamilton Couper, "Account of, and Directions for Erecting a Sugar Establishment," *Southern Agriculturist* 4 (May 1831): 227.

143. Ibid.

144. May 2006 interview with H. Alan McEwen of H. Alan McEwen (Boiler Repairs) Ltd, Boilermakers & Industrial Boiler Plant Engineers, Farling Top Boilerworks, Keighley, West Yorkshire, England.

145. Originally called Belle Isle Plantation, this plantation was renamed Black Out during World War II, when the owners painted the buildings black. The plantation is now part of the Tom Yawkey Wildlife Center, which is owned and managed by the South Carolina Department of Natural Resources. The chimney at Black Out was in an advanced state of deterioration until 2010. At that time Yawkey Center project leader Jamie Dozier arranged to have the chimney restored by Dennis Babich, who repointed most of the masonry and replaced the missing bricks from the crown. The Yawkey Foundation funded the restoration.

146. The conclusion that most rice chimneys had square stacks and square flues is based on examinations of extant chimneys and the bases of chimney remains.

147. Clifton, ed., *Life and Labor,* 108.

148. The original height of a chimney in which the crown and part of the upper stack is gone is an estimate.

149. Robinson, "Aiken's Rice Estate." 201–3.

150. Ibid. Robinson was mistaken when he said that both openings were manholes. One opening was a manhole, and the other was the smoke-duct entrance.

151. Farey, *A Treatise on the Steam Engine,* 583.

152. The acceleration of the rising air is based on Bernoulli's Principle, which states that, when a liquid or gas flowing through a conduit enters a region where the conduit's diameter is reduced, its speed increases.

153. Walter Pickles, *Our Grimy Heritage: A Fully Illustrated Study of the Factory Chimney in Britain* (Fontwell, Sussex, U.K.: Centaur, 1971).

154. Ibid., 16.

155. Gregorie and Surles, "Minute Book," 46.

156. James Haynes to Charles Manigault, April 22, 1847, in Clifton, ed., *Life and Labor,* 55.

157. Frederick Law Olmsted, *The Cotton Kingdom: A Traveller's Observations on Cotton and Slavery in the American Slave States*, edited, with an introduction, by Arthur M. Schlesinger (New York: Knopf, 1953), 186.

158. Charles Heyward, Heyward Family Papers (vol. 1855–62), South Caroliniana Library.

159. Doar, *Rice and Rice Planting,* 30–31.

160. Anne Simons Deas, *Reflections of the Ball Family of South Carolina and the Comingtee Plantation* (Charleston: South Carolina Historical Society, 1909), 160–61.

161. Allston, "Rice," 347.

162. Gregorie and Surles, "Minute Book," 113.

163. John H. Tucker, "Newly Invented Rice-Pounding Mill: The Kam in the Place of the Lifter," *Southern Agriculturist* n.s. 3 (August 1843): 305.

164. *The Charleston Directory for 1859* (Charleston, S.C., 1859), 1875.

165. P. McKinlay to Estate of H. Middleton, May 1857, Middleton Papers, box 5, folder 14, Middleton Place Plantation.

166. Illustration to "Rice Culture in the South," *Harper's Weekly* 28 (July 5, 1884): 436.

167. "A New Invention—McKinlay's Portable Rice Mill," *Rural Carolinian* 10, no. 4 (1873): 539.

168. A Mill to Free Rice from Its Hulls," *Scientific American* (July 9, 1887): 18.

169. *The Genuine "Engelberg" Rice and Coffee Hulling and Polishing Machines* [advertising pamphlet] (Syracuse, N.Y.: Engelberg Huller Co., n.d.), 8.

170. This description of the operation of the Engelberg huller is taken from D. H. Grist, *Rice*, 6th ed. (London: Longman, 1985), 430.

171. The only identifying mark on the huller is "Engelberg Huller Company, of Syracuse, New York." Because no patent number, inventor, or patent date was found on it, we could not locate a copy of the patent.

172. Grist, *Rice*, 430–31.

CHAPTER 7. City Mills and Large Plantation Toll Mills

1. Lucas, *Memorandum*, 6.

2. This account of Waverly Mills is taken from Alberta Morel Lachicotte, *Georgetown Rice Plantations* (Columbia: State Printing, 1955), 34–35.

3. Alberta Lachicotte Quattlebaum, "Supplementary Information," undated document in MS 0182.00, South Carolina Historical Society.

4. Interview by Richard Dwight Porcher, Jr., with Alberta Lachicotte Quattlebaum, July 1, 2010, Waverly Plantation, Georgetown County, South Carolina.

5. *Savannah Georgian,* March 27, 1829.

6. Will [W. W. Gordon] to Nellie, January 28, 1855, Gordon Papers, Southern Historical Collection.

7. The accession label in the Charleston Museum states that the pestles were shod in Russian steel.

8. Neither the Delaware Historic Society, the Delaware Public Archives (Dover), the Hagley Museum Library, nor the University of Delaware special collections has records of Hadley's company, Daniel I. Hadley & Associates, Inc..

9. Daniel L. Hadley to Mrs. C. B. Provost, chair of the Georgetown Historical Commission, June 18, 1969, archives of the Rice Museum, Georgetown, South Carolina.

10. Drayton, *A View of South-Carolina,* 123–24.

11. Georgetown Rice Milling Company (1880), pamphlet, Georgetown County Library, Georgetown, South Carolina.

CHAPTER 8. The Golden Age of Rice, 1800–1860

1. Dale E. Swan, "The Structure and Profitability of the Antebellum Rice Industry: 1859," (doctoral dissertation, University of North Carolina, Chapel Hill, 1972). Swan included Georgia and South Carolina in his study, but not North Carolina.

2. Dusinberre, *Them Dark Days*, 388.

3. Dusinberre claims that the figures in table 4 greatly exaggerate the decline of rice production in Colleton and Beaufort Counties because 1849 was a year of extraordinary high production, so the apparent decline is therefore largely factitious, and because the table makes clear that rice production was increasing in many parts of the Lowcountry.

4. Richard Dwight Porcher, Jr., and Sarah Fick, *The Story of Sea Island Cotton* (Charleston: Wyrick & Company, 2005), 318.

5. House, *Planter Management and Capitalism,* 23.

6. Smith, *Slavery and Rice Culture,* 34.

7. Clifton, ed., *Life and Labor,* xxxvii.

8. Production totals are from Smith, *Slavery and Rice Culture,* 30–44.

9. Lawrence S. Rowland, "'Alone on the River': The Rise and Fall of the Savannah River Rice Plantations of St. Peter's Parish, South Carolina," *South Carolina Historical Magazine* 88 (July 1987): 131, 139.

10. Alston, *Rice Planter and Sportsman.*

11. Dusinberre, *Them Dark Days,* 389.

12. For the values of the mill and threshing barn, see Clifton, ed., *Life and Labor,* xx and xxiv.

13. Lander, "Charleston," 330–51.

CHAPTER 9. The Last Days of Rice Planting

1. Charles Heyward, Heyward Family Papers (1855–1862), South Caroliniana Library.

2. Memo dated June 12, Manigault Plantation Records, Southern Historical Collection.

3. Betty L. Mitchell, *Edmund Ruffin: A Biography* (Bloomington: Indiana University Press, 1981), 211.

4. Erskine Clarke, *Dwelling Place: A Plantation Epic* (New Haven: Yale University Press, 2005), 443.

5. Barney (Edward Barnwell Heyward) to Tat (Catherine Maria Clinch), Combahee, November 5, 1867, quoted in Margaret Belser Hollis and Allen H. Stokes, eds., *Twilight on the South Carolina Rice Fields: Letters of the Heyward Family, 1862–1871* (Columbia: University of South Carolina Press, 2010), 293.

6. Myrta L. Avary, *Dixie after the War* (New York: Doubleday, Page, 1906), 161.

7. Clifton, ed., *Life and Labor,* 357–59.

8. Elizabeth Waties Allston Pringle, *Chronicles of Chicora Wood* (New York: Scribners, 1922), 284.

9. Edward Barnwell Heyward, Combahee, to Allen in Charleston, July 16, 1866, Heyward Family Papers, South Caroliniana Library.

10. *Progressive Farmer* 1 (April 14, 1886): 4.

11. Frances Butler Leigh, *Ten Years on a Georgia Plantation since the War* (London: Bentley, 1883), 22, 236–37.

12. John Girardeau Legaré, *The Darien Journal of John Girardeau Legaré, Ricegrower,* edited by Buddy Sullivan (Darien: Darien Printing & Graphics, 1997), 55.

13. United States Department of Agriculture, *Annual Report of the United States Commissioner of Agriculture* (Washington, D.C.: U.S. Government Printing Office, 1871), 43.

14. Heyward, *Seed from Madagascar,* 211.

15. Leigh, *Ten Years on a Georgia Plantation,* 269.

16. Rosser H. Taylor, "Fertilizers and Farming in the Southeast, 1840–1950: Part 1: 1840–1900," *North Carolina Historical Review* 30 (July 1954): 311–12.

17. Leigh, *Ten Years on a Georgia Plantation,* 124.

18. Contracts with these provisions are found in the Sparkman Family Papers and the Manigault Plantation Records, Southern Historical Collection.

19. Porcher and Fick, *Sea Island Cotton,* 138.

20. Malcolm Bell, Jr., *Major Butler's Legacy: Five Generation of a Slaveholding Family* (Athens: University of Georgia Press, 1987), 396, 400.

21. Clifton, ed., *Life and Labor,* 364–65.

22. George C. Rogers, Jr., *The History of Georgetown County, South Carolina* (Columbia: University of South Carolina Press, 1970), 487.

23. Pringle, *A Woman Rice Planter,* 446.

24. Peter A. Coclanis, "Bitter Harvest: The South Carolina Low Country in Historical Perspective," *Journal of Economic History* 45 (June 1985): 257.

25. James M. Clifton, "Twilight Comes to the Rice Kingdom: Postbellum Rice Culture on the South Atlantic Coast," *Georgia Historical Quarterly* 62 (Summer 1978): 149.

26. Arthur H. Cole, "The American Rice-Growing Industry: A Study of Comparative Advantage," *Quarterly Journal of Economics* 41 (August 1927): 607–8.

27. Our primary source for the history of the Gulf Coast rice industry is Edward Hake Phillips, "The Gulf Coast Rice Industry," *Agricultural History* 25 (April 1951): 91–96.

28. United States Bureau of the Census, *The Twelfth Census of the United States, 1900,* 10 vols. (Washington: U.S. Government Printing Office, 1901–02), 6: *Agriculture,* pt. 2: *Crops and Irrigation,* 56.

29. Roswell Trimble, *News and Courier,* September 29, 1940.

30. Robert B. Outland, *Tapping the Pines* (Baton Rouge: Louisiana State University Press, 2002), 5–6; James H. Tuten, *Lowcountry Time and Tide: The Fall of the South Carolina Rice Industry* (Columbia: University of South Carolina Press, 2010), 56.

31. Tuten, *Time and Tide,* 54.

32. Federal Writers' Project, Savannah, *Savannah River Plantations* (Savannah: Georgia Historical Society, 1947), 89, 91.

33. "Anne Simons Deas' Diary," Anne Simons Deas Papers, South Caroliniana Library.

34. *Wilmington Morning Star,* October 14 and 15, 1893.

35. Heyward, *Seed from Madagascar,* 245–46.

36. Charles Spalding Wylly, "These Memories" (1916) pamphlet, Local History Room, Brunswick Public Library, Brunswick, Georgia.

37. Tuten, *Time and Tide,* 68–72.

Bibliography

Archival Sources

Brunswick, Georgia. Brunswick Public Library: Charles Spalding Wylly Papers.

Chapel Hill, North Carolina. Southern Historical Collection, University of North Carolina: Gordon Papers, James Hamilton Couper Plantation Records (1818–54), John Berkeley Grimball Diary, Josiah Smith Letterbook, Manigault Plantation Records, Nathaniel Heyward Manuscript (1802), Sparkman Family Papers.

Charleston, South Carolina. Charleston Library Society: Hinson Collection.

Charleston, South Carolina. Drayton Hall, A Historic Site of the National Trust for Historic Preservation: Charles Drayton's Diaries, 1784–1820.

Charleston, South Carolina. Middleton Place Manuscript Collection, 1783–1899: Middleton Papers.

Charleston, South Carolina. South Carolina Historical Society: Ball Family Papers, Charles Manigault Letterbook, Henry Augustus Middleton Papers, John Sparkman Plantation Book, Lucas Family Papers, Manigault Family Papers, Mathurin Guerin Gibbs Plantation Register, Pinckney Lowndes Papers in the Ravenel Collection, Minute Book, Agricultural Society of South Carolina, Peter Manigault Letterbook.

Charleston, South Carolina. Special Collections, Robert Scott Small Library, College of Charleston: Henry Laurens account book and ledger.

Columbia, South Carolina. South Caroliniana Library, University of South Carolina: Anne Simons Deas Papers, Allston Family Papers, Ball Family Papers, Heyward Family Papers (1855–62), Jonathan Lucas Papers, Papers of Henry Laurens, Peter Manigault Letterbook, Williams Middleton Papers.

Columbia, South Carolina. South Carolina Department of Archives and History.

Crowley, Louisiana. Arcadia Parish Library: Freeland Archives.

Durham, North Carolina. Duke University Library: Louis Manigault Manuscripts.

Georgetown, South Carolina. Georgetown County Library.

Georgetown, South Carolina. The Rice Museum.

Lafayette, Louisiana. Southwestern Louisiana University: Records of the Rice Miller's Association.

Savannah, Georgia. Georgia Historical Society Library and Archives: Daniel Blake Financial Records, 1830–32, John McIntosh, Jr., Papers.

Published Primary Sources

Allston, Robert F. W. "Best Mode of Curing and Milling Rice." In *Reports Submitted to the Winyah and All-Saints Agricultural Society on the 20th April, 1848*. Charleston: Printed by Burges and James, 1848.

Alston, J. Motte. *Rice Planter and Sportsman: The Recollections of J. Motte Alston, 1821–1909.* Edited by Arney R. Childs. Columbia: University of South Carolina Press, 1953.

American, An. *American Husbandry.* 2 volumes. London: Printed for J. Bew, 1775.

Barker, S. W. "Implements and Machinery Adapted to Rice Culture." In *Address Delivered before the Agricultural Society of South Carolina at Their Seventy-sixth Annual Meeting, January 12, 1871.* Charleston: Walker, Evans & Cogswell, 1872.

Barnwell, Joseph W., ed. "Diary of Timothy Ford, 1785–1786." *South Carolina Historical and Genealogical Magazine* 13 (October 1912): 181–204.

Bartram, William. *Travels through North & South Carolina, Georgia, East & West Florida, the Cherokee Country, the Extensive Territories of the Muscogulges, or Creek Confederacy, and the Country of the Choctaws.* 1791. Reprint, New York: Penguin, 1988.

Black-River Planter, A. "Observations on the Bearded Rice." *Southern Agriculturist, and Register of Rural Affairs* 3 (November 1830): 578–79.

Blith, Walter. *The English Improver Improved; or, The Survey of Husbandry Surveyed.* London: Printed for John Wright, 1653.

Bryan, J. "On the Culture of Rice." *Southern Agriculturist, and Register of Rural Affairs* 5 (October 1832): 528–33.

Candler, Allen D., ed. *Colonial Records of the State of Georgia.* 25 vols. Atlanta, 1904–1916.

Carew, Richard, of Antony. *The Survey of Cornwall.* 1602. Reprint, edited by F. E. Halliday. London: Melrose, 1953.

Castiglioni, Luigi. *Viaggio: Travels in the United States of North America 1785–87.* Translated and edited by Antonio Pace. Syracuse: Syracuse University Press, 1983.

Catesby, Mark. *The Natural History of Carolina, Florida, and the Bahama Islands,* 2nd British edition. 2 vols. London: Printed for C. Marsh, T. Wilcox, and B. Stichall, 1754. First published 1731–43.

The Charleston Directory for 1859. Charleston, S.C., 1859.

Clifton, James M. "A Half-Century of a Georgia Rice Plantation." *North Carolina Historical Review* 47 (October 1970): 388–415.

———. "The Ante-Bellum Rice Planter as Revealed in the Letterbook of Charles Manigault, 1846–1848." Part 1. *South Carolina Historical Magazine* 74 (July 1973): 119–27. Part 2. 74 (October 1973): 300–310.

———. "Golden Grains of White Rice: Rice Planting on the Lower Cape Fear." *North Carolina Historical Review* 4 (October 1973): 365–93.

———. "Jehossee Island: The Antebellum South's Largest Rice Plantation." *Agricultural History* 58 (January 1985): 56–65.

———, ed. *Life and Labor on Argyle Island: Letters and Documents of a Savannah River Rice Plantation, 1833–1867.* Savannah: Beehive Press, 1978.

Committee, The. "Report on Cradling Rice." *Southern Agriculturist, Horticulturist, and Register of Rural Affairs* 5 (October 1845): 388–89.

Cooper, Thomas, and David J. McCord, eds. *Statutes at Large of South Carolina.* 10 vols. [covering 1682–1838]. Columbia: A. S. Johnston, 1836–41.

Couper, J. Hamilton. "Account of, and Directions for Erecting a Sugar Establishment." *Southern Agriculturist, and Register of Rural Affairs* 4 (May 1831): 225–32; (June 1831): 281–88.

Coulter, E. Merton, ed. *The Journal of William Stephens, 1743–1745.* Athens: University of Georgia Press, 1959.

Courtenay, William A. "Early Crops and Commerce." In *Year Book of the City of Charleston, 1883*. Charleston, n.d.

Darby, James C. "On Planting and Managing a Rice Crop." *Southern Agriculturist, and Register of Rural Affairs* 2 (June 1829): 247–54.

Deas, Anne Simons. *Reflections of the Ball Family of South Carolina and the Comingtee Plantation*. Charleston: South Carolina Historical Society, 1909.

De Brahm, William Gerard. *De Brahm's Report on the Genral Survey of the Southern District of North America*. Edited by Louis De Vorsey, Jr. Columbia: University of South Carolina Press, 1971.

Doar, David. *Rice and Rice Planting in the South Carolina Low Country*. Contributions from the Charleston Museum, no. 8. Charleston: Charleston Museum, 1936.

"Dotterer's Patent Rice-Sowing Machine." *Rural Carolinian* 1 (1870): 375.

Drayton, John. *A View of South-Carolina, as Respects Her Natural and Civil Concerns*. Charleston: W. P. Young, 1802.

Easterby, J. H., ed. *The South Carolina Rice Plantation as Revealed in the Papers of Robert F. W. Allston*. Chicago: University of Chicago Press, 1945.

Editor. "Management of a Southern Plantation." *DeBow's Review* 22 (January 1857): 38–44.

Editor, The. "Observations on Strong & Moody's Patent Rice Mill." *Southern Cabinet* 1 (December 1840): 736–40.

Emmons, Calvin. "Dear Sir." *American Farmer* 1, no. 37 (February 5, 1840): 293.

Forster, A. M. "Rice Culture in South Carolina." *Proceedings of the Elliott Society* 2 (December 1860): 40–46.

Friend to Improvement, A. "On Rice Threshing Machines." *Southern Agriculturist, and Register of Rural Affairs* 2 (September 1829): 404–5.

Gallardo, José Miguel. "The Spaniards and the English Settlement in Charles Town." *South Carolina Historical and Genealogical Magazine* 37 (April, July 1936): 91–99.

Glen, James. *A Description of South Carolina*. London: Printed for R. and J. Dodsley, 1761.

Gregorie, Anne K., and Flora B. Surles. "Minute Book, South Carolina Agricultural Society: 1825–1860." W.P.A. Project 65–33–118, 1936.

Hewatt, Alexander. *An Historical Account of the Rise and Progress of the Colonies of South Carolina and Georgia*. 2 vols. London: Printed for Alexander Donaldson, 1779.

Heyward, Duncan Clinch. *Seed from Madagascar*. Chapel Hill: University of North Carolina Press, 1937.

Hollis, Margaret Belser, and Allen H. Stokes, eds. *Twilight on the South Carolina Rice Fields: Letters of the Heyward Family, 1862–1871*. Columbia: University of South Carolina Press, 2010.

House, Albert Virgil, ed. *Planter Management and Capitalism in Ante-Bellum Georgia: The Journal of Hugh Fraser Grant, Ricegrower*. New York: Columbia University Press, 1954.

Irving, John B. *A Day on Cooper River*. 2nd ed. Edited by Louisa Cheves Stoney. Columbia: R. L. Bryan, 1932.

J.G. "On Rice Mills." *Southern Agriculturist, and Register of Rural Affairs* 2 (November 1829): 504–8.

Jefferson, Thomas. *Jefferson's Memorandum Books: Accounts, with Legal Records and Miscellany, 1767–1826*. Edited by James A. Bear, Jr., and Lucia C. Stanton. Vol. 2. Princeton, N.J.: Princeton University Press, 1997.

———. *The Papers of Thomas Jefferson.* Vol. 26: May 11–August 31, 1793. Edited by John Catanzariti, Eugene R. Sheridan, J. Jefferson Looney, Elizabeth Peters Blazejewski, and Jean-Yves M. Le Saux. Princeton: Princeton University Press, 1995.

Jones, George Fenwick, and Renate Wilson, trans. and eds. *Detailed Reports on the Salzburger Emigrants Who Settled in America.* Vol. 18: 1744–1745. Camden, Maine: Picton Press, 1995.

"Journal of an Officer's [Lord Adam Gordon's] Travels in America and the West Indies, 1764–1765." In *Travels in the American Colonies.* Edited by Newton D. Mereness. 367–456. New York: Macmillan, 1916.

Kemble, Frances Anne. *Journal of a Residence on a Georgian Plantation in 1838–1839.* Edited by John A. Scott. New York: Knopf, 1961.

King, Roswell, Jr. "Queries on the Culture of Rice." *Southern Agriculturist, and Register of Rural Affairs* 1 (September 1828): 409–13.

Laurens, Henry. *Papers of Henry Laurens.* Vol. 8: October 10, 1771–April 19, 1773. Edited by George C. Rogers, Jr., and David R. Chesnutt. Columbia: Published for the South Carolina Historical Society by the University of South Carolina Press, 1980.

———. *Papers of Henry Laurens.* Vol. 9: April 19, 1773–December 12, 1774. Edited by George C. Rogers, Jr., and David R. Chesnutt. Columbia: Published for the South Carolina Historical Society by the University of South Carolina Press, 1981.

———. *Papers of Henry Laurens.* Vol. 11: January 5, 1776–November 1, 1777. Edited by David R. Chesnutt and C. James Taylor. Columbia: Published for the South Carolina Historical Society by the University of South Carolina Press, 1988.

———. *Papers of Henry Laurens.* Vol. 16: September 1, 1782–December 17, 1792. Edited by David R. Chesnutt and C. James Taylor. Columbia: Published for the South Carolina Historical Society by the University of South Carolina Press, 2003.

Lawson, John. *A New Voyage to Carolina.* 1709. New edition, edited by Hugh Talmadge Lefler. Chapel Hill: University of North Carolina Press, 1967.

Lawton, William M. *An Essay on Rice and Its Culture, Read before the Agricultural Congress, Convened at Alabama, December 5, 1871.* Charleston, S.C.: Printed by Walker, Evans & Cogswell, 1871.

Legaré, John D. "Account of an Agricultural Excursion Made into the South of Georgia in the Winter of 1832." *Southern Agriculturist, and Register of Rural Affairs* 6 (March 1833): 138–47.

Legaré, John Girardeau. *The Darien Journal of John Girardeau Legaré, Ricegrower, 1877–1932.* Edited by Buddy Sullivan. Darien: Darien Printing & Graphics, 1997.

Leigh, Frances Butler. *Ten Years on a Georgia Plantation since the War.* London: Bentley, 1883.

Lofton, Alexander Lucas. *The Lucases of Haddrells Point.* Book 1: 1785–1835. Mount Pleasant, S.C: Privately published by the author, 1998.

Marsh-Planter, A. "Account of the Culture and Product of a Field of Bearded Rice." *Southern Agriculturist, and Register of Rural Affairs* 3 (June 1830): 292–93.

Mayrant, William. "On the Cultivation of Bearded Rice." *Southern Agriculturist, and Register of Rural Affairs* 2 (February 1829): 74–79.

McCord, David J. *Statutes at Large of South Carolina,* Vol. 7. Columbia: A. S. Johnston, 1840.

McCrady, Edward, Jr. *The History of South Carolina under the Royal Government, 1719–1776.* New York: Macmillan, 1899.

Meredith, Hugh. *An Account of the Cape Fear Country, 1731.* Edited by Earl Gregg Swem. Perth Amboy, N.J.: Charles F. Heartman, 1922.

"A Mill to Free Rice from Its Hulls." *Scientific American* (July 9, 1887): 18.

Molony, Thomas J., Jr. "Rice Cutting Machines." *Rural Carolinian* 2 (1871): 183–84.

Moore, Alexander. "Daniel Axtell's Account Book and the Economy of Early South Carolina." *South Carolina Historical Magazine* 95 (October 1994): 280–301.

Mortimer, John. *The Whole Art of Husbandry; Or, The Way of Managing and Improving of Land.* 2nd ed. London, 1708.

Munnerlyn, Charles. "Queries on the Culture of Rice; by William Washington, with Answers, by Charles Munnerlyn." *Southern Agriculturist, and Register of Rural Affairs* 1 (May 1828): 215–23.

Nairne, Thomas. "A Letter from South Carolina." 1710. Republished in *Selling a New World: Two Colonial South Carolina Promotional Pamphlets.* Edited by Jack P. Greene. Columbia: University of South Carolina Press, 1989.

"A New Invention—McKinlay's Portable Rice Mill." *Rural Carolinian* 10, no. 4 (1873): 539–41.

Norris, John. "Profitable Advice for Rich and Poor." 1712. Republished in *Selling a New World: Two Colonial South Carolina Promotional Pamphlets.* Edited by Jack P. Greene. Columbia: University of South Carolina Press, 1989.

Observer, An. "Protecting Crops from Birds." *Southern Agriculturist, and Register of Rural Affairs* 6 (August 1833): 424–27.

Olmsted, Frederick Law. *The Cotton Kingdom: A Traveller's Observations on Cotton and Slavery in the American Slave States.* Edited, with an introduction, by Arthur M. Schlesinger. New York: Knopf, 1953.

Phillips, Ulrich B., ed. *Plantation and Frontier Documents: 1645–1863.* Cleveland: A. H. Clark, 1909.

———. *Life and Labor in the Old South.* Boston: Little, Brown, 1929.

———, ed. "Some Letters of Joseph Habersham." *Georgia Historical Quarterly* 10 (June 1926): 144–63.

Pinckney, C. C. "On the Advantage of Using Animal Power and Machinery in the Culture of Rice." *Southern Agriculturist, and Register of Rural Affairs* 10 (April 1837): 171–74.

Pinckney, Thomas. "Thomas Pinckney's Letter on the Water Culture of Rice." In *Report of the Committee Appointed by the South Carolina Agricultural Society,* 16–24. Charleston: Printed by Archibald E. Miller, 1823.

———. "Letter from T. Pinckney to Jas. Gregorie, Chairman, &c." *Southern Agriculturist, and Register of Rural Affairs* 2 (April 1829): 148–49.

Porcher, Frederick A. "Historical and Social Sketch of Craven County, South Carolina." In *A Contribution to the History of the Huguenots of South Carolina.* N.p.: Republished for private circulation by T. Gaillard Thomas, 1887.

Pringle, Elizabeth Waties Allston (pseud. Patience Pennington). *A Woman Rice Planter.* New York: Macmillan, 1913.

———. *Chronicles of Chicora Wood.* New York: Scribners, 1922.

Ravenel, Edmund. "The Limestone Springs of St. John's Berkeley, and Their Probable Availability for Increasing the Quantity of Fresh Water in Cooper River." *Proceedings of the Elliott Society of Natural History* (September 1860): 28–32.

Ravenel, Theodore D. "The Last Days of Rice Planting." In David Doar, *Rice and Rice Planting in the South Carolina Low Country.* Contributions from the Charleston Museum, no. 8. Charleston: Charleston Museum, 1936.

Richards, T. Addison. "The Rice Lands of the South." *Harper's New Monthly Magazine* 19 (November 1859): 721–38.

Robinson, Solon. "Governor Aiken's Extensive Rice Estate." *DeBow's Review* 1 (August 1850): 201–3.

Rowland, Charles E. "Queries on the Culture of Rice." *Southern Agriculturist, and Register of Rural Affairs* 1 (August 1828): 352–57.

Rowland, Lawrence S. "Alone on the River: The Rise and Fall of the Savannah River Rice Plantations of St. Peter's Parish, South Carolina." *South Carolina Historical Magazine* 88 (July 1987): 121–50.

Salmon, William. *Botanologia: The English Herbal.* London: H. Rhodes and J. Taylor, 1710.

Saunders, William L., ed. *The Colonial Records of North Carolina.* 10 vols. Raleigh: Printed by P. M. Hale, 1886–90.

Shields, David S. "George Ogilvie's *Carolina or the Planter 1776*: an Introduction & Edition." *Southern Literary Journal,* special issue (1986).

South Carolina Commons House of Assembly. *Journal of the Commons House of Assembly, November 12, 1754–September 23, 1755.* Edited by Terry W. Lipscomb. Columbia: Published for the South Carolina Department of Archives and History by the University of South Carolina Press, 1986.

Spalding, Thomas. "On the Culture, Harvesting and Threshing of Rice, and on the Rust in Cotton." *Southern Agriculturist, and Register of Rural Affairs* 8 (April 1835): 167–74.

Smyth, J. F. D. *A Tour of the United States of America.* 2 vols. Dublin: T. Henshall, 1784.

Stewart, John. "Letters from John Stewart to William Dunlop." *South Carolina Historical Magazine* 32 (January 1931): 1–33.

———. "Letters from John Stewart to William Dunlop." *South Carolina Historical Magazine* 32 (April 1931): 81–114.

Tucker, John H. "Newly Invented Rice-Pounding Mill: The Kam in the Place of the Lifter." *Southern Agriculturist, and Register of Rural Affairs* n.s. 3 (August 1843): 305–6.

United States Bureau of the Census. The *Twelfth Census of the United States, 1900.* 10 vols. Washington, D.C.: U.S. Government Printing Office, 1901–2. 6: *Agriculture,* pt. 2: *Crops and Irrigation.*

United States Department of Agriculture. *Annual Report of the United States Commissioner of Agriculture.* Washington, D.C.: U.S. Government Printing Office, 1871.

W. "On Rice Threshers." *Southern Agriculturist, and Register of Rural Affairs* 7 (November 1834): 579–82.

Ward, Joshua John. "Letter from Col. Ward, on the Big Grain Rice." *Proceedings of the Agricultural Convention and of the State Agricultural Society of South Carolina,* 56–57. Columbia: Summer & Carroll, 1846.

Washington, William. "On the Advantage of Planting the Bearded Rice." *Southern Agriculturist, and Register of Rural Affairs* 2 (April 1929): 19–20.

Watson, Elkanah. *Men and Times of the Revolution.* Edited by Winslow C. Watson. New York, 1856.

Webber, Mabel L., ed. "Col. Senf's Account of the Santee Canal." *South Carolina Historical and Genealogical Magazine* 28 (January 1927): 8–21.

Wilkinson, R. A. "Production of Rice in Louisiana." *DeBow's Review* 6 (July 1848): 53–59.

Newspapers and Magazines

American Farmer
Charleston City Gazette
Charleston Morning Post, and Daily Advertiser

Charleston News and Courier
Crowley Daily Signal
DeBow's Review
Frank Leslie's Illustrated Newspaper
Frank Leslie's Popular Monthly
Gentleman's Magazine
Georgia Gazette
Georgian
Harper's New Monthly Magazine
Harper's Weekly
Progressive Farmer
Savannah Georgian
Scientific American
South Carolina Gazette
Southern Agriculturist
State Gazette of South Carolina
Wilmington Morning Star

Maps and Plats

Drayton, Charles. Map of Drayton Hall Rice Fields, circa 1790. Drayton Hall Historic National Trust.

Gaillard, J. P. Map of Richmond, The Farm, Hampstead, Johns Run, Tower Hill, and part of Bluford Plantations, owned by the Est. of Robert Marion, Situate in St. Stephen Parish, Berkeley County, S.C., 1928, 1929.

General Plan of the Canal and its Environs between Santee and Cooper Rivers in the State of South Carolina. Commenced in the Year 1793 and finished in the Year 1800, by Christian Senf Colonel Engineer and Director in Chief of the Canal. Original located in the South Carolina Historical Society.

Hardwick, J., "Plan of Calais, a Plantation belonging to the Rev'd Mr Hugh Fraser. Situated in All Saints Parish Georgetown, District State of So. Carolina, Having Such Courses, Distances, Bounds, and Marks as Are Represented and Expressed in the PLAN from an Actual Survey taken in July 1796."

Hardwick, John. "Plan of Limerick, a Plantation Belonging to Elias Ball, Esqr., Situated on the Head Branch of the Eastern Branch of Cooper River, in St. John's, St James, and St. Stephens Parishes, Charleston District, and State of South Carolina. Map dated 1797.

Palmer, J. O. "A Plan Exhibiting the Shape, Marks, Buttings, and Boundaries of the Woodboo Plantation, Situate in St. Johns Parish- Berkeley County and Charleston District. Belonging to Stephen Mazyck." Certified March 22, 1806.

Secondary Sources

Allston, Robert F. W. "Rice." *Commercial Review of the South and West* 1 (1846): 320–57.

Austin, Amory. *Rice: Its Cultivation, Production, and Distribution in the United States and Foreign Countries*. U. S. Department of Agriculture Miscellaneous Series, report no. 6. Washington, D.C.: U.S. Government Printing Office, 1893.

Avary, Myrta L. *Dixie after the War.* New York: Doubleday, Page, 1906.

Bagwell, James. *Rice Gold: James Hamilton Couper and Plantation Life on the Georgia Coast.* Macon: Mercer University Press, 2000.

Bathe, Greville, and Dorothy Bathe. *Oliver Evans: A Chronicle of Early American Engineering.* New York: Arno Press, 1972.

Bell, Malcolm, Jr. *Major Butler's Legacy: Five Generation of a Slaveholding Family.* Athens: University of Georgia Press, 1987.

Bullard, Mary R. *Cumberland Island: A History.* Athens: University of Georgia Press, 2003.

Carney, Judith A. "Landscapes of Technology Transfer: Rice Cultivation and African Continuities." *Technology and Culture* 37 (1996): 5–35.

———. *Black Rice.* Cambridge, Mass.: Harvard University Press, 2001.

Catling, David H. *Rice in Deep Water.* International Rice Research Institute. London: Macmillan, 1992.

Chang, Te-Tzu. "The Origin, Evolution, Cultivation, Dissemination, and Diversification of Asian and African Rices." *Euphytica* 25 (1976): 425–41.

———. "Rice." In *Evolution of Crop Plants.* Edited by N. W. Simmonds. 31–37. London: Longman, 1976.

Clarke, Erskine. *Dwelling Place: A Plantation Epic.* New Haven: Yale University Press, 2005.

Clifton, James M. "A Half-Century of a Georgia Rice Plantation." *North Carolina Historical Review* 47 (October 1970): 388–415.

———. "Golden Grains of White: Rice Planting on the Lower Cape Fear." *North Carolina Historical Review* 50 (October 1973): 365–93.

———. "Twilight Comes to the Rice Kingdom: Postbellum Rice Culture on the South Atlantic Coast." *Georgia Historical Quarterly* 62 (Summer 1978): 146–54.

———. "Jehossee Island: The Antebellum South's Largest Rice Plantation." *Agricultural History* 58 (January 1985): 56–65.

Cobley, Leslie S. *The Botany of Tropical Crops.* London: Longman, 1977.

Coclanis, Peter A. "Bitter Harvest: The South Carolina Low Country in Historical Perspective." *Journal of Economic History* 45 (June 1985): 251–59.

Cole, Arthur H. "The American Rice-Growing Industry: A Study of Comparative Advantage." *Quarterly Journal of Economics* 41 (August 1927): 595–643.

Croft, Terrell, and E. J. Tangerman. *Steam Engine Principles and Practice,* 2nd edition. New York: McGraw-Hill, 1939.

Davis, Robert Scott, Jr. "As Good as the French: The Rise and Decline of the Georgia Buhr Stone Industry." *Georgia Historical Quarterly* 77 (Fall 1993): 560–668.

Day, Geoffrey. *Tide Mills in England and Wales.* Woodbridge, Suffolk, U.K.: Published by the Friends of Woodbridge Tide Mill, 1994.

De Brahm, John G. W. *History of the Province of Georgia.* Wormsloe, Ga., 1849.

De Brahm, William Gerard. *De Brahm's Report on the General Survey of the Southern District of America.* Edited by Louis De Vorsey, Jr. Columbia: University of South Carolina Press, 1971.

Dedrick, B. W. *Practical Milling.* Chicago: National Miller, 1924. Reprint, Newton, N.C.: Society for the Preservation of Old Mills, 1989.

Dethloff, Henry C. *A History of the American Rice Industry.* College Station: Texas A&M Press, 1988.

Doar, David. *Rice and Rice Planting in the South Carolina Low Country.* Contributions from the Charleston Museum, no. 8. Charleston: Charleston Museum, 1936.

Dusinberre, William. *Them Dark Days: Slavery in the American Rice Swamps.* Athens: University of Georgia Press, 2000.

Edelson, S. Max. *Plantation Enterprise in Colonial South Carolina.* Cambridge, Mass.: Harvard University Press, 2006.

Ellis, Rev. William. *History of Madagascar.* London: Fisher, 1838.

Eltis, David, Philip Morgan, and David Richardson. "Agency and Diaspora in the Atlantic History: Reassessing the African Contribution to Rice Cultivation in the Americas." *American History Review* 112 (December 2007): 1329–58.

Farey, John. *A Treatise on the Steam Engine, Historical, Practical, and Descriptive.* London: Longman, Rees, Orme, Brown & Green, 1827.

Federal Writers' Project, Savannah, *Savannah River Plantations.* Savannah: Georgia Historical Society, 1947.

Ferguson, Eugene S. "The Origins of the Steam Engine." *Scientific American* 210 (January 1964): 98–107.

Gray, Lewis Cecil. *History of Agriculture in the Southern United States to 1860.* 2 vols. Gloucester, Mass.: Peter Smith, 1958.

Grist, D. H. *Rice.* 6th ed. London: Longman, 1986.

Groening, Richard I., Jr. "The Rice Landscape in South Carolina: Valuation, Technology, and Historical Periodization." Master's thesis, University of South Carolina, 1991.

Handler, Jerome S., and Michael L. Tuite, Jr. *The Atlantic Slave Trade and Slave Life in the Americas: A Visual Record.* http://hitchcock.itc.virginia.edu/slavery/ (accessed May 16, 2013).

Haskell, Jennie. "Rice: In the Fields and Mills of South Carolina." *Frank Leslie's Popular Monthly* 8 (February 1879): 161–69.

Heitzler, Michael J. *Goose Creek: A Definitive History.* Vol 1: *Planters, Politicians and Patriots.* Charleston: History Press, 2005.

Higman, B. W. *Slave Population and Economy in Jamaica, 1807–1834.* Cambridge: Cambridge University Press, 1976.

Hilliard, Sam B. "The Tidewater Rice Plantation: An Ingenious Adaptation to Nature." *Geoscience and Man* 12 (June 20, 1975): 57–66.

Hodge, P. R. *The Steam Engine.* New York: Appleton, 1840.

Howell, Charles, and Allan Keller. *The Mill at Philipsburg Manor, Upper Mills, and a Brief History of Milling.* New York: Sleepy Hollow Restorations, 1977.

Hunter, Louis C. *A History of Industrial Power in the United States, 1780–1930.* Vol. 2: *Steam Power.* Charlottesville: University Press of Virginia, 1985.

Hurt, R. Douglas. *American Farm Tools.* Manhattan, Kans.: Sunflower University Press, 1982.

Jobey, George. "Millstones and Millstone Quarries in Northumberland." *Archaeologia Aeliana* 5th ser., 14 (1986): 49–80.

Jones, Charles C. *The History of Georgia.* Boston: Houghton, Mifflin, 1883.

Kelley, Joe, and Dan Tufford. "Plant Succession in Cooper River, South Carolina, Tidal Former Rice Fields: Ecological and Human Use Implications." In *The Golden Seed: Writings on the History and Culture of Carolina Gold Rice.* Edited by David S. Shields. 111–28. Charleston: Douglas W. Bostick, 2010.

King, Edward. *The Great South.* Hartford: American Publishing, 1879.

Knapp, S. A. *Rice Culture.* U. S. Department of Agriculture, Farmers' Bulletin 417. Washington, D.C.: U.S. Government Printing Office, 1910.

Lachicotte, Alberta Morel. *Georgetown Rice Plantations.* Columbia: State Printing, 1955.

Lander, Ernest M., Jr. "Charleston: Manufacturing Center of the Old South." *Journal of Southern History* 26 (August 1960): 330–51.

Lees, William B. "The Historical Development of Limerick Plantation, a Tidewater Rice Plantation of Berkeley County, South Carolina, 1683–1945." *South Carolina Historical Magazine* 82 (January 1981): 44–62.

Leonard, Carroll M., and Vladimir L. Maleev. *Heat Power Fundamentals*. New York: Pitman, 1949.

Littlefield, Daniel. C. *Rice and Slaves: Ethnicity and the Slave Trade in Colonial South Carolina*. Baton Rouge: Louisiana State University Press, 1981.

Lucas, William Dollard. *A Lucas Memorandum*. Edited by David Henry Lucas. N.p.: Privately published, n.d.

Malone, Dumas. *Jefferson and the Rights of Man*. Vol. 2 of *Jefferson and His Time*. 6 Vols. Boston: Little, Brown, 1951.

McClung, Anna M., and Robert Fjellstrom. "Using Molecular Genetics As A Tool to Identify and Refine 'Carolina Gold.'" Paper presented at the Carolina Gold Rice Symposium, sponsored by the Carolina Gold Rice Foundation, Charleston, South Carolina, August 2005.

McLean, Beth Bailey, ed. *The Story of Rice: The World's Most Popular Food*. New Orleans: Rice Millers Association, 1934.

Mitchell, Betty L. *Edmund Ruffin: A Biography*. Bloomington: Indiana University Press, 1981.

Morgan, Philip D. "Work and Culture: The Task System and the World of Lowcountry Blacks, 1700 to 1880." *William and Mary Quarterly* 39 (October 1982): 563–99.

Officer, Lawrence H., and Samuel H. Williamson. "Purchasing Power of Money in the United States from 1774 to 2010." MeasuringWorth, 2009http://www.measuringworth.com/ppowerus/result.php (accessed May 16, 2013).

Outland, Robert B. *Tapping the Pines*. Baton Rouge: Louisiana State University Press, 2002.

Parrish, Lydia. *Slave Songs of the Georgia Sea Islands*. 1942. Reprint, as a Brown Thrasher Book. Athens: University of Georgia Press, 1992.

Phillips, Edward Hake. "The Gulf Coast Rice Industry." *Agricultural History* 25 (April 1951): 91–96.

Pickles, Walter. *Our Grimy Heritage: A Fully Illustrated Study of the Factory Chimney in Britain*. Fontwell, Sussex, U.K.: Centaur, 1971.

Porcher, Richard Dwight, Jr. "Rice Culture in South Carolina: A Brief History, the Role of the Huguenots, and Preservation of its Legacy." *Transactions of the Huguenot Society* 92 (1987): 11–22.

———. *A Teacher's Field Guide to the Natural History of the Bluff Plantation Wildlife Sanctuary*. New Orleans: Kathleen O'Brien Foundation, 1985.

Porcher, Richard Dwight, Jr., and Sarah Fick. *The Story of Sea Island Cotton*. Charleston: Wyrick, 2005.

Purseglove, J. W. *Tropical Crops: Monocotyledons*. New York: Wiley, 1972.

Pursell, Carroll W., Jr. *Early Stationary Steam Engines in America*. Washington, D.C.: Smithsonian Institution Press, 1969.

Ramsay, David. *History of South Carolina, from Its First Settlement in 1670 to the Year 1808*. 2 vols. 1809. Reprint, Newberry: W. J. Duffie, 1858.

Rivers, William J. *A Sketch of the History of South Carolina*. Charleston: McCarter, 1856.

Rogers, George C., Jr. *The History of Georgetown County, South Carolina*. Columbia: University of South Carolina Press, 1970.

Rowland, Lawrence S. "'Alone on the River': The Rise and Fall of the Savannah River Rice Plantations of St. Peter's Parish, South Carolina." *South Carolina Historical Magazine* 88 (July 1987): 121–50.

Salley, A. S., Jr. "The Introduction of Rice into South Carolina." *Bulletin of the Historical Commission of South Carolina*. No. 6. Columbia: State Company, 1919.

Second, G. "Evolutionary Relationships in the Sativa Group of *Oryza* Based on Isozyme Data." *Genetics, Selection, Evolution* 17 (1985): 89–114.

———. "Isozymes and Phylogenetic Relationships in *Oryza*." In *Rice Genetics: Proceedings of the International Rice Genetics Symposium, 27–31, May 1985*, 27–39. Manila: International Rice Research Institute, 1986.

Smith, Julia Floyd. *Slavery and Rice Culture in Low Country Georgia, 1750–1860*. Knoxville: University of Tennessee Press, 1985.

Swan, Dale E. "The Structure and Profitability of the Antebellum Rice Industry: 1859." Doctoral dissertation, University of North Carolina, Chapel Hill, 1972.

Taylor, Rosser H. "Fertilizers and Farming in the Southeast, 1840–1950: Part I: 1840–1900." *North Carolina Historical Review* 30 (July 1954): 311–12.

Tucker, D. Gordon. "Millstones North and South of the Scottish Border." *Industrial Archeology Review* 6 (Autumn 1982): 186–93.

———. "Millstone Making in Scotland." *Society of Antiquaries of Scotland* 114 (1984): 539–56.

———. "Millstone Making in the Peak District of Derbyshire: The Quarries and the Technology." *Industrial Archeology Review* 8 (Autumn 1985):42–58.

Tuten, James H. *Lowcountry Time and Tide: The Fall of the South Carolina Rice Industry*. Columbia: University of South Carolina Press, 2010.

Waring, Joseph Ioor. *A History of Medicine in South Carolina, 1825–1900*. Charleston: South Carolina Medical Association, 1967.

Watson, Andrew M. *Agricultural Innovation in the Early Islamic World*. Cambridge: Cambridge University Press, 1993.

Wood, Peter. *Black Majority: Negroes in Colonial South Carolina from 1670 through the Stono Rebellion*. New York: Norton, 1974.

Woosley, Theodore Dwight. *The First Century of the Republic: A Review of American Progress*. New York: Harper, 1876.

Index

Abbott, David, 307
ACE Basin, 75
African rice (*Oryza glaberrima*): spread through West Africa, 7–9; primary center of domestication, 9
African slaves: anarchy and rebellion, 63, 299; blacksmiths, 257–58; carpenters, 67, 85, 86, 137, 186, 257–58; contributions to rice culture, xxi, 5, 31, 62, 65, 157, 315, 317; desire for freedom, 63, 298, 318; destruction of property, 111, 299; disruption of plantation work, 60, 299; escaping and forming maroon colonies, 298; excellence of workmanship, 258; first to plant rice in the Lowcountry, 13, 31; fled to British camps, 63; freehold" property, 36; harvesting rice, 105, 108, 113; house slaves, 96; left no written records, 12, 30, 42, 320; mechanics, 150, 257–58; milling rice, 160, 165; old slave song, 96; provision ground system, 36, 37; refused to work, 298; served apprenticeships in the city, 258; "slave pharmacology," 97; trunk minders, 257; volunteered or forced into the American armies, 63; watermen, 257
African-type mortar and pestle, 1, 4, 153, 156–60, 162, 166, 186, 208, 300
Agricultural Society of South Carolina, 108, 124, 154, 225, 259
Aiken, Gov. William, Jr. (1806–1887), 39, 40, 142, 255
Albemarle, Duke of (1608–1670), Lords Proprietor, 14
Albemarle Point, 14, 35
aleurone layer of rice seed, 156–58, 218
Allston, Adèle Petigru (1810–1896), 302
Allston, Benjamin (1833–1900), 38, 39, 105
Allston, Charles Petigru (1848–1922), 138
Allston, Capt. John H., 78
Allston, Joseph Blyth (1833–1904), 279
Allston, Joseph Waties (d. 1834), 279
Allston, Robert Francis Withers (1801–1864), 19, 21, 22, 26, 27, 52, 65, 79, 105–6, 127–28, 138, 144, 167, 169, 180, 208–10, 218–19, 233, 259, 279
Allston, William Allen (1834–1878), 279
Alston, Col. William (1756–1839), 25
Alston, Jacob Motte (1821–1909), 37, 103, 189, 294
American, An (pseud., real name unknown), 52, 179
American Farmer, 23
American rice industry, 272
American rice production, 290, 310
animal-powered rice mills, 19, 170, 172–73, 175–77, 186–87, 200, 211
antebellum period or era, 2, 3, 13, 75, 290, 306
antebellum South, 93, 297
Argyle Island, Ga., 24, 40, 60, 61, 96, 117, 198, 292
Arkansas, 4, 106, 150, 309
Arkansas rice production, 309
artesian springs, 40, 54, 55, 314
artesian wells, 307
Asian rice (*Oryza sativa*): center of domestication, 10; *Indica* race, 10; *Japonica* race, 10; *Javanica* race, 10; spread into Madagascar, 11; spread into sub-Saharan West Africa, 10, 11
Atlantic Coast, 1, 3, 4, 28, 320

Atlantic Ocean, 58
Atlantic world, 3, 295
Audley, Erasmus, 118
Austin, Amory, 209, 214
Axtell, Daniel, 18, 47, 186–87
Axtell, Lady Rebecca, 18, 186

Bagwell, James, 115
Baillie, George, 64
Ball, Elias (1752–1810), 192, 316
Barbadian planters, 17, 35, 36, 331n.5
Barbados, 14, 15, 31, 34–36, 42, 58
Barker, Sanford William (1807–1891), 151
Bartram, William (1739–1823), 187
basil, 266–67
Bear Island Wildlife Management Area, 104
beaters in threshers, 111, 124–25, 127, 129, 132, 134–35, 137, 142, 149–50
Beaufort Area, 297
Beaufort County, S.C., 57, 91
Beaufort, S.C., 311, 320
Bennett & Company, C. S., 285
Bennett, Gov. Thomas (1781–1865), 283
Bennett, M. W. L. (d. 1874), 285
Berber traders, 9
beriberi, 157
bill of lading, 14, 17
bill of scanting:; of Daniel Blake, 129, 132; of Elias Ball, 193; of George Veitch, 181–82; of Henry Laurens, 166; of Peter Villeponteaux, 172
Black Majority (Wood), 13
Black Oak Church, St. John's Berkeley, 314
Black Rice (Carney), 13
black rice hypothesis, 13
Black River Planter, A (pseud., real name unknown), 25
Blake, Daniel, 59, 129, 132
Blith, Walter (d. 1652), 41
bobolinks, 98
boilers for steam engines: able to produce high pressure, 241; brick setting, 243, 245, 254; bridge wall, 245; cast-iron cap to seal manhole, 243; copper boilers, 232–33, 241–42; double boilers, 243; explosions, 246–47; feedwater drawn from wells, 247; feedwater force pump, function of, 247; foundation at Chicora Wood, 139; fuel, 132, 142, 149, 156, 241, 248; furnace, function of, 139, 229, 245, 254; glass-tube gauge, 245; made of black iron sheets, 242; made of cast iron, 242; made of wrought iron sheets, 242; mounted on incline for drainage, 245; plain cylindrical boilers, 241; single boiler, 243; try cocks, 245–46
Boltzius, Johann Martin (1703–1765), 186–87
Bonneau Ferry, 45
Boulton, Matthew (1728–1809), 226, 241
Boulton & Watt, 228, 250
Bowman, J., 202, 216, 233, 242, 343n.72
Bowman, T. B., 265, 271
Boyle, Mr., 175–76
British capture Charleston, 63
British capture Savannah, 62, 63
British Colonial America, 3
British Navigation Acts, 62
British North America, 93, 320
British tariff on American milled rice, 220
broken rice, 117, 123, 126, 153, 158, 160, 166–67, 177, 179–80, 219, 222, 224, 263, 269, 281, 288, 309
Brunswick, Ga., 298–99
brushing or polishing rice, 17, 154, 156, 158, 218–19, 221, 259, 263, 266–71, 274, 279–80, 286, 306, 309
Bryan, J., 118
Burmese rice, 306
Butler, General M. C., 300
Butler, Major Pierce (1744–1822), 104, 305
Butler, Pierce (1810–1867), 39, 95, 96, 128, 302, 305
Butler's Island, Ga., 39, 95, 96, 102, 104, 222–23, 302, 303, 305
Butler, William, 82
Butts, Jehiel, 112, 134, 149
Butts, Oliver J., 266–67, 288
Butts thresher, operation of, 111, 134–37, 138, 142, 149, 295, 338n.41, 339n.52

Calcasieu Parish, La., 307

cams in rice mills, 177, 259, 269
canals for rice fields, 38, 40, 41, 48, 50, 53, 62–64, 69, 83, 84, 87–93, 97, 101–2, 108, 153, 294–95, 299, 300, 307
carborundum shelling stones, 309
Carew, Richard (1665–1726), 72
Caribbean, 15, 36, 62
Caribbean markets for rice, 17, 62
Carney, Judith A., 8, 9, 13, 41, 42
Carolina Gold rice, xvii, 5, 12, 20–28, 103, 301, 307, 309–10, 330.n.37
Carolina Gold rice, history of: 1928–29 plat documents Maham's rice fields, 22; DNA analysis, 20, 25; introduction credited to Henry Laurens, 23; introduction credited to Henry Woodward, 20; Joshua John Ward's family connection to Maham, 23; Long Grain, 26, 27; Madagascar source questioned, 20; Robert Allston credits Hezekiah Maham with introducing Carolina Gold, 21, 22
Carolina Lowcountry, 57, 306
Carolina; or, The Planter (Ogilvie), 72
Carolina Plantation Society, 42
Carolina rice, 1, 3, 5, 6, 20, 28
Carolina rice, origin of culture: African origin (black rice hypothesis), 13; planter achievement, 12, 315; planter appropriation of slave crop, 13
Carolina rice, source of seeds: on board the *William & Ralph*, 14, 15, 31; Dubois seed, 15, 16; early undocumented sources, 16; Lords Proprietors, 14, 15; Thurber's seed from Madagascar, 19, 20
Carolina (ship), 14, 35
Castigioni, Luigi (1757–1832), 169
Catesby, Mark (1679–1749), 33, 47, 166, 168
Catling, David H., 8
Caw Caw County Park, Charleston County, 43
chaff fans for cleaning rice, 119, 132, 139–40, 142, 144, 149, 281
Champney's Island, Ga., 303
Chang, Te-Tzu, 7, 8
Charles II, 14, 328n.1

Charles Pinckney National Historic Site, 168
Charleston City Gazette, 21, 22, 173
Charleston, S.C., 2, 3, 59–61, 63, 75, 80, 93, 112, 118, 122, 124, 129, 134, 138, 170, 187, 204, 214, 223–24, 226, 228, 232–34, 250, 260, 263, 265–66, 268–69, 278, 283, 286, 288–89, 292, 295–96, 300–1, 309, 311–13
Charles Town, S.C., 3, 14, 16, 19, 20, 35, 36, 62, 63, 154, 320
Chicora Wood threshing barn, operation of: Butts thresher used, 138, 142; chaff fans, 139–40, 144; elevator belt to move rice, 140, 144; feed house for rice sheaves, 142; hopper to chaff-fans, 142; rake operation and placement in frame, 142; reconstruction and operation of rakes, 144; rolling screen placement and operation, 140, 144; sheaves lifted by conveyor, 142; spiral conveyor, 140, 142; straw ejected from barn, 142
Chicora Wood storage barn: hopper function, 148; operation of, 146–47; sloops and schooners, 144
chimneys for steam engines, early history, England: Farey, John (1791–1851), 249; Great Britain and the Industrial Revolution, 249; Savery, Thomas, 249
city mills and plantation toll mills: Bennett's Rice Mill, 283–85; Chisolm's Rice Mill, 25, 283, 295; city mills built by Jonathan Lucas I, 278; Georgetown Rice Mill, 288–89; Greenwich Mill on Shem Creek, 204; Lucas and Norton, 278; Savannah Rice Mill, operation of, 281–82; Waverly Rice Mill, 278–81; West Point Rice Mill, 282–83
Civil War, 36, 40, 53, 64, 65, 96, 102, 107, 111, 129, 134, 138, 142, 148–49, 150–51, 198, 202, 232–33, 239, 245–46, 257, 263, 266–67, 272, 275–76, 280, 283, 288, 290, 293–95, 297–98, 306, 312, 314
Clifton, James M. (1931–2008), 292, 306
Clinch, Catherine (Tat) Maria (1828–1870), 300, 349n.5
Clinch, Duncan Lamont (1787–1849), 59
Clinton, Gen. Sir Henry (1730–1795), 63

Coclanis, Peter A., 306
Cohen, Solomon, 173
Cole, Arthur H., 307
Colleton County, S.C., 61, 69, 272
Colleton, James (b. ca. 1706), 31
Colleton, Sir John, Lords Proprietor (1608–1666), 34
Colonel's Island, Ga., 299
Colonial Assembly, 162, 164
colonial era, 57
Columbian steam engine, 229–30
Committee on Foreign Mills, 225
Conakry, Guinea, West Africa, 9
Confederacy, 297, 299, 300
Confederate Army, 297–98
Continental Army, 63
Cooper, Lord Ashley, Lords Proprietor (1621–1683), 16
Cornwallis, Gen. Lord Charles (1738–1805), 23
cotton gin, animal powered, 176
Couper, James Hamilton (1794–1866), 93, 128, 221–23, 250, 292, 335n.91
Couper, John (1759–1850), 93
Courtenay, William Ashmead (1831–1909), 61, 62
Cowper, Basil, 170
Crisp, William, 20
Croft, Childermas, 202
Crow Island, Santee Delta, 294, 318–19
cultivation of rice, 8
culture of rice, 8, 16, 33, 52, 80, 268
Cumberland Island, Ga., 55
cypress, 29, 30, 34, 35, 62, 69, 70, 73, 158, 198, 208, 258, 317

Daddy Aleck, 302
Daniel I. Hadley & Associates, 348.n.8
Daniell, William, 247
Darby, James C., 39
Darien, Georgia, 3, 158, 299, 303, 320
Day on Cooper River, A (Irving), 313
deadly rice fields: *Aëdes aegypti* mosquito, 93, 96; *Anopheles* mosquito, 93, 95; Asiatic cholera, 96; diseases brought into Lowcountry, 93; high mortality rate of slaves, 95, 96; inadequate nutrition, 97; influenza, 96; malaria, 93, 96; planters' summer residences, 96; poor medical treatment, 97; poor sanitation, 97; pneumonia, 96; rice fields breeding mosquitos, 95; smallpox, 93; tuberculosis, 96; typhus, 93; working in burning sun, 96; working in cold water, 96; yellow fever, 93, 96
Deans, Robert, 163, 165–66
Deas, Anne Simons (1845–1928), 142, 256, 311, 316
De Bow's Review, 335n.86
De Brahm, William Gerard (1718–1799), 51, 52, 59, 332n.35
Deforest, J., 221–23
DeNeale, Mr., 112
Description of South Carolina, A (Glen), 51
Dethloff, Henry Clay (b. 1934), 18
Doar, David, 13, 69, 89, 99, 101, 117, 150, 209, 258
domestication of rice, 7
"Door of no Return," 315
Dotterer, Thomas Davis, Jr. (1832–1894), 80, 234
Drayton, Charles (1743–1820), 118, 169, 175–77, 182, 184, 186, 190, 211, 337n.15
Drayton, John (1767–1822), 82, 170–71, 176–77, 204, 217–18, 286, 318, 334n.79
drill plough, 79, 80
drills to move rice, 214
Dunlop, William, 16, 17, 31, 36
Dupont, Gideon, Sr. (1712–1788), 52
Dupre, Lewis, 154, 211–12, 219
Dusinberre, William (b. 1930), 290
"Dutch floodgate," 93
Dyer, Aaron, 124

East Florida, 1, 3, 60,
East India, 11, 16
Ebenezer, Ga., 186
economic downturn of 1819, 294
Edelson, S. Max, 12, 13, 31
editor, 219, 223
elevator belts to move rice, 140, 144, 214
Elliott, Stephen, 59

Ellis, Rev. William (1794–1872), 158
Eltis, David, 328n.2
Emancipation, 40, 296, 301, 304
Engelberg Huller Company, Syracuse, N.Y., 271–74
European Americans, 2–4, 13
Europeans, 2, 4, 7, 34, 93
Evans, Peter, 126
Evans, Oliver (1755–1819), 204, 214, 217, 226, 229–31, 236, 277

fire prevention in mills and barns, 149–50
Fitch, John (1743–1798), 231
FitzSimons, Christopher (1762–1825), 214
Fjellstrom, Robert, 20, 21
Flails, 4, 111–16, 118, 120–21, 124–26, 132, 136, 149–50, 152, 166, 300, 316
floodgates for rice fields: construction of a floodgate, 89; diamond gate, 90–93; Hopeton lock system, 93; paddle gates, 91, 92; single-gate floodgate, 88; two-gate floodgate, 88–91, 93
flow culture, steps of: dry growth, 81; harvest flow, 80; insect control, 81; layby-flow, 81; long-water flow, 81; point flow, 81; sprout flow, 80, 81; stretch flow, 80, 81; weed control, 81
Ford, Louise P., 87
Ford, Timothy, 25, 167–68
Forster, A. M., 77
Fort Sumter, Charleston Harbor, 296
foundries and manufacturing companies in Charleston: Cameron and Barkley, 340n.68; Cameron and McDermid, 232; Cameron and Mustard, 258; Eagle Foundry, 232; Eason and Dotterer, 232, 234, 248, 295; James M. Eason & Brothers, 262, 268; Phoenix Iron Works, 226, 232, 263; Smith and Porter, 232; Taylor Iron Works Manufacturing, 239; Thurston & Johns, 138; Vulcan Iron Works, 232; William S. Henerey, 80
freedmen: chose not to work, 302–3; contracts to work, 4, 301, 304–5; crop sharing contracts, 304; desire for land and autonomy, 305; end of slavery, 320; foreman of labor, 305; lived off the land, 304; looting and burning plantations, 299; military supervision, 304; moved to the cities to work, 303; quality of labor force, 301; "two-day" tenant system, 304; wintertime field maintenance, 303
Freedmen's Bureau, 304–5
Freeman, James, 19
freshets, 28, 32, 47, 50, 51, 63, 75, 87, 88, 96, 101, 309
Friend to Improvement, A (pseud., real name unknown), 125

Gaillard, John Palmer, Sr. (1874–1962), 22
Gaillard, Theodore, 202
Georgetown Historic Ricefields Association, Inc., xxiii
Georgetown Rice Milling Company, 266, 288–89
Georgetown, S.C., 2, 3, 112, 136, 173, 263, 285, 295, 299, 302
Georgia, 1, 3, 4, 24, 30, 37, 53, 58–60, 63, 64, 69, 82, 89, 92, 95, 101, 103, 106, 129, 146, 158, 186, 202, 214, 221–23, 233, 239, 248, 250, 252, 259, 262, 275, 290, 292–93, 296, 299, 300, 302–3, 305, 309–10, 312
Georgia colony, 58
Georgia Gazette, 170, 202
Georgia tidewater rice production, 59
Gervais, John Lewis (1741–1798), 166
Ghana empire, 9
Gibbs, John Ernest, xxiv
Gibbs, Mathurin Guerin (1788–1849), 53
Gignilliat, William, 59
Glen, James (1701–1777), 51, 166–69, 177, 179
Gold Coast, 327n.1
Golden Age of Rice: 1800–1860: advent of steam-powered threshing barns and mills, 294–95; earnings at Gowrie for first fourteen years, 292; increase in production before Civil War, 292; production totals for clean, milled rice, 290–91; tidal rice-field expansion, 293–94
Gondwanaland, 7, 327n.1

Goose Creek Men, 31, 32, 34–36, 58, 157, 313, 315
Goose Creek Men, list of, 331n.5
Gordon, Lord Adam (1726–1801), 62
Gordon II, William W., 281
Goree Island, Senegal, West Africa, 315
Grand Prairie, Ark., 309
Grant, Hugh Fraser (1811–1873), 59, 293
Grant, Governor James, 60
Gray, Lewis Cecil (1881–1952), 112
Great Britain, 1, 224, 249–50, 252, 256–57
Greene, Gen. Nathanael (1742–1786), 5
Gregorie, John P., 312
Grist, Donald Honey (b. 1891), 275, 348n.170
Groehing, Richard I., Jr., 332n.24
Guendalos Company, Georgetown, S.C., 306
Guerard, Peter Jacob (d. 1714), 17, 160–62, 340n.10
Guinea, West Africa, 9, 41

Habersham, John, 170
Habersham, Joseph (1751–1815), 170
Habersham, Robert (1783–1870), 247
Habersham, Robert & Son, 136
Hadley, Daniel I., xv, 285–88
half-moon bank, 101
Hall, Capt. Fayrer, 15, 16, 20
Hall, Shapter & Tupper, 281
Hamilton, Gen. James, 127
Hamilton, James, 93
Hamilton, J., Jr., 126
hand-mill (or hand mill) for rice, 166–70
Harper's Weekly, 146, 281
Harrison, Gen. George P., 305
Hell Hole Swamp, Berkeley County, S.C., 53
Henry Ford Museum, Dearborn, Michigan, 233, 239, 283
Hewatt, Alexander (1739–1824), 33, 34, 93, 95, 179, 335n.92
Heyward, Arthur, 59
Heyward, Charles (1802–1866), 258, 298
Heyward, Daniel, 123, 180
Heyward, Duncan Clinch (1864–1943), 12, 20, 53, 115, 145, 303, 306, 311–12

Heyward, Edward (Barney) Barnwell (1826–1871), 300, 302
Heyward, Nathaniel, 75
Heyward, W. M., 126
High Hills of the Santee, 25
Hill, Max L., III, 337n.13
Hilton Head Island, 57
History of Georgia, The (Jones), 333n.50
History of South Carolina . . . (Ramsay), 332n.36
Historical Account of the Rise and Progress of the Colonies of South Carolina and Georgia, An (Hewatt), 33, 34, 93, 95, 179, 335n.92
History of the American Rice Industry, 1685–1985, A (Dethloff), 18
Hobcaw Barony, Georgetown County, S.C., 56, 344n.91
Holland, 23, 93, 220
Holmes, Samuel, 172
Hopeton lock, 93
hopper, 80, 118–24, 130, 132, 134, 137, 139–40, 142, 144, 148–49, 169, 185, 205–8, 216–17, 222, 266, 271, 273–74, 276–78, 281, 283, 287–88, 339n.59
hopper-boy for drying rice, operation of, 276–78, 287
Horn of Africa, 11
Horne, Robert, 14
House, Albert Virgil, 106
Huger, Daniel Elliott (1779–1854), 126
Huger, T. Pinckney, 59
Hunter, Lewis C. (1898–1984), 229
hurricane damage to rice crops, 20, 75, 96, 101, 108, 280, 290, 299, 306, 311
Hurricane Hugo, 319
hurricanes of 1822, 290, 319; 1850, 292; 1851, 292; 1852, 292; 1854, 292; 1893, 311; 1894, 311; 1898, 311; 1906, 311; 1910, 311; 1911, 311; 1959, 285; 1989, 319
Hurt, R. Douglas, 119
Hutchinson Island, Ga., 60, 202

Improved Water Rice Machine: brushes removed the polish, 218; fully automated, 217; John Drayton's account of

its operation, 217; operated by only three workers, 217; Robert Allston's description, 218; rolling screen separated unhulled and hulled rice, 219
Indian slave trade, 35
Indian subcontinent, 9
Indian-type mortar and pestle, 157
indigo, 4, 17, 63, 153
Industrial Revolution, 1, 105, 226, 232, 249, 293
inland-swamp rice culture:; acreage not documented, 332n.41; dam construction and function, 28, 46; end of inland swamp cultivation, 53; fields abandoned after the Revolution, 101; flanking canals, 50; Gideon Dupont credited for beginning water culture, 52; hoeing to control weeds, 52, 61; installation of trunks, 42–48; John Stewart, first mention of possible water control in 1690, 17, 31; labor by slaves, 48; overflows, 46, 50; Petigru disputes Ramsay's credit to Dupont, 52; plantations shared same watercourse, 50; reached complex level by 1744, 48; reservoir function, 42, 48; small-stream floodplains, 48; threat from freshets, 32, 47, 50, 51, 63, 75, 87, 88, 96, 101; variable water supply, 50; water culture to control weeds, 51–53, 82
Irving, John Beaufain (1800–1881), 313
Islam, 8–10
isozyme analysis, 7, 327n.3
Italy, 24, 41, 162, 164
Izard, John, 75
Izard, Ralph (1742–1804), 24

Japan, 10
Japanese Kiushu rice, 308
Jefferson Embargo of 1812, 290, 294
Jefferson, Thomas, 24, 121, 220
jetties, 311
J. G. (pseud., real name unknown), 223–24
Jehossee Island, 39, 40, 142, 254–55
Johns Island, S.C., 16
Johnson, President Andrew, 300
Johnstone, McKewn, 62

Jolof empire, 9
Jones, Charles Colcock, Jr. (1831–1893), 59
Judd, William Robert, xvii, xviii, xx, xxi

Kemble, Frances Anne (1809–1893), 38, 39, 95, 331n.15
King, Roswell (1765–1844), 223
King, Roswell, Jr. (1796–1854), 102, 223
king post, 175, 177, 182, 222–23
Knapp, Seaman Asahel (1831–1911), 210, 308, 343n.84
Knight, Samuel, 154
Kogar, Joseph, 118, 337n.14
Kogar wind fan, 118–20, 206

LaBruce, James Louis, 306
Lachicotte, Philip Rossignol (1824–1896), 265, 279, 306
Lachicotte, P. R., and Sons, 279, 306
Lake Chad, Africa, 9
Lander, Ernest McPherson, Jr. (b. 1915), 232
Latrobe, Benjamin Henry Boneval (1764–1820), 204, 217, 343n.74
Laurasia, 327n.1
Laurens, Henry (1724–1792), 21, 23, 24, 60, 63, 121, 154, 165–68, 191–92, 214, 217, 316
Lawson, John (1674?–1711), 19, 157
Lawton, William M., 26, 27
Lebby, William (d. 1882), 239, 282, 347n.141
Lees, William B., 332n.39
Legaré, John D., 223
Legaré, John Girardeau (1852–1932), 303
Leigh, Frances Butler (1838–1910), 302–3
Liberia, West Africa, 9
Life and Labor on Argyle Island (Clifton), 332n.20
Littlefield, Daniel C., 13
London, England, 14–16, 20, 23, 33, 64, 121, 165, 213, 220
longleaf pines, 32–35, 58, 96, 310
Lords Proprietors, 14, 16, 17, 30, 35, 157, 164, 315, 328n.1
Louisiana, 4, 106, 150, 169, 209, 250, 285–86, 306–9
Louisiana rice industry, 209, 306, 308–9

Lowcountry, xvii, xviii, xix, xx, xxi, 1, 4, 5, 13, 16, 20, 30–32, 34, 36, 37, 41, 42, 47, 50, 53, 57, 60, 61, 63, 65, 78, 87, 93, 95, 96, 106–7, 110, 157–58, 164, 188, 192, 210–11, 272, 292, 296–98, 306, 311–13, 315, 318, 319–20
lowland culture of rice, 8
Lowndes, Richard I'on (1847–1889), 288
loyalists, 59, 60, 63
Lucas, Henry Edward (1822–1900), 294, 318
Lucas, John (1824–1909), 162, 171–73, 175–76
Lucas, Jonathan, I (1754–1821), xviii, xxi, 2, 154, 162, 184, 188, 190, 192, 202, 204–6, 208–9, 213–14, 216–20, 232–33, 250, 278, 283, 286, 316, 318
Lucas, Jonathan, Jr. (1775–1832), 154, 204, 216, 220–21, 224, 340n.12
Lucas, Jonathan, III (1800–1853), 282, 340n.12
Lucas, William (1789–1878), 241, 340n.12
Lucas, William Dollard (b. 1920–), 278
Lynch, Thomas, Sr. (ca. 1727–1776), 55

Madagascar, 11, 12, 15, 16, 19–21, 31, 327n.1
Maham, Col. Hezekiah (1739–1789), 5, 21–23
Maham Cemetery, St. Stephen's Parish, S.C., 22
Maham Swamp, 21
malaria, 37, 93, 96, 318, 320
Malay Archipelago, 10
Mali empire, 9
Mande-speaking people, 9
Manigault, Charles (1789–1874), 40, 75, 84, 85, 97, 100, 113, 116–18, 123, 134, 136–37, 144, 146, 149, 180–81, 197–98, 206, 224, 247–48, 254, 257, 292–93
Manigault, Gabriel (1704–1781), 165
Manigault, Louis (1828–1899), 69, 75, 77, 117–18, 128, 144, 149, 248, 254, 298, 300, 305
Manigault, Peter (1731–1773), 51, 160
manual-powered rice mills, 157–70
market rice, 140, 144, 160, 180, 222
market sieve for cleaning rice, 166, 179
Marsh Planter, A (pseud., real name unknown), 26
Mayrant, William, 25, 26
Mazÿck, Benjamin, 331n.5

Mazÿck, Stephen (1749–1808), 55
McClung, Anna M., 20, 21, 25, 330n.28
McCrady, Edward, Jr. (1833–1903), 333n.53
McEwen, H. Alan, xxiii, 250, 256, 347n.144
McLean, Beth Bailey (1892–1976), 210
McLearn, Mr., 198
Mead, Oliver Middleton, 312
Meade, Capt. Samuel, 19
mechanical reapers, 307
Middle Passage, 4, 158
Middleton, Henry Augustus (1793–1887), 148, 262
Middleton, Williams, 150, 233
Midway District, Ga., 59
mill engines, steam: beam eliminated, 231; Cameron.McDermid.Mustard engine, 234, 239; conversion of water power to steam, 233; diversity of engine design, 235; flywheel, 231, 235–36, 239; function and operation of steam chest, 235–39; horizontal cylinder, 231–32; method of mounting engine, 241; operation of fly-ball governor, 235–36; slide valve operation, 236; valve stem function, 236
Miller, Charles, 14, 15
milling rice in Louisiana, 209, 307
millrace, xix, 187–90, 192, 194, 197, 199–1, 205, 233, 316, 344n.101
millrace flooring, 194, 199
millstones: action of rice between the millstones, 209–10; adjusted for different lengths of rice, 340n.4; bed stone, 169, 185, 205–6, 210, 213, 280; brought with English settlers, 209; buhrstones (millstones) made from French buhr stone, 213; dressing stones, 211; Georgia buhr stone, 214; grooves cut into working surface, 210; Lafayette Burr Millstone Company, Ga., 214; La Ferté-sous-Jouarre quarry, France, 213; leveling rods, 281; Millstone Grit, 213; Peak stone from Derbyshire quarry, England, 205, 211, 213, 274; quarry in Northumberland, England, 210–11; runner stone, 169, 185, 205–8, 210–11, 280; standard size of millstones, 210; sweeper swept the rice

from the millstone, 208; weight of millstones, 211; West Point Rice Mill, 283; for wheat and corn, 213; "Winnowing Screen Pendulum," 154, 211

Mississippi, 150, 285

Molony, Thomas J., Jr., 106

Moreland Hunt Club, Santee Delta, 319

Morgan, Philip D. (b. 1949), xviii, 37, 328n.2, 331n.10

mortars, 38, 96, 118, 157–59, 162–63, 165–66, 170–73, 175, 177, 181–82, 184, 186, 192, 197, 201, 206, 208–10, 217, 219, 223–24, 259, 262–63, 266, 271, 283, 287, 295, 309, 316, 345n.103

mortar and pestle use by slaves: broken rice, 160; fatigued and debilitated slaves, 160; milling a large crop, 160; made of native wood by slaves, 158, 208; partially hulled, 160; planters not satisfied with results, 160; pointed end of pestle removed the hulls, 158; removed hulls and outer bran layer, 158; screens removed broken rice, 158; technology brought from Africa, 1, 3, 4, 157; two-step method with pestle, 158–59, 166; winnowing with fanner basket, 158

Mortimer, John (1656?–1736), 41, 45

Moultrie, John, 60

Mr. Timothy's Gazette, 181

Munnerlyn, Charles, 106, 346n.132

Murray, A. B., 285

Muslim empires, 9

Myrick, Mr., 67

Nairne, Thomas (d. 1715), 32, 170,

Napoleonic Wars, 293

Native Americans, 3, 157

Natural History of Carolina, Florida, and the Bahama Islands . . . (Catesby), 166

naval-stores industry, 58, 310

Naylor, Thomas, 232, 241, 344n.91

Nesbit, Robert, 79, 80

Newcomen, Thomas, 226

Newcomen engine, 226, 228, 241, 345n.118

New Orleans, La., 274, 306, 309

New Voyage to Carolina . . . , A (Lawson), 329n.20

Nile Valley, 10

Norris, John, 19, 32, 33

North Carolina, 1, 3, 4, 19, 24, 30, 34, 53, 58, 59, 63, 69, 82, 124, 248, 271, 275, 302, 311, 317

North, the, xxi, 4, 235, 247

oak smackers, 98

Observer, An (pseud., real name unknown), 99

obstacles to overcome to get rice to market: bobolinks, 98; burrowing animals, 101; insect damage, 80, 100; maintenance of hydraulic system, 101; rats, 100; red or volunteer rice, 79, 99; river waves eroded outer banks, 101; saltwater intrusion, 102; slaves' health, 160; soil compaction, 102; soil fertility, 101; storms and hurricanes, 75, 88, 101, 292; water weevil, 100; waterfowl damage to crops, 99; weeds, 47, 51–53, 61, 68, 79, 80–82, 97, 153, 299, 300

Ogilvie, George, 72, 73

Oglethorpe, James Edward (1696–1785), 58

Old South, the, 4

Olmsted, Frederick Law (1822–1903), 258

Oryza barthii Chev., 8

Oryza glaberrima Steudel, 7

Oryza longistaminata Chev. & Roer, 8

Oryza nivara Sharma & Shastry, 9

Oryza rufipogon Griffiths, 9

Oryza sativa L., 7, 99

Our Grimy Heritage . . . (Pickles), 256

overseers, in general, 2, 24, 26, 27, 37–39, 51, 63, 97, 100, 149–50, 160, 202, 255, 294–95, 297–98, 302

overseers, by name: Belflowers, 105; Cooper, Jesse T., 40; Haynes, James, 68, 85, 123, 137, 149, 180, 257; King, Roswell (1765–1844), 223; King, Roswell, Jr. (1796–1854), 102, 223; Maham, Col. Hezekiah, 21; Pittman, 105; Sellers, B. T., 150; Skinner, K. Washington, 84, 206; Venters, Leonard F., 248

overshot water wheel, 192, 198, 286

oxen working a rice mill, 65, 67, 124, 170–71, 173, 175, 343n.73

paddy rice, 219, 272, 274, 340n.5
Pelleu, Marion J., 87
Pangaea, 327n.1
Papers of Henry Laurens, 24
Parishes, S.C.: Middle St. John's, 54, 55, 314; St. Andrew's, 334n.79; St. Bartholomew's, 19; St. James, Goose Creek, 52; St. John's, Berkeley, 21, 188; St. Luke's, 173; St. Peter's, 59, 60, 293; St. Stephen's, 5, 21, 22
Parliament, England, 57
Parrish, Lydia (1871–1953), 158,
patents of rice processing machines, holders of:; Brotherhood, Fred, 269, 288; Butts, Jehiel, 112, 134; Butts, Oliver J., 266; Chapman, William C., 325; Dupre, Lewis, 154, 211; Emmons, Calvin, 125; Emmons, William, 128; Engelberg, Evaristo Conrado, 271–72; Habanards, 274; Hort, Benjamin S., 112; Hort, Elias B., 112; Lachicotte, Philip R., 279; Lachicotte, Philip R., and T. B. Bowman, 265; Lockfaw, John A., 269; Lucas, Jonathan, Jr., 154, 220; Ludlow, 129; Mathewes, William, 112, 132; McKinlay, Peter, 260, 262, 268; Napier, 259; Nourse, Asa, 154; Norton, John L., 154; Ravenel, John, 154, 223; Read, Jacob, 154; Taylor, John F., 263
patriots, 59, 63
Peak millstones, 205, 211, 213, 274, 344n.91
Pedington, Adam, 154
Pennington, Patience, 335n.83. *See also* Pringle, Elizabeth Waties Allston
persistence of earlier technologies, 111
pestles, xix, 1, 4, 136, 153, 156–67, 170–73, 175–77, 181–82, 184, 186–87, 189, 192, 197, 201, 204, 206, 208–9, 217–24, 259–60, 262–63, 265–66, 269, 271–72, 281–83, 287–88, 295, 300, 309, 316
pestle shod in Russian steel in West Point Rice Mill, 283
Petirgu, Capt. Thomas, U.S.N. (1793–1857), 26

Petigru, James Louis (1789–1863), 52
Phillips, Ulrich Bonnell (1877–1934), 331n.16, 333n.53, 341n.33
phosphate industry of S.C., 303
Pickles, Walter, 256
Pinckney, Charles (1757–1824), 108–9
Pinckney, Charles Cotesworth (1746–1825), 108
Pinckney, Elise M., 233
Pinckney, Thomas (1750–1828), 24, 82, 121
plantations:; Barnhill, xxiv; Black Out, 252, 254, 347n.145; Bluff, 48, 50; Brewton, 312; Brookgreen, 23, 26; Buck Hall, Cooper River, 142, 339n.53; Butler's Island, 39, 95, 96, 102, 104, 222–23, 302–3, 305; Causton, 306; Cedar Hill, xxiv, 234, 247; Cedar Spring, 314; Chicora Wood, xxiv, 24, 122, 128, 130, 132, 137, 138–48, 275, 302; Cockfield, xxiv, 24, 122–23, 272, 274; Colerain, 292; Comingtee, 311, 316; Dean Hall, 321; Drayton Hall, xxiii, 42, 43, 53; East Hermitage, 75, 85, 117, 247, 300; Eldorado, 148; Estherville, xxiv, 57, 62; Fairfield (South Santee River), 234; Fairfield (Waccamaw River), xxiv, 234, 252; Fairlawn (Charleston County), 234; Goodwill, 53, 298; Gowrie, 75, 96, 117, 136, 144, 149, 198, 257, 292–93, 300–1, 305–6; Hobonny, 150, 213, 233, 312; Hoover, 330n.37; Hope, xxiv, 254; Hopeton, 92, 93, 221–23, 250, 292; Indianfield, 54; Jericho, 53; Kinloch, xxiv, 25, 70, 122, 148, 150, 272, 275; Laurel Hill (Georgetown County), 252, 254, 257; Laurel Springs, 312; Lavington, xxiv, 25; Limerick, 53, 188; Magnolia, 98; Mansfield, xxiv, 25, 136–37, 234, 275; Maybank, 299; Mepkin, 23, 24, 63, 165, 191, 192, 217, 316; **Middleburg**, xix, xxiv, 118, 192, 209, 225, 233–34, 243, 245, 257, 275, 315–17; Middleton, xxiv; Millbrook (Santee River), 316; Mulberry Castle, 315; Myrtle Grove, 312; Nemours, xxiv, 243; Newington, 18, 186–87; Nieuport, 91, 150, 312; Nightingale, xxiv; Old Combahee, xxiv, 233; Old Town, Louisville, Ga., 214; Peach Island, 202; Plumfield, 330n.37;

Pooshee, 314; Popular Grove, 45; Proctor, 306; Rice Hope (North Santee), xxiv, 126; Richmond, 21, 67; Rochelle, xxiv, 211; Rose Hill (Combahee River), xxiv, 194, 258, 298, 312; Silk Hope (Cooper River), 292; Social Bluff, 299; Social Hall, xxiv, 69, 243; South Mulberry, 340n.69; Sunnyside, 294; Tibwin, 190; Turnbridge, xxiv, 272; Tweedside, 292; Wantoot, 55, 188; Waverly, xxiv, 279; Weehaw, 322; Weymouth, xxiv; Willtown, xxiv, 252, 322; Woodboo, 54, 55; Woodbourne, 103, 189, 294; Woodville, 299

plantation enterprise, beginning of: Barbados model for Carolina rice plantations, 34, 36; Goose Creek Men's role, 31, 32, 34, 36; Proprietors' plans to introduce crops, 14, 15; provision ground system of slave labor, 36; reasons why Goose Creek Men created early plantation enterprise, 36

Plantation Enterprise in Colonial South Carolina (Edelson), 12

planter elite, xxi, 3, 34, 42, 64, 317

Pleistocene epoch, 57, 333n.45

polishing rice, 279, 306

Porcher, Eva, 311

Porcher, Francis Yonge (1789–1862), 108

Porcher, Frederick Adolphus (1809–1888), 21

Porcher, Richard Dwight, Jr., xvii, xviii, xxii, xxiii, xxiv, 22, 285, 313

Popple, William, 18

portable steam engines, 151, 226, 239–41, 248

Port Royal, S.C., 15, 298

Portuguese mariners, 8, 41

Portuguese traders, 11, 35

Potter, James, 198

Potter, John, 59, 100, 292

pounding rice, xix, 96, 165, 181, 186, 208

prairie land, Southwest Louisiana, 307–8

pre–Civil War, xviii, xxi

Prevost, Mrs. C. B. (1904–1982), 348

Pringle, Elizabeth Waties Allston (1845–1921), 84, 85, 142, 306. *See also* Pennington, Patience

providence rice: John Stewart's role, 30–34; moisture strictly from rainfall, 30; nature of the rice plant, 30, 31; planters appropriated slaves' knowledge, 31; slaves and planters first experimented with growing rice, 31; slaves' knowledge of rice growing, 31; small-stream floodplains, 32

Province of Carolina, 14, 18, 20, 328n.1

pumps worked by wind, 191

Purrysburg, S.C., 60

Pursell, Carroll W., Jr., 231, 233

Quattlebaum, Alberta Lachicotte, xxiv, 280, 320

rakes for cleaning rice, 130, 132, 134, 137–44, 148–49, 216, 258

Ramsay, David (1749–1815), 52, 112, 329n.10

Randolph, Edward, 18

Ravenel, Edmund (1797–1871), 55

Ravenel, John (1793–1862), 154, 223–24, 345n.112

Ravenel, Theodore DuBose (1863–1944), 79, 312

Reconstruction, 297, 301, 304

red rice, 79, 99, 100, 153, 336n.8

Reid, John, 214

Reid, Peter, 214

reservoir rice, types of: 1000-acre rice field, 56, 57; artesian springs, 40, 54, 55, 314; barrier island ridges and swales, 57; inland ridges and swales, 57; inland swamps, 5, 41, 48, 50, 51, 53, 60, 95, 101, 313; reclaimed salt marshes, 45, 53, 54

Revolutionary War effects on rice industry, 59, 60, 62

rice birds, 98

Rice (Grist), 348n.170

Rice and Rice Planting in the Carolina Low Country (Doar), 13

Rice and Slaves . . . (Littlefield), 13

rice chimneys: batter of stack, 252; flue, function of, 252; for steam mill on sugar plantation, 250; function of the chimney, 249–52; height of chimneys, 254; made of metal, 280; manholes, function of, 255; masonry crown, function of, 253–54;

rice chimneys: (*cont.*)
    octagonal design of stack, 252; openings in the stack, 256; origin from Great Britain, 252; oversiller, function of, 253–54, 257; planters' "signatures," 256; put-log holes, 256; round design of stack, 252; square design of stack, 252; star-shaped design of stack, 252, 257; underground duct to boiler, function of, 254–55; variation in chimney design, 253

rice culture, reasons for end of: alteration of fresh/saltwater interface, 300, 311; city mills shut down, 312; competition from low-cost Burmese rice, 306; competition from the Southwest, 306; cost of producing rice rose dramatically, 306; destruction of the infrastructure, 299, 300, 311; difficulty with labor contracts, 303–4; free labor system unsuitable, 303, 305; hurricanes, 309, 311; Irish and Chinese failed as laborers, 303; lack of experienced labor force, 303; loss of planters' sons in the Civil War, 300; neglect of the rice fields during the Civil War, 299, 300; newer labor-saving machines too expensive, 2, 112, 309; shortage of capital, 299

rice culture, spread of from Cooper River epicenter: north to Santee Delta and Winyah Bay, 57; south to Georgia starting in 1752, 58, 59; to Cape Fear River tidal swamps, 58; to Lower Cape Fear around 1720, 58; to Savannah River, St. Peter's Parish, S.C., 59, 60; to St. Marys and St. Johns Rivers in East Florida, 60

rice documented as export crop:; Commons House of Assembly in 1698, 17; Commons House of Assembly to John Thurber in 1715, 19; Commons House of Assembly to Samuel Meade in 1715, 19; Council of Trade to James Stanhope in 1715, 19; Daniel Axtell's 1701 account book, 18; "Duty of Rice," 1698, 17; Edward Randolph to the Lords Commissioners in 1700, 18; His Majesty's Commissioners in 1699, 18; Logan's Bill of Lading in 1695, 17; quitrents paid in rice in 1696, 17; Richard Splatt's invoice to William Crisp in 1726, 19, 20; William Salmon's *Botanologia: The English Herbal* in 1710, 19; William Thronburgh to William Popple in 1699, 18

rice factors, 26, 224, 259, 301

rice flats: archival plans missing, 85; care and handling of flats, 85; caulking flats, 85; chine-log flat, 85; flat house to store flats, 85; function, 83; plank-built flats, 85; unique to rice fields, 83

rice fruit, morphology of: aleurone layer, 156; bran, 156–57; "brown rice," 155; embryo, 156; endosperm, 156; germ, 156; hulls, 155–56; rice polish, 156; rough rice (= paddy rice), 156

rice, harvesting: "rice bird" reaper, 106; under the task system, 37–40; with cradle scythe, 107–8; with sickle, 105–7

Rice Kingdom, xxi, 1–6, 20, 24, 25, 47, 48, 53, 57, 58, 61, 62, 65, 78, 79, 92, 93, 105, 107, 111–13, 121, 148, 154, 157–60, 162–63, 164, 166, 169–70, 172, 177, 184, 186–88, 192, 198, 202, 209–10, 214, 220, 222, 226, 232–34, 239, 243, 245–46, 250, 254, 256–57, 271–72, 274, 280, 282, 285, 294–97, 299–1, 303, 305–10, 312, 314, 317–18, 320, 327n.1

Rice Millers Association, La., 210

rice mills, types of: Brotherhood rice pounding and hulling mill, 269–70; Carolina Rice Mill-1850 model, operation of, 286–88; cog mill, 173–77; Deans rice-pounding machine, 165–66; Deforest rice mill, 221–23; drum mill, 173–76; Engelberg disk huller, 274–75; Engelberg huller and polisher, 274; Engelberg steel huller, 272; Guerard Pendulum Engine, 160–62; Habarnards Rice Huller, 274; Italian pestle mill, 162–65; Lachicotte and Bowman machine for cleaning rice, 265–66; Lockfaw rice mill, 269–71; Lucas Jr. rice cleaner, 220–21; Lucas I Improved water rice machine, 216–20; Lucas Water Rice Machine, 202–8. *See also* in main Index for more detail; McKinlay "Crank Pestle Mortar," 262; McKinlay portable rice mill, 268–69;

McKinlay rotary rice cleaner, 262–63; mortar and pestle, 157–60; Napier's wire cards, 259; Oliver J. Butts brushing machine, 266–67; pecker mill, 171–73; Ravenel rice-hulling machine, 223–24; Salzburger rice mill, 186–87; spring mill, 162; stamps, 186–87; stamp machines, 51; Strong and Moody's patent rice mill, 345n.106; Taylor machine for cleaning rice, 263–64; Veitch pestle mill, 181–84; water-powered mill, 184–224; Williams cam pestle mill, 259–60; wooden rotary quern, 166–70

Rice Museum, Crowley, La., 285

Rice Museum, Georgetown, xxiii, 285

Rice Planter and Sportsman (Alston), 331n.11

rice plant, nature of, 155–57

rice production in the Rice Kingdom, 310

rice scourer, 309

rice, types of in the Rice Kingdom: bearded, 25, 26; Carolina Gold, 20–25; common white, 26, 27; Long Grain Carolina Gold, 26, 27; white, 19, 20, 156

Richardson, David, 328n.2

rivers and waterways: Abbapoola Creek, 16; Altamaha Delta, 61; Altamaha River, 29, 30, 59, 92, 93, 95, 102, 128, 221–23, 250, 292, 303, 312, 317; Ashepoo River, 58, 69, 75, 243; Ashley River, 14, 35, 53, 186, 278, 282–83; Atkinson Creek, Santee Delta, 243; Back River, Berkeley County, 313; Back River, Jasper County, 45; Big Duck Creek, Santee Delta, 294; Black River, 61; Bull Creek, 103, 294; Cape Fear River, 4, 58; Charleston Harbor, 30, 296, 313; Combahee River, 43, 53, 58, 61, 75, 87, 90, 91, 145, 150, 194, 199, 243, 258, 298, 302, 306, 311–12, 316–17; Cooper River, 29, 30, 32, 34, 35, 44, 47, 57, 58, 67, 102–3, 142, 151–52, 165, 192, 197, 217, 225, 233, 258–59, 278, 292, 296, 313–15; Cooper River, Eastern Branch, 118, 234, 243, 316; Cooper River, Western Branch, 23, 31, 48, 117, 311, 316; Coosawhatchie River, 58; Dorchester Creek, 18, 186; Gambia River, West Africa, 9; Goose Creek, 17, 35, 36, 42, 108, 160, 313; Great Pee Dee River, 29, 57, 128, 278, 302; Hazzard's Creek, 344n.91; Huger Creek, 53; Lake Moultrie, 55, 188, 314; Lower Cape Fear, 58, 63, 311; Mississippi River, 306; Mississippi River valley, 169; New River, 58; Nicholson Creek, 53, 314; Niger River, West Africa, 9; North Santee River, 39, 69, 70, 150, 204, 211, 216, 243, 317–18; Ogeechee River, 29, 30, 59, 103, 129, 146, 292, 310, 317; Old Town Creek, 14; Pamlico Sound, 19; Port Royal, 15, 298; Quenby Creek, 317; Sabine River, La., 307, 309; Sampit River, 57, 288; Santee Delta, 30, 57, 102, 243, 274, 294, 313, 317; Santee River, xvii, 19, 21, 69, 70, 82, 103, 150, 202, 211, 243, 306, 316–18; Satilla River, 59, 292; Savannah River, 59, 61, 63, 84, 103, 117, 128, 170, 198, 209, 281, 292, 293, 305–6, 311; Senegal River, West Africa, 9; Shem Creek, 190, 204; South Newport River, Ga., 59; South Santee, 148, 202, 234, 317; Saint Johns River, 4, 60, 69; Saint Marys River, 59, 60, 292; Tar River, N.C., 19; Tibwin Creek, 190; Turkey Creek, 314; Vermilion River, La., 307; Waccamaw River, 57, 103, 138, 189, 234, 257, 278–79, 294, 306; Wadboo Creek, 31, 329n.12; Wambaw Creek, 209; Wateree River, 19, 53, 298; Winyah Bay, 21, 22, 30, 56, 57, 61, 317; Wright River, 58

rivers and wetlands, types of: black-water rivers, 28–30, 69, 102, 315; brown-water rivers, 28, 30, 69, 102; cypress/mixed hardwood swamps, 29, 30, 62, 103; salt marsh, 30, 45, 53, 54, 293; salt-point, 29, 30, 56; salt-wedge estuary, 29, 30; small-stream floodplain, 30, 32, 33, 46, 48; vertically homogenous estuary, 29, 30

Rivers, William James (1822–1909), 329n.6

Robinson, Solon, 254–55

Rogers, George C., Jr. (1922–1997), 340n.2

rolling screens: at Chicora Wood threshing barn, 122, 144; at Cockfield Plantation, 122; at Kinloch Plantation, 122; hexagonal, 122, 181; operation of, 123; octagonal, 122–23; round, 122–23; used in milling, 177; used in threshing, 113, 121, 123

rotary quern, 166–70, 204

rough rice shipped to England, 220, 224
Roundhouse Museum, Savannah, Ga., 239
Rowland, Charles E., 212
Rowland, Lawrence S., 293
Ruffin, Edmund (1794–1865), 102, 298
Rural Carolinian, 268
Rutledge, Frederick, 137
Rutledge, John Jr. (1776–1819), 60

St. Augustine, Fl., 14–16, 34, 63
St. Catherine's Island, Ga., 299
St. Mary's, Ga., 299
St. Simons Island, Ga., 298, 302
Salley, Alexander Samuel, Jr. (1871–1961), 16, 328n.4
salt marshes, 30, 45, 53, 54, 293, 342n.61
Salzburgers, Ebenezer, Ga., 186–87
Sanders, Joshua, 61
Sandford, Robert, 15, 16
Santee Canal, 55, 188, 333n.42
Santee Coastal Preserve, 319
Santee Delta, 30, 102, 243, 274, 294, 313, 317
Santee Gun Club, 319
Sapelo Island, Ga., 109, 223,
Savage, Jack, 75
Savannah, Ga., 2, 3, 60, 62, 63, 93, 126–27, 170, 202, 224, 232, 239, 252, 262, 281, 286, 295, 305, 311, 320
Savannah Wildlife Refuge, S.C., 103
Savery, Thomas (ca. 1650–1715), 249
Schencking, Barnard, 17, 36
schooners, 40, 144–46, 211, 276, 281, 285–86, 295, 339n.55
Schulze, Richard, 330n.37
Scientific American, 271
screens for cleaning rice, 158, 177
Screven, John H., 233
Seabrook, Archibald Hamilton, Jr. (1856–1941), 148
Sea Island cotton, xx, 292–94, 301, 304, 318,
Second Continental Congress, 23
Second, G., 7
Seed from Madagascar (Heyward), 20
seed barn to store seed rice, 337n.13
seed rice, preparation of, 117–18

Sellers, B. T., 150
Sellers, John, 122
Sellers, N. & D., 121
Senegal, West Africa, 8, 9, 315
separators for different size rice, 129
shaking rice straw, 119, 121, 142, 150
sheaves: curing, 106, 109–10; Devonshire hook to haul sheaves, 108–9; "mow-burned," 110; stacked in ricks, 109; transporting to barnyard, 108–10
Sherman, Gen. William Tecumseh, 300
Sherman Reservation, 300
Sherman's army, 233, 299
Shields, David S., xvii-xviii
short-staple cotton, 292
Sierra Leone, West Africa, 9
sieves for cleaning rice, 148, 177–81
Slavery and Rice Culture . . . (Smith), 95
slave trade, 4, 35, 48, 64, 159, 293, 294
sloops, 144, 146, 276, 286, 339n.55
Smith, James, 122
Smith, Joseph, 122
Smith, Josiah, 51
Smith, Julia Floyd, 95
Smith, Thomas, second Landgrave (1670–1738), 61
Smyth, J. F. D. (John Ferdinand Dalziel), 167–69
Songhay empire, 9
Sothel, Gov. Seth (d. 1694), 17
South Carolina, 1, 3, 5, 16, 18, 19, 23, 24, 29, 37, 39, 48, 51, 52, 58–60, 69, 78, 82, 89, 90, 93, 96, 98, 101–4, 112, 150, 166, 187, 211, 220, 232–34, 243, 248, 258, 275, 278, 283, 285, 292–93, 295, 299, 300, 302, 303, 311–12, 319
South Carolina Agricultural Society, 23
South Carolina Commons House of Assembly, 181
South Carolina Gazette, 61, 172–73, 181
South Carolina Rice Plantation, The (Easterby), 336n.12
Southern Agriculturist, 108, 223, 250
Southern Cabinet, 219, 345n.106
South, the, 3, 4, 214, 232, 239, 289, 301, 314

Southwest rice industry, 99, 210, 214, 267, 285–86, 300, 306–10
sowing rice seeds: broadcast, 78; covered seed sowing in trenches, 78; date for sowing, 78; drill plough, 79; open-trench sowing with clayed seeds, 78, 79; pressing into ground with heels, 78
Spain, 10, 14, 34, 60,
Spalding, Thomas (1774–1851), 109, 121, 223
Sparkman, James Ritchie (1815–1897), 38, 39
spiral conveyor to move rice, 140, 142, 146, 281
Splatt, Richard, 19, 20
spokes in rice mills to lift pestles, 164, 177, 184, 192, 201, 206, 259, 283, 342n.53
Sri Lanka, 10, 41
State Agricultural Society, 26
Statutes at Large of South Carolina, 332n.31, 334n.60
steam engines, operation and history of: condensing, low-pressure engine, 226; operation of, 225–26; cylinder cutoff, 230, 232; expansive nature of steam, 229; feedwater, 226; high-pressure, noncondensing, vertical cylinder, Columbian engine, operation of, 226, 229–31; horizontal cylinder, 231; mill engine 231–39. *See also* mill engines in main Index; Newcomen engine, 226, 228, 241, 345n.118; reciprocating motion, 228, 239; rotary motion, 228, 231, 239, 260; walking-beam engine, 228, 283, 285
steam thresher for wheat, 307
Stephens, William, 186–87, 259
Stewart, John, 16, 30–34, 36, 42, 51
Stoke Rice Mill, 197, 233, 258, 275, 316
Stono Rebellion of 1739, 62
storage barn for rice, 140–42, 144–48, 339n.59
storm tower, 319
Stuttgart, Arkansas, 309
sugar, 4, 34, 37, 42, 93, 250
Swan, Dale E., 290
Swinton, Wm., 61

tabby, 90
task system for plantation slaves: basic task unit, 37; freedom and mobility, 40; full-task, 37; gang work, 40; half-task, 37, 39; nonhands, 37; not suited to some types of work, 40; no work on Good Friday, Christmas day, or Sundays, 39; only half-tasks on Saturday, 39; preserved slaves' African culture, 40; quarter acre of work, 38, 39; quarter-task, 37; slave free time to plant gardens, 38; three-quarter hand, 38; weeding spurred development of task system, 51
Texas Gulf Coast rice industry, 309
Them Dark Days . . . (Dusinberre), 335n.97
thiamine, vitamin, 157
Thornburgh, William, 18
"thrashing" rice, 127, 336n.1
threshers for rice, mechanical, evolution and types of: Andrew Meikle thresher, 120–21; Bernard thresher, 124–25; Calvin Emmons thresher, 125–28; Chicora Wood thresher, 138–44; flail, use of and output in bushels, 112–16; Invincible thresher, 150–51; Jehiel Butts thresher, 134–37; Ludlow thresher, 129; Mathewes thresher, 132–34; patents of South Carolinians lost in U.S. Patent Office fire, 112; planters praise for Butts thresher, 134–35; portable threshers, 151; post-war threshers, 150–52; Scotch thresher, 121, 125–26; Sinclair's portable thresher, 151; threshing barn, construction and design of, 130–32, 138–44; William Emmons thresher, 128–29
threshing barns, 2, 26, 40, 85, 120, 122, 129–30, 132, 136–42, 146, 148–51, 217, 231–33, 239, 241, 248, 250, 254, 256, 275, 293–95, 300, 317–19
threshing floor, 113, 115, 117
threshing yard, 84, 105, 150, 152, 295
tidal rice fields, abandoned, legacy of: abandoned fields under bank today, 104; climax forest composition unknown, 103; ecological dynamics of abandoned fields with breached banks, 103; ecological value of abandoned fields today, 103; recreational value of abandoned fields today, 103; reversion back to swamp forest, 102;

tidal rice fields, abandoned, legacy of: *(cont.)* state policy against repairing banks of abandoned fields, 103

tidal rice culture, history of: deltas and floodplains converted to rice fields, 61; increased yields over inland rice, 61; July 16, 1737, earliest recorded mention of tidal swamp suited for fields, 61; London provided the capital after the Revolution, 64; main type of culture after the Revolution, 64; McKewn Johnstone credited with the first tidal crop of rice, 62; more profitable than inland swamp culture after the Revolution, 64; origin contributed to many individuals, 62; tidal freshwater swamp converted to rice fields, 65

tidal rice fields, construction of: boundary surveyed and marked, 65; check banks established, 68; obstacles to overcome, 65; permanent outer bank constructed, 66; quarter drains constructed, 68; removal of trees and stumps, 67; smaller individual fields constructed, 67; temporary outer bank, 67; tide-trunks installed, 67

Tidewater Atlantic Coast, 4

Tidyman, Philip (d. 1850), 204, 216, 318,

Timbuktu, 9

Timmons, George, 154

tobacco, 4, 14, 35, 37, 64, 100

toll rice mills, 224–25, 276, 278, 281, 286, 316–17

Tories, 63

Tory privateers, 63

Tour of the United States of America, A (Smyth), 167

Treatise on Steam Engines . . . , A (Farey), 249

Treaty of Paris of 1783, 23, 60

Trimble, Roswell, 309

trunks, types of: Essex trunk, England, 45; lever-gate trunks, 42, 44, 45, 48, 50; lift-gate trunks, 42–45, 54, 72; plug trunks, 42, 47, 69, 70, 72; swing-gate trunks, 45, 46, 54; tide trunks, construction of, 67, 68, 73; tide trunks, operation of, 73–75; tide trunk, ACE Basin design, 75; tide trunk, original design, 75; tide trunk, Santee design, 75; tide trunk, Savannah design, 75; trunk-minder, role of, 73–75

Tucker, D. Gordon, 344.n.92

two-step milling of rice with mortar and pestle, 158–59, 166

Ultisols, 37, 331n.9

Union forces destruction along Georgia coast, 298

upland culture of rice, 8, 28, 100

upland rice, 8, 24, 31, 186, 271

Upper Guinea Coast, 41

U.S. Department of Agriculture, 330n.37

U.S. Patent Office fire in 1836, 112, 154, 220, 223, 336n.3

U.S. Patent Office, Washington, 111, 125

Veitch, George, 181–82, 184, 187

View of South-Carolina, A (Drayton), xxi, 170, 204, 217

Village Museum, McClellanville, S.C., xxiii, 209, 274

Villeponteaux, Peter, 172–73

Virginia, 35, 107, 121, 213, 328n.1

Vitruvius, 184

W (pseud., real name unknown), 125, 129–30, 132

Wadboo Barony, 31, 34, 329n.12

walking-beam steam engine, 228, 283, 285

Ward, Joshua John (1800–1853), 23, 26

Ward, Samuel Mortimer, 306

Waring, Dr. John Beamer, 341n.25

Waring, Joseph Ioor (1897–1977), 335n.96

War of Austrian Succession, 62

War of Jenkins Ear, 62

Washington, N.C., 19

Washington, William, 25,

Washo Reserve, Santee Coastal Preserve, 102

water control for crops, history of: Chinese early as 722–481 B.C.E., 41; English perfected drainage and flooding by late 1600s, 41; Incas between 1300 and 1150 B.C.E., 41; piedmont of Northern Italy since 15th century, 41; pre-conquest Aztecs, South America, 41; Sri Lanka about 300 B.C.E., 41; tide mills in England in 1006, 41; West-African agriculture mid-15th century, 41

water culture of rice: met with limited success, 82; only two floodings used, 82

water power for mills, history of: bed stone, 169, 185, 205–6, 210, 213; horizontal water wheel, 222–23; runner stone, 169, 185, 205–8; vertical water wheel, 184, 223

water power for rice mills, source of: canals, 188; flowing streams, 188; pump-fed reservoirs, 191; rain-fed reservoirs, 188; saltwater tides, 190; tidal freshwater rice fields, 188

water-powered rice mills: breastwork to reduce drag on water wheel, 199; canted millrace gate, 194; cast iron hub to strengthen water wheel, 197; dimensions of the mill house, 197; debris rack for millrace, 197; dimensions of the millrace, 192; millrace floor of wood, 194; millrace gate design and operation, 194; millrace gate operated by lever or windlass, 197; millrace walls, 194; operation of water-powered mill, 201; pestle rice mill, 192; pit or driving wheel's function, 194; plantations for sale advertised with water-powered mills, 202; reverse gearing, 200; run by undershot water wheel, 188–90, 198; water wheel anchored to millrace walls, 194; water wheel construction, 197

Water Rice Machine of Jonathan Lucas I: detailed operation of millstones, 204; distance between the stones adjusted, 205; first to use millstones to remove hulls, 192, 202, 204; J. K. Bowman, Peach Island Plantation, 202; John Drayton's sketch of Tidyman's rice mill, 204; Latrobe's diagram of Water Rice Machine, 217, 343n.74; mortars and pestles removed bran, 206; operation of Water Rice Machine, 205; revolutionized milling of rice, 202

Watkins, Jabez Bunting (1845–1921), 307
Watkins Syndicate, 307
Watson, Andrew M., 10, 11
Watson, Elkanah, 170

Watt, James, 226, 228–29, 241
Waverly Rice Mill, Waccamaw River, 138, 278–81
Wayne, Gen. Anthony (1745–1796), 59
West Africa, 1, 4, 7–11, 41, 48, 315
West, Gov. Joseph (d. 1691), 14
West Indies, 14, 34–36, 58, 60, 250
Weston, Anthony, 148
Weston, Plowden, 85, 149–50, 204
West Point Rice Mill, 228, 278, 282–83, 285, 295, 312
wheat and corn, milling of, 158, 169, 184, 209, 213–14, 344n.103
Wilkinson, R. A., 169–70
William and Ralph (ship), 14, 15, 31
Williamson, John, 60
Williams, S. K., 259
Willoughby, Lord Francis, 5th Baron, 14
Wilmington, N.C., 2, 3, 124, 271, 320
wind fans, 111, 118, 121–22, 127, 129, 132, 134, 137, 150–51, 205–6, 208, 217, 219, 274, 283, 286–87
winnowing rice manually: fanner basket, 111, 115–16, 118, 150; straw shaken to remove embedded rice, 115; winnowing house, 111, 116, 118, 120, 132; winnowing platform, 116
wire-sieves, 179–80, 219
Woman Rice Planter, A (Pringle), 335n.83
wooden mill for milling rice, 166–79
Wood, Peter H., 13
Woodward, Henry (1746?–1685), 15, 16, 20, 31
World War I, 312
World War II, 233
Wright, Charles, 60
Wright, Jermyn, 60
Wylly, Charles Spalding, 312

Yangtze Valley, China, 10w d
Yellow River Valley, China, 10
Yemassee War, 1715, 19
Yorkshire, England, 23, 213
Young, Arthur, 121
Young Mill-Wright and Miller's Guide, The (Evans), 277
Young, Thomas, 126

About the Authors

RICHARD DWIGHT PORCHER, JR., is professor emeritus at the Citadel and adjunct professor of biological sciences at Clemson University, where he established the Wade T. Batson Endowment in field botany. Porcher is the author of *Wildflowers of the Carolina Lowcountry and Lower Pee Dee*, and the coauthor of *A Guide to the Wildflowers of South Carolina* (both by the University of South Carolina Press) and *The Story of Sea Island Cotton*. Porcher serves as a trustee of the South Carolina Nature Conservancy and on the board of directors of the Charleston Library Society, the Waring Library, and the Carolina Gold Rice Foundation. He is the 2008 recipient of the South Carolina Environmental Awareness Award.

WILLIAM ROBERT JUDD, a self-taught draftsman/artist, archaeologist, and historian, is retired from the U.S. Space and Naval Warfare Systems Command (SPARWAR) and lives with his family on James Island, South Carolina.